Praise for the previous edition of this book

"This book does a superb job with one of the most complex, politicized, and profound issues of our day. It's even-handed and accessible, skipping the jargon and rhetoric to give the reader a solid understanding of the climate challenge and the most promising solutions."

—**Eileen Claussen**, President,
Center for Climate and Energy Solutions

"Scientifically up-to-date and clearly written, this courageous book cuts through mystery and controversy to explain climate change for readers who prefer facts."

—**Richard Somerville**, Distinguished Professor,
Scripps Institution of Oceanography

"Much has been published on climate change in recent years, but nothing as clear, succinct, comprehensive, and readable as Bob Henson's handy volume. Henson sheds much light on the heat in a balanced, meticulous fashion. His assessment of a staggering number of recent studies and on-going research is invaluable. This will be the first go-to climate change book on my shelf."

—**David Laskin**, Author of *Braving the Elements*
and *The Children's Blizzard*

"Bob Henson is one of the world's clearest and most engaging writers on the atmospheric sciences."

—**Keay Davidson**, Author of *Carl Sagan: A Life*

Praise for Robert Henson's *Weather on the Air*

"The history, the personalities, the science, the challenges, the beauty, and the warts of weathercasting: It's all here."
—Ray Ban, retired executive vice president of programming and meteorology, The Weather Channel

"Supplanting baseball, talking about the weather has become our national pastime. This book is a must-read for anyone seeking to understand the evolution of our attraction to, and dependence on, up-to-the-minute news about the weather."
—Edward Maibach, Center for Climate Change Communication, George Mason University

"Henson clearly charts the internal storms as well as the fair skies of a profession that has been surprisingly slow to accept the growing scientific consensus of humankind's contribution to climate change."
—Joe Witte, former NBC/Today meteorologist

"From green screens to greenhouse gases, a thorough and very readable history of broadcast meteorology."
—Greg Carbin, NOAA/NWS Storm Prediction Center

". . . lovingly written and meticulously researched."
—Bob Swanson, *Weatherwise* Magazine

THE THINKING PERSON'S GUIDE TO
CLIMATE CHANGE

THE THINKING PERSON'S GUIDE TO
CLIMATE CHANGE
ROBERT HENSON

AMERICAN METEOROLOGICAL SOCIETY

Cover Images:
robas/E+/Getty Images
Ralph Lee Hopkins/National Geographic/Getty Images
Patti McConville/Photographer's Choice RF/Getty Images

Published by the American Meteorological Society
45 Beacon Street, Boston, Massachusetts 02108

For more AMS Books, see http://bookstore.ametsoc.org.

The mission of the American Meteorological Society is to advance the atmospheric and related sciences, technologies, applications, and services for the benefit of society. Founded in 1919, the AMS has a membership of more than 13,000 and represents the premier scientific and professional society serving the atmospheric and related sciences. Additional information regarding society activities and membership can be found at www.ametsoc.org.

Library of Congress Cataloging-in-Publication data is available and can be found online at www.ametsoc.org.

MIX
Paper from
responsible sources
FSC® C005010

Acknowledgments

My deep gratitude goes to the experts who have graciously provided interviews and text reviews for this and previous versions of this material, which originated as *The Rough Guide to Climate Change*. Listed here with their affiliations at the time of their input, they are: **Sim Aberson** (NOAA Atlantic Oceanographic and Meteorological Laboratory); **Lisa Alexander** (Monash University); **David Anderson** and **Connie Woodhouse** (NOAA Paleoclimatology Program); **Myles Allen, Ian Curtis, Tina Jardine, Lenny Smith,** and **David Stainforth** (Oxford University); **Caspar Ammann, Aiguo Dai, James Done, John Fasullo, Andrew Gettelman, Mary Hayden, Michael Henry, Greg Holland, Aixue Hu, Jim Hurrell, Mari Jones, Jeff Kiehl, Joanie Kleypas, David Lawrence, Sam Levis, Linda Mearns, Jerry Meehl, Andrew Monaghan, Susi Moser, Bette Otto-Bliesner, David Schneider, Britt Stephens, Claudia Tebaldi, Tom Wigley, Steve Yeager,** and **Jeff Yin** (NCAR); **Vicki Arroyo, Elliot Diringer, Judi Greenwald, Katie Mandes, Namrata Rastogi, Truman Semans,** and **John Woody** (Pew Center on Global Climate Change); **Mustafa Babiker** (Arab Planning Institute); **Richard Baker** (U.K. Central Science Laboratory); **Dan Becker** (Sierra Club); **Lance Bosart** (University at Albany, State University of New York); **Lori Bruhwiler** (NOAA Earth System Research Laboratory); **Harold Brooks** (NOAA Storm Prediction Center); **Shui Bin** (University of Maryland); **Andy Challinor** (University of Reading); **John Christy** (University of Alabama in Huntsville); **Hugh Cobb** (NOAA National Hurricane Center); **Judith Curry** (Georgia Institute of Technology); **Andrew Derocher** (University of Alberta); **Lisa Dilling, Roger Pielke Jr., Konrad Steffen,** and **Eric Wolf** (University of Colorado Boulder); **Simon Donner** (University of British Columbia); **Nikolai Dotzek** (Deutsches Zentrum für Luft- und Raumfahrt); **David Easterling, Tom Karl,** and **Thomas Peterson** (NOAA National Climatic Data Center); **Kerry Emanuel** (Massachusetts Institute of Technology); **Ben Felzer** (Marine Biological Laboratory); **Jeff Fiedler** (UCAR); **Piers Forster** (University of Leeds); **Hayley Fowler** (Newcastle University); **Greg Franta** and **Joel Swisher** (Rocky Mountain Institute); **Jonathan Gregory** (University of Reading/Met Office Hadley Centre); **James Hansen** and **Cynthia Rosenzweig** (NASA Goddard Institute for Space Studies); **Eric Haxthausen** (Environmental Defense); **Katharine Hayhoe** (Texas Tech University); **Paul Higgins** (American Meteorological Society); **Paul Holper** (CSIRO); **Brian Huber** (Smithsonian National Museum of Natural History); **Peter Huybers** (Harvard University); **Tim Johns, Richard Jones,** and **Jason Lowe** (Met Office Hadley Centre); **Phil Jones** and **Nathan Gillett** (University of East Anglia); **David Karoly** (University of Oklahoma); **David Keith** (University of Calgary); **Georg Kaser** and **Thomas Mölg** (University of Innsbruck); **Paul Kench** (University of Auckland); **Eric Klinenberg** (New York University); **Chuck Kutscher** (U.S. National Renewable Energy Laboratory); **Beate Liepert** (Lamont-Doherty Earth Observatory); **Diana Liverman** (University of Arizona); **L. Scott Mills** (North Carolina State University); **Glenn Milne** (Durham University); **Andy Parsons** (Natural Environment

Research Council); **Greg Pasternack** (University of California, Davis); **Rick Piltz** (ClimateScienceWatch.org); **John Church** and **James Risbey** (CSIRO); **Alan Robock** (Rutgers, The State University of New Jersey); **Jason Rohr** (University of South Florida); **Vladimir Romanovsky** (University of Alaska Fairbanks); **William Ruddiman** (University of Virginia); **Stephen Schneider** (Stanford University); **Peter Schultz** (U.S. Climate Change Science Program); **Richard Seager** (Lamont-Doherty Earth Observatory); **Glenn Sheehan** (Barrow Arctic Science Consortium); **Scott Sheridan** (Kent State University); **Thomas Smith** (NOAA); **Susan Solomon** (NOAA/IPCC); **Eric Steig** (University of Washington); **Peter Stott** (Met Office Hadley Centre); **Jeremy Symons** (National Wildlife Federation); **David Thompson** (Colorado State University); **Lonnie Thompson** (Ohio State University); **Jeff Trapp** (Purdue University); **David Travis** (University of Wisconsin–Whitewater); **Isabella Velicogna** (University of California, Irvine); **Chris Walker** (Swiss Re); **Chris West** (U.K. Climate Impacts Programme); **S. Joseph Wright** (Smithsonian Tropical Research Institute); and **Gary Yohe** (Wesleyan University).

Additional thanks go to the friends, colleagues, and loved ones who read large portions of the book and provided much-needed moral support along the way, including Joe Barsugli, Brad Bradford, Andrew Freedman, Zhenya Gallon, Roger Heape, Theo Horesh, Richard Ordway, John Perry, Catherine Shea, Wes Simmons, Stephan Sylvan, and Ann Thurlow. Matt Kelsch gets special kudos for providing nonstop encouragement and a careful and thoughtful read of this most recent version. And every writer should have an agent and supportive friend like Robert Shepard in her or his corner.

The enthusiasm of others for this project smoothed the way on many occasions. I'm grateful to Mark Ellingham and Andrew Lockett at Rough Guides for originally championing the book and to Sarah Jane Shangraw and Ken Heideman for bringing it to the AMS. Special thanks go to Duncan Clark, who provided editing, designing, and illustration for the first and second editions of the original book.

No one could ask for a better springboard for writing about climate change than working at the University Corporation for Atmospheric Research, which operates the National Center for Atmospheric Research. It's impossible to adequately convey my appreciation for the opportunity I've had as a writer and editor at UCAR and NCAR to learn about climate change from many dozens of scientists over more than two decades. This book couldn't have been written without periods of leave made possible by Lucy Warner, Jack Fellows, Rachael Drummond, and Scott Rayder, along with the support of my colleagues in UCAR Communications. The opinions expressed herein are my own (as are any errors). The content does not necessarily reflect the views of my UCAR and NCAR colleagues, the institution as a whole, or the American Meteorological Society and its members.

The choices we make in the next few years will shape the planet we bequeath to the youth of this century. This book is dedicated to them, including those in my own circle of life—David, Chloe, Ava, and Donovan Henson; Renée BeauSejour; and Natasha Kelsch Mather.

Contents

Part 4: Debates and Solutions
From spats and spin to protecting the planet

Part 5: What Can You Do?
Reducing your footprint and working for action

Foreword

Society has reached a critical point with climate change. Carbon dioxide concentrations in the atmosphere are now more than 40 percent higher than they were at the start of the industrial revolution and they are set to increase rapidly as we burn fossil fuel at faster and faster rates. As a direct result, we are pushing Earth's climate outside the range that has existed since human civilization began.

The choices we make over the next several years will help set the direction for civilization (and Earth's climate) for decades and centuries to come.

Yet the significance of our near-term choices for society's future prospects is almost entirely lost on us. Policy makers, members of the media, and the public—all of us—face a raft of pressing concerns and priorities that keep climate change largely outside our immediate consciousness. How we handle our day-to-day stresses has clear implications for our immediate happiness and well-being, whereas our individual climate choices have only a small influence on a seemly remote and distant problem. Nevertheless, individual actions add up with climate change and ultimately they could matter much more than the issues that exert their importance more transparently and forcefully.

The good news is that a great deal is understood about climate change and how people are disturbing the climate system. We know, based on multiple independent lines of evidence, that climate is changing and that human activities are causing climate to change. Furthermore, the options for climate change risk management are numerous, well developed, and reasonably straightforward to think through. Those options imply we could manage climate change risks and realize new opportunities with relatively minor, but sustained, efforts.

To those seeing and hearing only the public discourse on climate change, these basic conclusions may seem surprising but, as outlined in this book, they're based on the comprehensive assessment of the scientific evidence.

The scientific conclusion that climate is changing is based on more than a dozen separate lines of evidence. They include: 1) temperatures at Earth's surface measured by thermometers, 2) temperatures throughout the atmosphere measured by satellites and balloon-borne instruments, 3) ocean temperatures (i.e., greater heat content), 4) glaciers throughout the world (the vast majority are melting), and 5) species migrating and undergoing changes in the timing of key life events. Taken together, these independent lines of evidence (and more) demonstrate that climate is changing.

Multiple independent lines of evidence also demonstrate that people are causing climate to change. Simple math and the growing chemical signature of fossil fuels in the atmosphere demonstrate that people are causing greenhouse concentrations to increase. The warming influence of those greenhouse gases is clear based on laboratory experiments, evidence from past changes in climate due to greenhouse gases, and the role of greenhouse gases on other planets (e.g., Venus is hotter than Mercury because of its greenhouse gases). The patterns of climate change over the last several decades also match the fingerprint of greenhouse gases but not the fingerprint of the usual suspects: the sun, volcanoes, aerosols, land-use patterns, and natural variability. These natural factors could be adding (or subtracting) a small amount to the changes we're seeing but there is no doubt that our greenhouse gas emissions are causing climate to change.

What are our options for addressing climate change? In a very general sense, climate policies fall into four broad categories:

* Efforts to reduce greenhouse gas emissions (often called mitigation)
* Approaches that increase society's capacity to cope with climate impacts (adaptation)
* Attempts to counteract some climate change impacts through additional, deliberate manipulation of the earth system (geoengineering or climate engineering)
* Efforts to better understand climate change, its implications, and society's options (knowledge base expansion)

Reducing emissions is a little like disease prevention (e.g., exercise, eat well, and don't smoke), which can help keep problems from arising. Adaptation is like managing illness (e.g., take medicine to cope with symptoms and alleviate problems). Geoengineering is a little like organ transplantation—an option of last resort that is probably best avoided but may be better than the alternative. Every category encompasses a family of more specific options, each of which has potential to reduce climate change risks, create new sources of risk, or confer additional benefits unrelated to climate change (co-benefits). The different categories can be used together in a wide range of combinations.

The conclusions that climate is changing, that people are causing climate to change, and that we possess a range of potentially beneficial risk management options are based on a comprehensive assessment of scientific evidence. Furthermore, the relevant subject matter experts (i.e., those who are familiar with the evidence) overwhelmingly agree.

Despite that broad expert agreement on these basics of climate change science, deep uncertainty remains over the societal consequences that will result. Predicting these consequences is exceedingly difficult because doing so requires integrating information from complex physical, natural, and social systems. Without a clear picture of what awaits, it is proving too difficult to recognize the seriousness of the risks and too easy to be complacent. The complexity also makes it too easy for pundits, politicians, and interest groups to downplay the issue or cloud the discussion.

The need to work through the extremely messy public debate on climate science falls on each of us. We must develop our capacity to consider climate science and policy in an informed and thoughtful way. That will make it easier to cut through facile arguments and misinformation.

Claims made in public discourse that are at odds with expert assessments generally aren't credible. It was true when some cigarette makers disputed the scientific evidence that revealed the health effects of smoking. It is true for those who cannot reconcile their spiritual beliefs with the scientific evidence that the earth is several billion years old. It's true for climate change now.

The suggestions that climate isn't changing or that people aren't causing climate to change are rejections of science. Those who downplay the potential for serious climate change impacts or who claim that serious consequences are certain to occur are also on shaky ground. The argument (routinely made in public discussions) that climate policy will lead to economic disaster is also firmly at odds with basic economics.

Conflict between science and ignorance is not new. Neither is our interest in having science prevail in that conflict: society's advancement depends on it. What may be new with climate change is the size and scale of the implications to civilization and, to a degree, the ease with which people can ignore scientific insights without knowing it.

Investing the time to learn a few basics can be the difference between being thoughtful and informed or misled and manipulated. That distinction is critical for charting a course for civilization that gives us the best chance to prosper while living in a climate system that is, as described in *The Thinking Person's Guide to Climate Change*, increasingly shaped by human activities.

—Paul Higgins
Director of Policy
American Meteorological Society
June 2014

Introduction

When strangers meet at a bus stop or in a coffee shop, weather is the universal icebreaker. Yesterday's sweltering heat, the storm predicted for this weekend: it's all fair game. Even longer-term climate shifts find their way into chitchat. "It used to snow harder when I was a kid" is a classic example—and one explained in part by the fact that any amount of snow looks more impressive from a child's height.

Today, however, such clichés have an edge to them, because we know that humans play a role in determining the course of climate. When we hear about Arctic tundra melting or a devastating hurricane, we're now forced to consider the fingerprints of humanity—and that's going well beyond small talk. Indeed, climate change can be as much a divider as weather has traditionally been a unifier. Weather has always seemed to transcend politics, but human-induced climate change is wedded to politics: it's an outgrowth of countless decisions made by local, regional and national governments, as well as individuals and corporations. Sadly, it's also become—particularly in the United States—a polarized subject, linked to other issues so frequently that it often serves as shorthand for one's entire world view.

It might come as a surprise, then, how much of the basic science behind global climate change is rock-solid and accepted by virtually all parties. Aside from a handful of skeptics and a few headline-grabbing controversies, most of the debate among experts these days revolves around interpretation. Just how warm will Earth get? Which computer projections for the year 2050 are likely to be the most accurate? How should we go about trying to reduce the blanket of greenhouse gases that's getting thicker each year? How can we best adapt to unavoidable changes? These are difficult questions—but they're about the *nature* of global climate change, not its mere existence.

Although more and more people recognize the risks of climate change, not everyone is convinced of the danger, in part thanks to a battlefield of rhetoric. By the early 2010s, the reverberations from worldwide economic turmoil, the polarization of U.S. politics, and the near-collapse of global climate negotiations made any grand solution to climate change seem further away than ever.

Still, if global warming is one of the most daunting challenges humanity has faced, it's also a unique opportunity. Fossil fuels do more than power our cars and homes. They fuel crisis—from instability in the oil-drenched Middle East to smog-choked skies across the great cities of the developing world. Even though unconventional sources of fossil fuel have proliferated in recent years, the total amount of oil, gas, and coal on Earth is undeniably finite, and we can't afford to burn it all without putting our planet's climate at serious risk. With any luck, the difficult steps needed to deal with global warming could hasten the world's inevitable transition to cleaner, sustainable forms of energy. And as many who have written on this subject point out, we may yet emerge from that transition with new ways of achieving progress on other tough issues. In that light, a warming world could affect our interconnected global society as much as it shapes the climate itself.

How this book works

Whether you're alarmed, skeptical, or simply curious about climate change, this book will help you sort through the many facets of this sprawling issue. **The Basics** lays out some key questions and answers, explains how global warming actually works, and examines the sources of the greenhouse gases heating up our planet. **The Symptoms** provides an in-depth look at how climate change is already affecting life on Earth and how these changes may play out in the future.

The Science describes how the global warm-up has been measured and puts the current climatic changes in the context of Earth's distant history and future. It also takes a look at the computer models that tell us what we can expect over the next century. Debates and Solutions surveys the global warming dialogue and explores the ways in which we might be able to eliminate or reduce the threat of climate change. These include political agreements, such as the Kyoto Protocol, as well as cleaner energy sources and sci-fi–esque geoengineering schemes. For solutions on an individual or family scale, turn to **What Can You Do?**, which provides tips on reducing your carbon footprint at home and on the road.

More online

The website associated with this book—**bookstore.ametsoc.org**—will bring you to a list of relevant books and research articles. You'll also find digital versions of key graphics that appear in the print edition.

1

The Basics

GLOBAL WARMING IN A NUTSHELL

Fragmented Arctic sea ice, September 2009. (Patrick Kelly/U.S. Coast Guard)

CHAPTER ONE

Climate Change: A Primer

KEY QUESTIONS AND ANSWERS

Before exploring the various aspects of climate change in depth, let's quickly answer some of the most frequently asked questions about the issue. The following pages will bring you up to speed with the current situation and the future outlook. For more information on each topic, follow the reference to the relevant chapter later in the book.

THE BIG PICTURE

Is the planet really warming up?

In a word, yes. Independent teams of scientists have laboriously combed through more than a century's worth of temperature records (in the case of England, closer to 300 years' worth). These analyses all point to a rise of around 0.8°C (1.4°F) in the average surface air temperature of Earth when comparing the period 2003–2012 to 1850–1900. The estimated linear trend from 1880 to 2012 is a bit higher, roughly 0.85°C (1.53°F). Chapter 10 explains how this average is calculated. The map on p. 4 shows how warming since the 1970s has played out regionally.

In recent decades, global temperatures have spiked dramatically, reaching a new high in 1998 and similar levels in 2010. Strong **El Niño** events

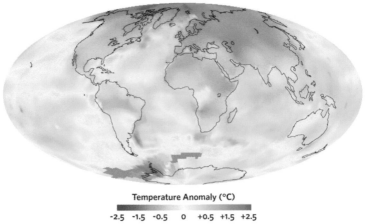

Temperature Anomaly (°C)

-2.5 -1.5 -0.5 0 +0.5 +1.5 +2.5

The change in surface temperature for the period 2000–2009 relative to 1951–1980 was calculated by NASA's Goddard Institute for Space Studies using data from surface weather stations for land areas and satellite-derived temperature estimates for the oceans. Gray areas denote places where temperatures were not recorded. The warmth was especially prominent at high northern latitudes, as predicted by computer models. (NASA Earth Observatory)

leading into both years (see chapter 7) helped boost global temperatures, but it's not as if readings were chilly in between. The first decade of the twenty-first century was the hottest on record—and quite possibly warmer than any others in the past millennium (see chapter 11).

Apart from what temperatures tell us, there's also a wealth of circumstantial evidence to bolster the case that Earth as a whole is warming up.

* **Ice** on land and at sea is melting dramatically in many areas outside of interior Antarctica and Greenland. The major glaciers remaining in Montana's Glacier National Park could be reduced to fragments as soon as the 2020s. Arctic sea ice has lost nearly three-quarters of its average summer volume since 1980, and the next several decades may see a point in summer when open water covers nearly all of the Arctic Ocean, perhaps for the first time in tens of thousands of years. The warmth is already heating up international face-offs over shipping, fishing, and oil-drilling rights in parts of the Arctic once written off as inaccessible.
* **The growing season** has lengthened across much of the Northern Hemisphere, with a boost from urban heat islands in some areas. The average blooming date of Japan's famed sakura (cherry blossoms) is now earlier than at any time in more than a thousand years of record

keeping, pushing the traditional festivals from April to March. Daffodils in the United Kingdom's Royal Botanic Gardens typically bloom more than two weeks sooner now than in the 1980s. In the region around Massachusetts' Walden Pond—where Henry David Thoreau kept landmark botanical records in the 1850s—the record-setting warm spring of 2012 coaxed the highbush blueberry into flowering on the first day of April, a full six weeks earlier than in Thoreau's time.

* **Mosquitoes, birds, and other creatures** are being pushed into new territories, driven to higher altitudes and latitudes by increasing warmth. Southern flying squirrels are meeting and mating with their northern relatives in Canada. Armadillos—once confined to Texas and Florida—have made their way as far north as Kansas and Indiana. Inuit in the Canadian Arctic report the arrival over the last few years of barn swallows, robins, black flies, and other previously unseen species. As we'll see later, however, not all fauna will migrate so successfully.

* **Many forms of marine life** are moving poleward even more dramatically. The most comprehensive review to date, published in 2013, found that marine species have shifted their ranges by roughly 10 times the average for land-based species.

But don't many experts claim that the science is uncertain?

There is plenty of uncertainty about details in the global warming picture: exactly how much it will warm, the locations where rainfall will increase or decrease, and so forth. Some of this uncertainty is due to the complexity of the processes involved, and some of it is simply because we don't know how individuals, corporations, and governments will change their greenhouse emissions over time. But there's near-unanimous agreement that global climate is already changing and that fossil fuels are at least partly to blame (see chapter 13).

The uncertainty that does exist has been played both ways in the political realm. Skeptics use it to argue for postponing action, while others point out that many facets of life require acting in the face of uncertainty—buying insurance against health or fire risks, for example.

Is a small temperature rise such a big deal?

A degree or so of warming may not sound like such a big deal, given that temperatures can vary by much more in a single day. However, little

Climate change or global warming?

The phrases that describe climate in transition have a history of their own. Early in the twentieth century, researchers preferred **climatic change** or **climate change** when writing about events such as ice ages. Both terms are nicely open-ended and still used often. They can describe past, present, or future shifts—both natural and human-produced—on global, regional, or local scales.

Once scientists began to recognize the specific global risk from human-produced greenhouse gases, they needed a term to describe it. In 1975, Wallace Broecker (Lamont-Doherty Earth Observatory) published a breakthrough paper in the journal *Science*, entitled "Climatic change: Are we on the brink of a pronounced global warming?" By the early 1980s, the phrase **global warming** was gaining currency among scientists. Meanwhile, the term **global change** emerged as a way to embrace all modes of large-scale tampering with the planet, including such emerging issues as the Antarctic ozone hole. When 1988's watershed events arrived (see chapter 13), the global warming label broke into headlines worldwide and became standard shorthand among media and the public.

Of course, the planet as a whole *is* warming, but many scientists avoid that term, preferring "global change" or, more specifically, **global climate change**. One of their concerns is that global warming could be interpreted as a uniform effect—an equal warming everywhere on the planet—whereas a few regions may in fact cool slightly, even as Earth, on average, warms up. Politicians hoping to downplay the reality of global warming gravitate toward "climate change" for entirely different reasons. U.S. political pollster and consultant Frank Luntz reportedly advised clients that "climate change" sounds less frightening to the layperson's ear than "global warming." Scary or not, a number of other surveys support the idea that "global warming" gets people's attention more quickly than the less ominous (though more comprehensive) "climate change." And a few activists and scientists now favor **global heating** or **global weirding**—phrases that imply humans are involved in what's happening.

increments make a big difference when they're in place day after day. The average annual temperature in Boston is only about 4°C (7°F) below that of Baltimore. Recent overall warming has been larger in certain locations, including the Arctic, where small changes can become amplified into bigger ones, and the planet as a whole could warm by 5°C (9°F) or more by the end of this century—hardly a small rise. That's more than half of the difference between current temperatures and the depths of the last ice age.

Any warming also serves as a base from which heat waves become that much worse, especially in big cities, where the **heat-island effect** comes into play. Like a thermodynamic echo chamber, the concrete canyons and oceans of pavement in a large urban area heat up more readily than a field or forest, and they keep cities warmer at night. During the most intense hot spells of summer, cities can be downright deadly, as evidenced by the hundreds who perished in Chicago in 1995 and the thousands who died in Paris in 2003 and Moscow in 2010 (see chapter 4).

How could humans change the whole world's climate?

Humans have transformed Earth's atmosphere by adding enormous quantities of carbon dioxide (CO_2) and other **greenhouse gases** to it over the last 150 years. As their name implies, these gases warm the atmosphere, though not literally in the same way a greenhouse does. The gases absorb heat that is radiated by Earth, but they release only part of that heat to space, which results in a warmer atmosphere (see p. 27).

The amount of greenhouse gases we add is staggering—in carbon dioxide alone, the total is more than 30 billion metric tons per year, which is more than four metric tons per person per year. And that gas goes into an atmosphere that's remarkably shallow. If you picture Earth as a soccer ball, the bulk of the atmosphere would be no thicker than a sheet of paper wrapped around that ball.

Even with these facts in mind, there's something inherently astounding about the idea that a few gases in the air could wreak havoc around the world. However, consider this: the eruption of a single major volcano—such as Krakatoa in 1883—can throw enough material into the atmosphere to cool the global climate by more than 1°C (1.8°F) for over a year. From that perspective, it's not so hard to understand how the millions of engines and furnaces spewing out greenhouse gases each day across the planet, year after year, could have a significant effect on climate. (If automobiles spat out chunks of charcoal every few blocks in proportion to the invisible carbon dioxide they emit, the impact would be more obvious.) Despite this, many people respond to the threat of global warming with an intuitive, almost instinctive denial.

When did we discover the issue?

Early in the twentieth century, the prevailing notion was that people could alter climates locally (for instance, by cutting down forests and plowing

The densest part of Earth's atmosphere extends only a few kilometers above the surface, as shown in this photo taken from the International Space Station. The orange sky near the bottom represents the troposphere, or "weather layer," where temperatures have been warming. Above this layer is the stratosphere, seen here as a light gray region, where temperatures have been cooling and ozone depletion has been a concern over recent decades. (NASA Earth Observatory)

virgin fields) but not globally. Of course, the ice ages and other wrenching climate shifts of the past were topics of research. But few considered them an immediate threat, and hardly anyone thought humans could trigger worldwide climate change. A few pioneering thinkers saw the potential global impact of fossil fuel use (see chapter 2), but their views were typically dismissed by colleagues.

Starting in 1958, precise measurements of carbon dioxide confirmed its steady increase in the atmosphere. The first computer models of global climate in the 1960s, and more complex ones thereafter, supported the idea floated by mavericks earlier in the century: that the addition of greenhouse gases would indeed warm the climate. Finally, global temperature itself began to rise sharply in the 1980s, which helped raise the issue's profile among the media and the public as well as among scientists.

Couldn't the changes have natural causes?

The dramatic changes in climate we've seen in the past 100 years are not proof in themselves that humans are involved. As contrarians are fond of pointing out, Earth's atmosphere has gone through countless temperature swings in its 4.54 billion years. These are the results of everything from cataclysmic volcanic eruptions to changes in solar output and cyclic variations in Earth's orbit (see chapter 11). The existence of climate upheavals in the past raises the question asked by naysayers as well as many people on the street: how can we be sure that the current warming isn't "natural"—that is, caused by something other than burning fossil fuels?

That query has been tackled directly over the last couple of decades by an increasing body of research, much of it compiled and assessed by the **Intergovernmental Panel on Climate Change (IPCC)**, a unique team that has drawn on the work of more than 1000 scientists over more than 20 years. We'll refer often throughout this book to the IPCC's work; see chapter 15 for more on the panel itself. As shown below, the IPCC's second, third, fourth, and fifth assessments have each issued progressively stronger statements on the likelihood of human involvement on the climate change of recent decades.

* **1995:** "The balance of evidence suggests a discernible human influence on global climate."
* **2001:** "There is new and stronger evidence that most of the warming observed over the last 50 years is attributable to human activities."
* **2007:** "Human-induced warming of the climate system is widespread."
* **2013:** "It is extremely likely that human influence has been the dominant cause of the observed warming since the mid-twentieth century."

To support claims like these, scientists call on results from two critical types of work: **detection** and **attribution** studies. Detection research is meant to establish only that an unusual change in climate has occurred. Attribution studies try to find the likelihood that particular factors, including human activities, are involved.

One way to attribute climate change to greenhouse gases is by looking at the signature of that change and comparing it to what you'd expect from non-greenhouse causes. For example, over the past several decades, Earth's surface air temperature has warmed most strongly near the poles and at night. That pattern is consistent with the projections of computer models that incorporate rises in greenhouse gases. However, the pattern agrees less well with the warming that might be produced by other causes, including natural variations in Earth's temperature and solar activity.

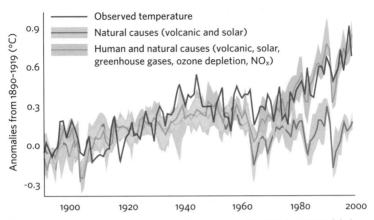

The orange and green shaded areas show the departure of twentieth-century global temperature from the 1890–1919 average, as produced by four computer model simulations. The orange and green lines show averages of the four models. The green runs factor in natural agents of climate change only; the orange runs include both human and natural factors. The blue line shows the temperature measured in the real world. (Parallel Climate Model/DOE/NCAR)

As computer models have grown more complex, they've been able to incorporate more components of climate. This allows scientists to tease out the ways in which individual processes helped shape the course of the last century's warm-up. One such study, conducted at the National Center for Atmospheric Research (NCAR), examined five different factors: volcanoes, sulfate aerosol pollution, solar activity, greenhouse gases, and ozone depletion. Each factor had a distinct influence. The eruption of Mount Pinatubo in 1991 helped cool global climate for several years. Sulfate pollution peaked in the middle of the twentieth century, between World War II and the advent of environmentalism, and it may have helped produce the slight drop in global temperatures observed from the 1940s through the 1970s. Small ups and downs in solar output probably shaped the early-century warming and perhaps the midcentury cooling as well. However, the sun can't account for the pronounced warming evident since the 1970s. The bottom line (see graphic) is that the model couldn't reproduce the most recent warming trend unless it included greenhouse gases.

Couldn't some undiscovered phenomenon be to blame?

Although many people would love to find a "natural" phenomenon to blame for our warming planet, such as the relationship between clouds and cosmic rays (see chapter 13), it is growing extremely unlikely that a suitable candidate will emerge. Even if it did, it would beg a difficult question: if some newly discovered factor—something we're not aware of now—can account for the climate change we've observed, then why aren't carbon dioxide and the other greenhouse gases producing the warming that basic physics tells us they should be?

And there's another catch. Any yet-unknown process could just as easily be a cooling as a warming agent, and if it were to suddenly wane, it could leave us with even greater warming than we imagined possible. Trusting in the unknown, then, is a double-edged sword. Most scientists in the trenches look instead to Occam's razor, the durable rule credited to the medieval English logician and friar William of Occam: one should not increase, beyond what is necessary, the number of entities required to explain anything.

How do rainforests fit into the picture?

As a whole, Earth's land-based ecosystems absorb more CO_2 than they release, but when deforestation is considered on its own, it's a major source of carbon dioxide. Though deforestation has eased significantly over the last few years, it still accounts for roughly 10%–15% of recent human-produced CO_2 emissions. When tropical forests are razed or burned to clear land, the trees, soils, and undergrowth release CO_2. Even if the land is eventually abandoned, allowing the forest to regrow, it would take decades for nature to reconcile the balance sheet through the growth of replacement trees that pull carbon dioxide out of the air.

Rainforests also cool the climate on a more local level: their canopy helps trap moisture in the atmosphere, providing a natural air-conditioning effect. When the rainforest has been slashed and burned over large areas, hotter and dryer conditions often set in, although the exact strength of this relationship is difficult to quantify. By contrast, in midlatitude and polar regions, it appears that forests actually tend to warm the climate, as they absorb more sunlight than bright, reflective snowfields do (see chapter 21).

Are hurricanes getting worse because of global warming?

There's tremendous variation in hurricane activity over time and from place to place. A number of studies since 2005 indicate that the number and/or strength of hurricanes have increased in various regions, especially since the 1970s. However, it's likely that some hurricanes at sea went unnoticed in the days before satellites and hurricane-hunter aircraft, and that complicates the assessment.

In the North Atlantic, there's no doubt that hurricane activity has stepped up since the mid-1990s. Here, ocean temperatures have risen through long-term warming and an apparent multidecadal cycle in Atlantic currents, and warm oceans provide the energy to drive hurricanes. The tropics as a whole are part of a global trend toward ocean warming that goes hand in hand with atmospheric warming. As for the future, computer models tend to point toward steady or decreasing numbers of hurricanes overall (although this is still a question mark, with research continuing). For those hurricanes that do form, models generally project an overall strengthening of winds and rainfall.

Trends aside, a catastrophic storm can strike in any year, and it's impossible to tie any single hurricane or other weather event directly to global warming. Take Hurricane Katrina, which ravaged New Orleans in 2005. Several hurricanes of comparable strength have been observed across the Atlantic over the last century, and the horrific damage caused to the city was the result not only of Katrina's strength, but also of the storm's track, the weakness of levees, and many other factors. That said, the waters of the Gulf of Mexico that fueled Katrina were at near-record warmth at the time. Likewise, several ingredients led to the enormous damage wrought by Hurricane Sandy in 2012—including warmer-than-usual waters off the mid-Atlantic coast, the storm's vast size, and its virtually unprecedented westward path into New Jersey—but scientists have debated the extent to which climate change might be involved. See chapter 8 for more on Katrina, Sandy, and other hurricanes and chapter 7 for details on oceanic changes.

Didn't we have "global cooling" a while back?

The planet did cool slightly from the 1940s to the 1970s, mainly in the Northern Hemisphere and most likely related to the postwar boom in industrial aerosol pollutants that bounce sunlight away from Earth. Despite a flurry of 1970s media reports on an imminent ice age (see chapter 13), there was never anything approaching a scientific consensus on the likelihood of further cooling, and it appears that greenhouse warming has long since eclipsed the midcentury cool spell.

After temperatures reached a new global high in 1998, the following decade remained warm, but with smaller ups and downs that continued into

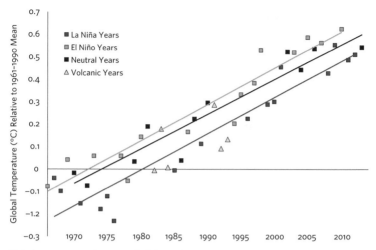

Global temperatures have trended above their twentieth-century averages since the late 1970s. The blue squares show years when the influence of La Niña predominated. Normally cooler than El Niño years (orange squares), La Niña years have been occurring more often this century, which appears to have helped temper long-term atmospheric warming. However, even La Niña years are now warmer than they used to be. (Dana Nutticelli/John Cook/SkepticalScience.com)

the 2010s. The absence of another show-stopping record led many skeptics and pundits to claim that global warming "stopped" in 1998. In truth, nobody expects the global average temperature to rise smoothly from one year to the next. Just as any spring in London, New York, or Beijing will see a few cold snaps, we can expect long-term warming to be punctuated by periods with little rise, or even a slight drop, in global temperature. There can be many factors driving such variability, but one big one appears to be the natural ebb and flow of heat between the oceans and atmosphere. During El Niño (see chapter 7), warmth held in the tropical Pacific Ocean flows into the atmosphere, and global temperatures typically rise by several tenths of a degree Celsius. The opposite occurs during La Niña, when the Pacific draws more heat than usual from the global atmosphere. The twenty-first century so far has seen an outsized number of La Niña years (see graph), which helps to explain the slowdown in atmospheric warming.

None of this changes the longer-term outlook, and the extra heat tucked into the ocean during the apparent hiatus could influence global temperature far down the line. As for individual record warm years, these will continue to be most likely during El Niño events.

Recent temperature trends have sometimes been ascribed to "global dimming." While a slight decrease in the amount of sunlight reaching Earth has been measured over the last few decades, it has not been enough to counteract the overall warming (see chapter 10).

What about the ozone hole?

There are a few links between ozone depletion and global warming, but for the most part, they're two separate issues. The world community has already taken steps to address the Antarctic ozone hole, which is expected to disappear by the end of this century (see chapter 2).

THE OUTLOOK

How hot will it get?

According to the 2013 IPCC Working Group I report, there's a wide range of possibilities for how far global average temperature is likely to climb above its 1986–2005 values. The spectrum extends anywhere from 0.4° to 2.6°C (0.7°–4.7°F) by 2046–2065 and from 0.3° to 4.8°C (0.5°–8.6°F) by 2081–2100. Importantly, this would be on top of the warming of about 0.6°C (1.1°F) that occurred from the 1850–1900 period to the 1986–2005 comparison period.

Why is there such a large spread in the above estimates? It reflects uncertainty about the quantities of greenhouse gases we'll add to the atmosphere in coming decades, and also about how the global system will respond to those gases. The lower-end projections assume huge, and perhaps unlikely, cuts in emissions that would require major economic shifts, whereas the higher-end values would push Earth more than 2°C above its preindustrial average, raising the odds of serious global impacts. See box [BELOW?] for more about these scenarios, or representative concentration pathways (RCPs). Even for what's known as a "midrange" scenario, which assumes that carbon dioxide emissions will level off by midcentury and drop thereafter, the IPCC's estimate of likely warming between 1986–2005 and 2081–2100 is 1.1°–2.6°C (2.0°–4.7°F). When you add in the 0.6°C of previous warming noted above, there's a substantial risk of pushing beyond the 2°C increase noted above. (There's nothing magic about that value, as we'll see in chapter 14, but it's a useful and often-cited benchmark.)

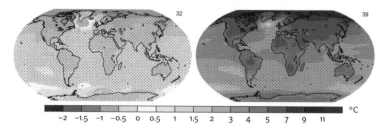

Change in annual mean surface temperature from the period 1986–2005 to 2081–2100 for two scenarios, the low-end RCP2.6 (left) and the high-end RCP8.5 (right). These represent the averages from more than 30 climate models. In the areas covered by dots, the indicated changes are large relative to natural variability and at least 90% of the models agree on the sign of the change. Crosshatched areas indicate where the change is small compared to natural variability. (IPCC)

Some parts of the planet, such as higher latitudes, will heat up more than others. The warming will also lead to a host of other concerns—from intensified rains to melting ice—that are liable to cause more havoc than the temperature rise itself. And the new climate will be a moving target. Barring major changes in policy and energy use, warming and its impacts will only continue to grow in the years after 2100.

Is global warming necessarily a bad thing?

Whether you view climate change as bad, good, or neutral depends on your perspective. Some regions—and some species—are likely to benefit, at least temporarily, while many others will suffer intense problems and upheavals. And some of the potential impacts, such as major sea level rise, increased flood and drought risk, intensified hurricanes, and many species being consigned to extinction, are bad news from almost any perspective. So while it may be a bit of a reach to think in terms of "saving the planet" from global warming, it's perfectly valid to think about preserving a climate that's sustaining to as many of Earth's residents as possible.

Perhaps the more pertinent question is whether the people and institutions responsible for producing greenhouse gases will bear the impacts of their choices, or whether others will—including those who had no say in the matter. Indeed, people in the poorest parts of the world, such as Africa, will generally be least equipped to deal with climate change, even if the changes there are no worse there than elsewhere. Yet such regions are releasing only a small fraction of the gases causing the changes.

Pathways to our future climate

Predicting the atmosphere's behavior decades into the future is a challenging task in itself. The job gets even tougher when it comes to human-induced climate change, because we don't know how quickly people, companies, and nations will act to cut down on greenhouse emissions. The IPCC has long relied on emission scenarios, which map out several different courses global emissions might take over the next few decades on the basis of how society might evolve. The 2001 and 2007 IPCC reports drew on a set on 40 scenarios grouped into four "families," distinguished by whether they emphasized globalization or regionalization and economic growth or environmental protection. As one might expect, the A1 scenario (rapid growth in a globalized economy) depicted the greatest rise in emissions, while the B2 scenario (slower growth in a more localized economy) showed the lowest rise.

Though they're very useful in capturing the flavors of our possible future, emission scenarios can also be restrictive. Each one specifies the amount of carbon emitted year by year. In reality, we could end up getting to the same atmospheric concentration of a greenhouse gas in several different ways (e.g., emitting a great deal early on and then tailing off versus a more steady growth in emissions). Moreover, recent studies have found that it's the total amount of carbon dioxide in the air that's most important to future climate, rather than how quickly it's emitted.

With this in mind, a new set of tools called **representative concentration pathways (RCPs)** was used in the 2013–2014 IPCC assessment. Each RCP is keyed to a particular concentration of greenhouse gas in the year 2100, which in turn would produce a particular amount of radiative forcing (the net energy kept from escaping Earth and heading into space, in watts per square meter). For RCP2.6, it's assumed that major cuts would be in place before midcentury, with CO_2 emissions plummeting thereafter. In contrast, RCP8.5 would allow CO_2 emissions to roughly triple over current levels by 2100—but with a huge risk of climate effects. The other two RCPs fall between these two extremes.

	Change in radiative forcing, year 2100–year 1750 (W/m^2)	CO_2 concentration in 2100, parts per million (ppm)	Concentration of CO_2 equivalent in 2100, including methane and nitrous oxide (ppm)	Likely range of global temperature increase by 2081–2100 over 1986–2005 (°C)
RCP2.6	2.6	421	475	0.3–1.7 (mean 1.0)
RCP4.5	4.5	538	630	1.1–2.6 (mean 1.8)
RCP6.0	6.0	670	800	1.4–3.1 (mean 2.2)
RCP8.5	8.5	936	1313	2.6–4.8 (mean 3.7)

How many people might be hurt or killed?

Quantifying the human cost of climate change is exceedingly difficult. The World Health Organization estimated that in the year 2000 alone, more than 150,000 people died as a result of direct and indirect climate change impacts. A more comprehensive study, released in 2009 by the Global Humanitarian Forum (a think tank headed by former United Nations chief Kofi Annan) found that disasters related to climate change kill some 300,000 people each year. Taking an even broader view, the humanitarian group DARA found that some five million people die each year from indoor and outdoor air pollution, heat-related infectious diseases, and other ills connected to climate change and combustion-related fuel use. The medical journal *Lancet*, building on IPCC findings, warned in the executive summary of a 2009 report that "climate change could be the biggest global health threat of the 21st century."

It's important to keep in mind that many of the projected impacts of global warming on society are the combined effects of climate change, development patterns, and population growth. (Some claim the last factor is the most important.) It's thus hard to separate out how much of the potential human suffering is due to each factor. Moreover, weather-related disasters kill thousands of people each year regardless of long-term changes in the climate.

Some of the most direct impacts of climate change will occur where land meets sea. In the decades to come, the warming of the planet and the resulting rise in sea level will likely begin to force people away from some coastlines. Prosperous low-lying cities such as Miami could be vulnerable to enormous damage. The implications are especially sobering for countries such as **Bangladesh**, where millions of impoverished people live on land that may be inundated before the century is out. Residents may not have the luxury of retreating from a slow rise in water level, as deadly surges from hurricanes and other oceanic storms will strike on top of the ever-rising sea.

Another concern is moisture—both too much and too little. In many areas, rain is falling in shorter but heavier deluges conducive to flooding, as evidenced by Pakistan's unprecedented monsoon rains and the resulting flood catastrophe in 2010. However, **drought** also seems to be becoming more intense. See chapter 5 for more on this seeming paradox. Changes in the timing of rainfall and runoff could complicate efforts to ensure clean water for growing populations, especially in the developing world.

Warming temperatures may also expand the range of vector-borne diseases such as **malaria** and dengue fever (see chapter 9), although other factors are critical and the science on this continues to evolve.

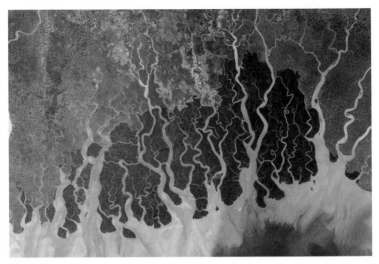

Much of the low-lying Sundarbans region of southwest Bangladesh and southeast India—one of the world's largest preserves of tidal mangrove forest—could be inundated this century by sea level rise, which will also threaten millions of nearby delta residents. (NASA image created by Jesse Allen, Earth Observatory, using data obtained from the University of Maryland's Global Land Cover Facility.)

Will agriculture suffer?

The effects on agriculture will vary based on where, when, and how farming and ranching is done. Global agricultural productivity is predicted to go *up* in some areas, at least in the near term, thanks both to the fertilizing effect of the extra CO_2 in the atmosphere and to now-barren regions becoming warm enough to bear crops (as well as the gradual improvements in crop technology that have boosted yields for decades). However, the rich world looks set to reap most of the benefits, as crop yields in much of the tropics, home to hundreds of millions of subsistence farmers, are likely to experience negative impacts. Unwanted plants could thrive in some areas. Periodic shortages of some staple foods, such as corn, may become more likely as ever-larger swaths of cropland are used to grow biofuels (see chapter 9) and as heat and drought become more intense. Once global temperatures rise beyond a certain point—perhaps as soon as the 2030s, according to the latest IPCC Working Group II assessment—the combined regional effects of climate change are expected to become a net drag on global crop production, one that only worsens over time.

How will the changes impact wildlife?

Because climate is expected to change quite rapidly from an evolutionary point of view, we can expect major shocks to some ecosystems—especially in the Arctic—and possibly a wholesale loss of species, especially given the other stresses on biodiversity now being exerted by increasing global population and development. A report in the journal *Ecology Letters* found that, to keep up with twenty-first-century warming, a wide array of species would need to evolve at rates up to 10,000 times more quickly than they did during past evolutionary shifts triggered by climate change.

According to a 2004 study in *Nature* led by Chris Thomas of the University of Leeds, climate change between now and 2050 may commit as many as 37% of all species to eventual extinction—a greater impact than that from global habitat loss due to human land use. Recent studies with more complex models suggest a wide range of possibilities as warming proceeds. The 2014 report from the IPCC's Working Group II concluded that "a large fraction of both terrestrial and freshwater species faces increased extinction risk under projected climate change during and beyond the 21st century."

How will the economy be affected?

There is a divergence of views regarding the impact of climate change—and indeed emissions cuts—on national and international economies in the coming century. The DARA study cited above estimated that climate change and carbon-related pollution are already trimming 1%–2% per year from global GDP. Entities ranging from the World Bank to major reinsurance companies are urging nations to take the economic implications of climate change seriously. The most widely cited research into this question was carried out on behalf of the U.K. government by the economist Nicholas Stern. The Stern Review, released in 2006, concluded that global warming could reduce global GDP by 20% within the current century; it recommended immediately spending 1% of global GDP each year on reducing emissions. Stern later concluded that his report underestimated the speed, scale, and likelihood of some serious climate impacts, and in a 2008 talk he advocated spending 2% of GDP on mitigation. Other economists questioned Stern's methodology, putting forth either higher or lower figures for the costs of avoiding and ignoring climate change. One of the biggest questions now on the table is how to handle the inevitable costs to the world's largest energy companies should billions of dollars of fossil fuel reserves be left in the ground to reduce the risk of major climate trouble.

In short, as with any other large-scale task, tackling climate change will not occur in an economic vacuum. The IPCC's Working Group III noted this in its 2014 report, stating that "climate policy intersects with other societal goals, creating the possibility of co-benefits or adverse side effects. These intersections, if well-managed, can strengthen the basis for undertaking climate action."

Will rising seas really put cities such as New York and London under water?

Not right away, but it may be only a matter of time. Depending on how much greenhouse gas is emitted this century, the IPCC's 2013 assessment concluded that sea levels by 2081–2100 could be anywhere from 0.26 to 0.81 m (10–32 in.) higher than they were in the period 1986–2005. Especially toward the high end, that's enough to cause major complications for millions of people who live and work near coastlines.

Unfortunately, sea level won't stop rising then, which means coastal areas will have to contend with ever-worsening problems. If emissions continue to rise unabated through this century, the Greenland and/or West Antarctic ice sheets could eventually be thrown into an unstoppable melting cycle that would raise sea level by more than 7 m (23 ft) each. This process would take some time to unfold—probably a few centuries, although nobody can pin it down at this point—but should it come to pass, many of the world's most beloved and historic cities would be hard-pressed to survive.

Will the Gulf Stream stop, freezing the United Kingdom and northern Europe?

The Gulf Stream and North Atlantic Current bring warm water (and with it warm air) from the tropical Atlantic to northern Europe. This helps keep the United Kingdom several degrees warmer than it would otherwise be. Although this system is unlikely to collapse entirely, there is a possibility that it could be diminished by climate change. One reason is that increasing rainfall and snowmelt across the Arctic and nearby land areas could send more freshwater into the North Atlantic, possibly pinching off part of the warm current. The Atlantic meridional overturning circulation—an index often used to gauge the northward flow of warm water—might weaken by anywhere from 12% to 54% during this century, on the basis of the highest-emission scenario considered in the 2013 IPCC assessment. That's probably not enough to offset global warming completely for the United Kingdom or northwest Europe, although it could certainly put a dent in it. In any case, the impacts would be much smaller and would take much longer to play out than the scenario dramatized in the film *The Day After Tomorrow* (see p. 345).

Will we reach a "tipping point"?

Some aspects of climate change are expected to unfold in a roughly linear fashion. Computer models, and the changes we've seen to date, suggest that many of the regional and seasonal fingerprints of our future climate will be roughly twice as strong for 2°C of global warming as for 1°C.

On top of these incremental changes, there's a real risk of even bigger trouble because of positive-feedback processes. Although "positive feedback" may sound like a good thing, in the physical world this process tends to amplify change, pushing conditions well beyond what one would otherwise expect. The challenge is to identify the points at which the most dangerous positive feedbacks will kick in. For instance, scientists consider it likely that the Greenland ice sheet will begin melting uncontrollably if global temperatures climb much more than 2°C (3.6°F) beyond preindustrial readings. Because of the implications for coastal areas, as noted above, this is a particularly worrisome threshold.

As each positive feedback has its own triggering mechanism, some would no doubt occur sooner than others. Thus, there is no single temperature agreed upon as a tipping point for Earth as a whole. However, scientists, governments, and activists have worked to identify useful targets that might focus attention and action. One goal adopted by the European Union, as well as many environmental groups, is to limit global temperature rise to 2°C (3.6°F) over preindustrial levels, but that ceiling looks increasingly unrealistic. We're already close to halfway there, and only the most optimistic of the latest IPCC projections keep us below the 2°C threshold by century's end.

Another approach is to set a stabilization level of greenhouse gases—a maximum concentration to be allowed in the atmosphere, such as 500 parts per million (ppm) of CO_2 as compared to 270–280 ppm in preindustrial times and around 400 ppm today. Still another concept is to limit the total amount of carbon added to the air over time. This perspective is especially useful in demonstrating that we likely have far more oil, coal, and gas available than we can afford to burn without hitting the 2°C point. You can learn more about these and other goals in chapter 14.

WHAT CAN WE DO ABOUT IT?

What would it take to stop climate change?

It isn't possible to simply "stop" climate change. The century-plus lifespan of atmospheric CO_2 means that the planet is already committed to a substantial amount of greenhouse warming. Even if we turned off every fuel-burning machine on Earth tomorrow, climate modelers tell us that the world would warm at least another 0.5°C (0.9°F) as the climate adjusts to greenhouse gases we've already emitted.

The best the world can do, therefore, is adapt to whatever warming can't be avoided while trying to limit the potential amount of additional warming. The latter will only be possible if and when changes in technology and lifestyle enable us to pull back far beyond our current emission levels, or unless we find some safe method to remove enormous amounts of carbon from the atmosphere, or both. That's a tall order, but if we're determined to reduce the risk of a wide range of climate impacts, we have no choice but to fulfill it.

Given the global nature of the climate problem, slashing emissions on the scale needed will almost certainly only be possible as part of an ambitious political agreement between the world's nations. Otherwise, there's a big risk: if one region or nation cuts back emissions on its own, that could actually bring down the price of fossil fuels on the open market and thus spur consumption elsewhere. We'll look at this conundrum in more detail in chapters 14 and 15.

Which countries are emitting the most greenhouse gases?

For many years the United States was in first place, with 30% of all of the human-produced greenhouse emissions to date and about 20% of yearly totals, despite having only a 5% share of the global population. However, China has now taken the lead. Its emissions are considerably lower per capita, but because of its growing population and affluence, China leapfrogged the United States as the world's leading greenhouse emitter around 2006. By 2012, China accounted for more than 28% of the planet's annual energy-related carbon dioxide emissions, according to the Netherlands Environmental Assessment Agency, while the U.S. share had dropped to around 15%. Of course, part of this reversal is due to China's hugely expanded role in manufacturing goods for other nations, including the United States.

As shown in chapter 15, the world's industrialized countries vary widely in how much they have increased or decreased their total emissions since 1990. Some of the decreases were due to efficiency gains and growth in alternative fuels, while others were due to struggling economies.

Has the Kyoto Protocol made a difference?

This United Nations–sponsored agreement among nations was the first to mandate country-by-country reductions in greenhouse gas emissions. Kyoto emerged from the UN Framework Convention on Climate Change, which was signed by nearly all nations at the 1992 mega-meeting popularly known as the Earth Summit. The framework pledges to stabilize greenhouse gas concentrations "at a level that would prevent dangerous anthropogenic interference with the climate system." To put teeth into that pledge, a new treaty was needed, one with binding targets for greenhouse gas reductions. That treaty was finalized in Kyoto, Japan, in 1997, after years of negotiations, and it went into force in 2005. Nearly all nations have now ratified the treaty, with the notable exceptions of the United States (which never ratified it) and Canada (which pulled out in 2011). Developing countries, including China and India, weren't mandated to reduce emissions, given that they'd contributed a relatively small share of the century-plus buildup of CO_2 (although that's no longer the case, as we will see).

Under Kyoto, industrialized nations pledged to cut their yearly emissions of carbon from 1990 levels, as measured in six greenhouse gases, by varying amounts that averaged 5.2% during the "first commitment period" (2008–2012). That equates to a 29% cut in the values that would have otherwise occurred. However, the protocol didn't become international law until global emission amounts had already climbed substantially from their 1990 values. Some countries and regions met or exceeded their Kyoto goals for the first commitment period—the nations that were part of the European Union when Kyoto was ratified trimmed their collective emissions by more than 12%—but other large nations fell woefully short. And the two biggest emitters of all, the United States and China, churned out more than enough extra greenhouse gas to erase the reductions made by all other countries during the Kyoto period. Worldwide, carbon dioxide emissions soared by more than 50% from 1992 to 2012, according to the Netherlands Environmental Assessment Agency.

At a 2012 meeting in Doha, Qatar, nations pledged to finalize a new global deal by 2015. Meanwhile, the Kyoto Protocol was formally extended into a second commitment period running from 2013 to 2020. This time,

What's a climate skeptic?

Any scientist worth her or his salt is a skeptic by nature. Researchers are constantly checking facts, testing hypotheses, and rooting out assumptions. That's why many top climate experts bristle when the term "climate skeptic" is applied to the small but vocal contingent—most of them from outside climate science—who critique the foundations of climate change understanding. Most of the mainstream researchers on global warming consider themselves skeptics in the truest sense, accepting nothing at face value. In their eyes, journalists and others have misappropriated the term "skeptics" to lend implicit credence to outlier views. In fact, the comprehensive website **SkepticalScience.com** uses the tagline "Getting skeptical about global warming skepticism." The name game works both ways, though. Because the root causes of climate change are so well understood, the label "climate change denier" is often applied to those who disagree with consensus viewpoints on the subject. Yet these "deniers" can range from out-and-out conspiracy theorists to working scientists who accept the basic premise of anthropogenic climate change but disagree on the nature or importance of possible outcomes. Throughout this book, and especially in Part 4 (Debates and solutions), we most often use the term "skeptic" to identify non-mainstream voices (keeping the above caveats in mind), but we also employ the term "contrarian" at times. In truth, there's no single term that fully describes this motley group, although an apt description might be "those unconvinced by the evidence."

the global target is more stringent: a cut of at least 18% relative to 1990 values. However, this second phase of Kyoto isn't legally binding, and the countries initially participating represent only about 15% of worldwide emissions.

What will the next global deal look like?

Even if every nation had lived up to its initial Kyoto goals, it would have made only make a tiny dent in the world's ever-increasing output of greenhouse gases. Reducing greenhouse gas emissions by a few percent over time is akin to overspending your household budget by a decreasing amount each year: your debt still piles up, if only at a slower pace. Hence, most people who are concerned about climate change are pinning their hopes on world leaders agreeing to much more ambitious emissions cuts.

At one point, hopes were high among activists for a strong post-Kyoto deal, especially following the election of U.S. president Barack Obama in 2009. However, the momentum faded during the run-up to a massive UN meeting that December in Copenhagen, where the outlines of a new global agreement were to be set. That meeting descended into near chaos, with countries far apart on a variety of issues and global economic turmoil overshadowing it all. The final outcome—mainly a set of voluntary national goals—struck many observers and participants as a huge disappointment. Although the next several annual meetings saw some modest progress, the negotiations remained years away from solidifying the next major global agreement on emission cuts, which organizers hope will go into effect in 2020 and include all UN nations.

Does the growth of China and India make a solution impossible?

Not necessarily. Although its growth in coal production is hugely worrisome, China is already making progress on vehicle fuel efficiency and other key standards. And because so much of the development in China and India is yet to come, there's a window of opportunity for those nations to adopt the most efficient technologies possible. At the same time, the sheer numbers in population and economic growth for these two countries are daunting indeed—all the more reason for prompt international collaboration on technology sharing and post-Kyoto diplomacy.

If oil runs out, does that solve the problem?

Hardly. Even if oil and/or gas resources were to "peak" in the next few years, which now seems unlikely, the question becomes what fuel sources the world would turn to: coal, nuclear, renewables, or some combination of the three. If the big winner were coal—or some other, less-proven fossil source such as shale or methane hydrates—it raises the potential for global warming far beyond anything in current projections.

Even if renewables win the day as the century unfolds, we're still left with the emissions from today's stocks of oil, gas, and coal, many of which would likely get burned between now and that eco-friendly transition. With this in mind, research has intensified on **sequestration**—how carbon might be safely stored underground. The idea appears promising, but big questions remain. There's also geoengineering, a set of grand schemes to pull carbon directly from the atmosphere or block sunlight through mirrors, soot, or artificially boosted clouds. All of these geoengineering ideas have huge unknowns, but scientists and governments are taking a closer look at them, if only as last-ditch alternatives to unstoppable warming (see chapter 16).

Won't nature take care of global warming in the long run?

Only in the *very* long run. Assuming that it takes a century or more for humanity to burn through whatever fossil fuels it's destined to emit, it would take hundreds of years before most of the CO_2 emissions were absorbed by Earth's oceans and ecosystems, and some of the remaining human-produced CO_2 would remain in the air for many thousands of years.

There are few analogies in the geological past for such a drastic change in global climate over such a short period, so it's impossible to know what will happen after the human-induced greenhouse effect wanes. All else being equal, cyclical changes in Earth's orbit around the sun can be expected to trigger an ice age sometime within the next 50,000 years, and other warmings and coolings are sure to follow, as discussed in chapter 11. In the meantime, we will have our hands full dealing with the next century and the serious climate changes that our way of life may help bring about.

CHAPTER TWO

The Greenhouse Effect

HOW GLOBAL WARMING WORKS

Imagine our planet suddenly stripped of its atmosphere—a barren hunk of rock floating in space. If this were the case, then Earth's near-ground temperature would soar by day but plummet by night. The average would be something close to a bone-chilling −18°C (0°F). In reality, though, Earth's surface temperature now averages a much more pleasant 14.5°C (58.1°F). Clearly, there's something in the air that keeps things tolerably warm for humans and other living things. But what?

One of the first people to contemplate Earth's energy balance was the French mathematician and physicist Joseph Fourier. His calculations in the 1820s were the first to show the stark temperature contrast between an airless Earth and the one we actually enjoy. Fourier knew that the energy reaching Earth as sunlight must be balanced by energy returning to space, some of it in a different form. And though he couldn't pin down the exact process, Fourier suspected that some of this outgoing energy is continually intercepted by the atmosphere, keeping us warmer than we would otherwise be.

Harking back to experiments by others on how a glass box traps heat, Fourier likened the atmosphere to a hothouse (or greenhouse), and voilà—

the concept of the **greenhouse effect** was born. It's been with us ever since, even though it's a flawed analogy. The atmosphere doesn't imprison the air the way a glass box does. Instead, it absorbs **infrared radiation** rising from Earth's sun-warmed surface. All else being equal, the more greenhouse gas there is, the less radiation can escape from Earth to space, and the warmer we get (but there are a few twists along the way, as we'll see).

The diagram opposite shows what happens, on average, to the sunlight that reaches our planet:

* **About 30%** gets reflected or scattered back to space by clouds, dust, or Earth's surface. (Light-colored features, such as snow and ice, reflect far more than they absorb, while dark features such as forests absorb more than they reflect.)
* **More than 20%** is absorbed in the atmosphere, mainly by clouds and water vapor.
* **Almost 50%** gets absorbed by Earth's surface—land, forests, pavement, oceans, and the rest.

The incoming radiation from the intensely hot sun is mostly in the visible part of the spectrum, which is why you shouldn't stare at the sun. Being much cooler than the sun, Earth emits far less energy, most of it at infrared wavelengths we can't see.

Some of Earth's outgoing radiation escapes through the atmosphere directly to space. Most of it, though, is absorbed en route by clouds and greenhouse gases (including water vapor), which in turn radiate some back to the surface and some out to space. Thus, Earth's energy budget is maintained in a happy balance between incoming radiation from the sun and a blend of outgoing radiation from a warm surface and a cooler atmosphere (an important temperature distinction, as we'll see below).

> " Remove for a single summernight the aqueous vapor from the air which overspreads this country, and you would assuredly destroy every plant capable of being destroyed by a freezing temperature." —*John Tyndall, 1863*

The air's two main components, nitrogen (78%) and oxygen (20%), are both ill-suited for absorbing radiation from Earth, in part because of their linear, two-atom (diatomic) structure. But some other gases have three or more atoms, and these branched molecules capture energy

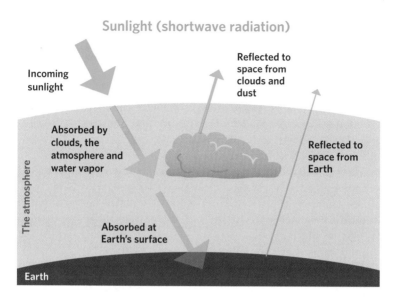

Sunlight (shortwave radiation)

Incoming sunlight

Reflected to space from clouds and dust

Absorbed by clouds, the atmosphere and water vapor

Reflected to space from Earth

The atmosphere

Absorbed at Earth's surface

Earth

Incoming and outgoing radiation: the diagram above shows sunlight entering the atmosphere. The light absorbed (as opposed to reflected) is given out as infrared radiation, as shown in the diagram below. The widths of the arrows in both diagrams reflect the warming power of each process, relative to the total sunlight coming in.

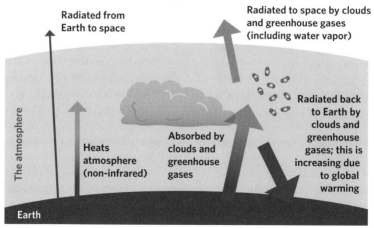

Infrared (longwave radiation)

Radiated from Earth to space

Radiated to space by clouds and greenhouse gases (including water vapor)

Heats atmosphere (non-infrared)

Absorbed by clouds and greenhouse gases

Radiated back to Earth by clouds and greenhouse gases; this is increasing due to global warming

The atmosphere

Earth

far out of proportion to their scant presence. These are the **greenhouse gases**, the ones that keep Earth inhabitable but appear to be making it hotter.

Most greenhouse gases are well mixed throughout the **troposphere**, the lowest 8–16 km (5–10 mi) of the atmosphere. (Water vapor is the big exception; it's much more concentrated near ground level.) As mountain climbers well know, the troposphere gets colder as you go up, and so these greenhouse gases are cooler than Earth's surface is. Thus, they radiate less energy to space than Earth itself would. That keeps more heat in the atmosphere and thus helps keep our planet livable.

That's all well and good, but the more greenhouse gas we add, the more our planet warms. As carbon dioxide and other greenhouse gases accumulate, the average elevation from which they radiate energy to space ascends to even higher, and colder, heights. This means less energy leaving the system, which in turn causes the atmosphere to heat up further. The diagram below shows how this works in more detail.

Once the added gases get the ball rolling through this process, a series of atmospheric readjustments follows. These are largely **positive feedbacks** that amplify the warming. More water evaporates from oceans and lakes, for instance, which roughly doubles the impact of a carbon dioxide increase. Melting sea ice reduces the amount of sunlight reflected to space. Some feedbacks are less certain: we don't know whether cloud patterns will change to enhance or diminish the overall warming. And, of course, we're not talking about a one-time shock to the system. The whole planet is constantly readjusting to the greenhouse gases we're adding.

A rogue's gallery of gases

In the 1860s, eminent Irish scientist **John Tyndall** became the first to explore and document the remarkable power of greenhouse gases. Tyndall put a number of gases to the test in his lab, throwing different wavelengths of light at each one to see what it absorbed. Almost as an afterthought, he tried coal gas and found it was a virtual sponge for infrared energy. Tyndall went on to explore carbon dioxide and water vapor, both of which are highly absorbent in certain parts of the infrared spectrum. The broader the absorption profile of a gas—that is, the more wavelengths it can absorb—the more powerful it is from a greenhouse perspective.

Here are the main greenhouse gases of concern. The diagram shows the prevalence of each gas and how much of an impact it's had in enhancing the overall greenhouse effect.

* **Carbon dioxide (CO₂)**, the chief offender, accounted for roughly 400 of every one million molecules in the air, or 400 **parts per million (ppm)**, by the year 2014, as measured on top of Hawaii's Mauna Loa. That number varies by about 8 ppm with the seasonal cycle: it first touched the 400-ppm threshold midway through 2013, then sank below it later in the year. The yearly average has been climbing by 1–3 ppm, or about 0.25%–0.75%, per year (see p. 39). The worldwide **emissions** of CO₂ are increasing at a faster rate—several percent per year, on average—but their impact on airborne CO₂ in percentage terms goes down when you consider the large amount of CO₂ already in the air.

Both a pollutant and a natural part of the atmosphere, carbon dioxide is produced when fossil fuels are burned, as well as when people and animals breathe and when plants decompose. Plants and the ocean soak up huge amounts of carbon dioxide, which is helping to keep CO₂ levels from increasing even more rapidly (see p. 44). Because the give-and-take among these processes is small compared to the atmospheric reservoir of CO₂, a typical molecule of carbon dioxide stays airborne for more than a century.

The prevalence of human-produced greenhouse gases and the best estimate of their relative importance in affecting Earth's radiative balance in the year 2011, according to the 2013 IPCC report. The percentages of radiative forcing depend on each gas's prevalence and its molecular ability to trap energy. The numbers total less than 100% because of rounding and other trace gases. The percentages do not account for the longevity of each gas; when considered over its longer lifetime, carbon dioxide's total radiative impact per molecule is larger than methane's, for example. Not shown are tropospheric ozone (produced indirectly) and water vapor. Radiative effects from airborne particles (aerosols) and clouds are also omitted.

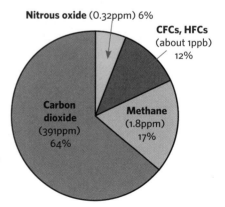

Atmospheric levels of CO_2 held fairly steady for centuries—at around 270–280 ppm—until the Industrial Revolution took off. In past geological eras, the amount of CO_2 has risen and fallen in sync with major climate changes. Although factors other than carbon dioxide can kick off some of these transitions, CO_2 is a critical factor in making them more powerful (see chapter 11).

❋ **Methane** emerges from a wide array of sources—including rice paddies, peat bogs, belching cows, and insect guts, as well as vehicles, homes, factories, wastewater, and landfills—making it far more challenging to analyze globally than CO_2. Methane is a greenhouse powerhouse: though it stays in the air for less than a decade on average, its effect on global warming is roughly 25 times that of carbon dioxide per molecule when both are considered over a century-long span. Methane's total an-

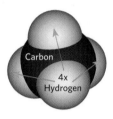

nual contribution to the greenhouse effect is thought to be roughly a third as big as carbon dioxide's, even though it makes up only around 2 ppm of the atmosphere.

After soaring in the last few decades, the amount of atmospheric methane increased more slowly and fitfully in the 1990s and through much of the following decade, implying that emissions themselves decreased. Some of the slowdown may have been related to more frequent dry conditions in the tropics, along with sagging economies across eastern Europe and Russia in the 1990s. Another possible factor is the increased capture and use of natural gas previously allowed to escape from oil fields. Starting in 2007, methane concentrations began to rise once more, apparently due to a combination of heavy tropical rains, Arctic warmth, and biomass burning. Expanded natural gas operations drop in North America related to hydraulic fracturing (fracking) may also be playing a role, although research is still in its early days. Studies thus far have produced wildly varying results on the percentage of methane lost to the atmosphere, largely through pipeline leaks. Estimates vary from as low as 0.5% to as high as 9%, with the actual leakage rates likely depending on local regulation and other factors.

❋ **Ozone**, like a versatile actor, takes on multiple roles, but in the greenhouse drama, it's only a supporting player. Instead of being emitted when fossil fuels are burned, ozone forms when sunlight hits other pollutants and triggers ozone-making reactions. Its presence can spike to unhealthy levels when stagnant air takes hold for days near ground level, where people, animals, and plants live and breathe. Fortunately, ozone survives only a few days in the troposphere. This makes it hard to assess its global concentration. The limited data that exist indicate an average of around 34 parts per billion (ppb), but the values range from less than 10 ppb in clean ground-level air to more than 100 ppb in polluted regions. There's apparently been little overall change in tropospheric ozone amounts since the 1980s, perhaps because of cleaner-burning cars and factories, but models hint at a global increase of about 30% since the Industrial Revolution.

Higher up, a protective natural layer of ozone warms the stratosphere while helping to shield Earth from ultraviolet light. The human-induced hole that appears

Venus: A cautionary tale

Scientists don't have the luxury of an extra Earth for experimentation. However, our nearest planetary neighbor serves as an example of what can happen when the greenhouse effect runs amok. Many scientists once saw Venus as a mysterious sibling, one whose climate might be mild enough to support human or humanlike life. Although the planet is named for the Roman goddess of love and beauty, it's actually a rather harsh place. That's because of carbon dioxide, which makes up less than 0.04% of Earth's atmosphere, but more than 96% of Venusian air. Factor in the planet's location—about 25% closer to the sun than Earth is—and you end up with a surface air temperature in the neighborhood of 460°C (860°F), hot enough to melt lead. (All that CO_2 is invisible, by the way; it's a dash of sulfuric acid that produces Venus's legendary cloak of haze.) By comparison, Mercury, with a far thinner atmosphere, is much colder than Venus on average, even though it's twice as close to the sun.

Carl Sagan, the late U.S. astronomer and science popularizer, built his early career on Venus, as it were, studying its sizzling atmosphere. In the early 1960s, Sagan drew on radio observations and simple mathematical modeling to explain how the planet's dense, superheated atmosphere could be produced by what he called a "runaway greenhouse effect." Sagan's findings led to his concern about the fate of Earth's own atmosphere: in 1984, he was the first witness in then-senator Al Gore's landmark congressional hearings on global warming.

each year in this layer of "good" ozone has led to record-low temperatures in the lower stratosphere.

* **Water vapor** isn't a very strong greenhouse gas, but it makes up for that weakness in sheer abundance. If you're on a warm tropical island with 100% relative humidity, that means the balmy breezes are wafting as much water vapor as they can—perhaps as much as 5 out of every 100 molecules in the air as a whole. When averaged globally, the concentration of water vapor is much less, and it varies greatly by location, altitude, and time of year, with colder air able to carry less of it. You can't see water vapor itself, only the haze, clouds, rain, and snow it produces.

Water vapor works to propagate itself and boost global warming through an interesting twist. As global temperatures rise, oceans and lakes release more water vapor, obeying a well-known law of thermodynamics. In turn, the added water vapor—up to 7% more for every degree Celsius of warming—adds to the warming cycle. This is one of several positive feedbacks critical to the unfolding of future climate change. A global increase in water vapor of roughly 0.1% per decade over land areas appears to have stalled over the last few years, in line with global temperatures leveling off, but both indices are expected to eventually resume their climb.

* **A few other gases**, which are extremely scant but extremely powerful, make up the rest of the greenhouse players. **Chlorofluorocarbons** (CFCs) and related compounds increased rapidly until they were identified as a key player in stratospheric ozone depletion. Under the Montreal Protocol, they've begun to level off. Along with helping to destroy "good" ozone, they are also powerful, long-lived greenhouse gases—another good reason we're phasing them out, though some replacements are just as risky for climate. **Nitrous oxide** is also an industrial and agricultural by-product, commonly referred to as laughing gas. It shows up at only about 320 ppb, but with about 300 times the molecule-for-molecule effect of CO_2 over its century-long lifespan in the atmosphere. Though its immediate effect on radiative balance is less than that of CFCs (as shown in the chart on p. 31), its long lifetime makes it far more important than CFCs when the effects are measured over a 100-year time frame, and its atmospheric concentration is increasing by almost 1% per year.

Greenhouse pioneers

Virtually no one at the peak of the Victorian age reckoned that burning coal or oil would tamper with our climate, but it was becoming clear that great changes in climate had occurred before. Chief among those were the ice ages, which coated North America and Eurasia with kilometer-thick sheets of ice as far south as modern-day Germany and Illinois. Fossils proved that this ice cover had persisted well into the era of early human civilization, some 12,000 years ago. What made the climate plummet into such frigidity and then recover? And could it happen again?

Svante Arrhenius

In the mid-1890s, the Swedish chemist **Svante Arrhenius** took a look at how carbon dioxide might be implicated in ice ages. His unexpected results provided our first real glimpse of a greenhouse future. Scientists already knew that volcanoes could spew vast amounts of greenhouse gas, so Arrhenius wondered if a long period of volcanic quiet might allow carbon dioxide levels to draw down and perhaps help plunge Earth into an ice age. He set out to calculate how much cooling might result from a halving of CO_2. When he included the role of water vapor feedback, Arrhenius came up with a global temperature drop of around 5°C (9°F).

Soon enough, a colleague inspired him to turn the question around: what if industrial emissions grew enough to someday double the amount of carbon dioxide in the air? Remote as that possibility seemed, Arrhenius crunched the numbers again and came up with a similar amount, this time with a plus sign: a warming of about 5°C (9°F).

> By the influence of the increasing percentage of carbonic acid [CO_2] in the atmosphere, we may hope to enjoy ages with more equable and better climates, especially as regards the colder regions of the earth."
>
> —*Svante Arrhenius, 1908*

Is the ozone hole linked to global warming?

The saga of ozone depletion in the stratosphere is conflated with global warming in the minds of many. Both topics came into public view during the 1980s, often lumped together under the heading of **global change**. There are several links between the two—just enough to cause confusion—but at heart they're two distinct issues.

Ozone, a greenhouse gas, is a pollutant at ground level, harmful when we breathe it. However, the ozone layer that sits within the **lower stratosphere** (especially at about 25-40 km/15-25 mi high) is a godsend. Even though the ozone makes up only a tiny fraction of the stratospheric air, it intercepts much of the **ultraviolet light** that can produce sunburns and skin cancer, damage our eyes, and cause other kinds of trouble for people and ecosystems.

A dramatic seasonal depletion in this layer of ozone was found over Antarctica in 1985. Shortly thereafter, scientists identified the three factors that conspire to form the ozone hole, which waxes and wanes during each Southern Hemisphere spring. The first ingredient is a special type of cloud, a **polar stratospheric cloud**, that only forms when winter temperatures fall below about −80°C/−112°F at high altitudes and latitudes. Also needed are CFCs, used since the 1920s in spray cans, air conditioners, and many other places. CFCs are heavier than clean air, but they mix easily through the atmosphere; once lofted into the stratosphere, they can remain there long enough to do damage. (CFCs are also greenhouse gases themselves; they account for roughly 10% of the human-enhanced greenhouse effect.)

The final protagonist is **sunlight**. As the six-month Antarctic night comes to an end each September, round-the-clock sunshine helps break down the CFCs. This releases **chlorine**, and the chlorine uses the surface of the polar stratospheric cloud to break down ozone into **oxygen**. A single molecule of chlorine can destroy many ozone molecules over a few weeks. During that time, about half of all the ozone through the depth of the Antarctic atmosphere typically vanishes, with near-complete ozone loss in parts of the lower stratosphere. By December of each year, though, the stratosphere has warmed up, the clouds disappear, and the ozone hole fills in.

Fortunately, the ozone hole has never extended much beyond Antarctica, although it has encroached on southern Chile. Southern Australia and New Zealand, while outside the hole per se, have seen ozone reductions of more than 10% at times. As far as the Arctic goes, its wintertime vortex is less stable than its Antarctic counterpart, which limits the growth of polar stratospheric clouds and helps keep a bona fide ozone hole from forming. Still, the springtime depletion over the Arctic can be as high as 60% in some years. A weaker but broader and more persistent **ozone depletion** exists, running about 3%–4% below pre-1970s values in northern midlatitudes and closer to 5%–6% in southern midlatitudes. Some of this worldwide depletion is likely due to the yearly dispersal of the Antarctic ozone hole and the mixing of that ozone-depleted air around the globe.

Because ozone absorbs sunlight and is also a greenhouse gas, its depletion can have a **cooling** effect. This helps explain why, even as Earth's surface air temperatures reach record highs, record lows are being notched up in the stratosphere (see p. 246). The resulting changes in air circulation over Antarctica sometimes extend to the surface, where the frozen interior has warmed far less than coastal areas (see p. 120).

Unlike global warming, there's an end to ozone depletion in sight, at least on paper. The 1987 **Montreal Protocol**, orchestrated by the United Nations and ratified with amazing speed, called for CFCs to be replaced by substitutes such as **halochlorofluorocarbons**, which have shorter lifetimes and are far less likely to break down in a way that damages ozone. Chlorine concentrations have stabilized in the stratosphere and may already be going down. According to the IPCC, the ozone layer should be rebuilding over the next several decades, although we can expect a few ups and downs along the road to recovery. Indeed, while stratospheric chlorine began decreasing in the late 1990s, and the annual ozone loss over Antarctica appears to have lessened slightly since then, the most extensive ozone hole on record occurred in 2006.

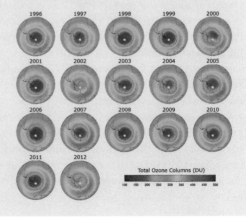

Satellite-derived measurements of total ozone above the South Pole in Dobson units, averaged for the period September through November. Deeper blues indicate a stronger ozone hole. The 2012 hole was the weakest in a decade. (European Space Agency)

Amazingly, this century-old calculation isn't too far off the mark. Scientists now use a **doubling of CO_2** as a reference point for comparing computer models and a benchmark for where our atmosphere might stand by around 2050 or a little afterward, if present trends continue. Most models have pegged the **equilibrium warming** from a doubling of CO_2 (the point, many years after the doubling, when temperatures have stabilized) at somewhere between 1.5° and 4.5°C (2.7°–8.1°F). The 2013 IPCC report reinforced this range as the most likely. All in all, Arrhenius's initial forecast was quite impressive for the pre-computer era.

In his own time, though, Arrhenius was a lone voice. He himself far underestimated the pace of global development, figuring it would take until the year 4000 or so for his projected doubling to occur. Even then, he figured, a little global warming might not be such a bad thing. Europe was just emerging from the hardships of the Little Ice Age, which had much of the continent shivering and crops withering from about 1300 to 1850. Arrhenius figured a little extra CO_2 might help prevent another such cold spell.

Another greenhouse pioneer came along in the 1930s, by which time the globe was noticeably heating up. **Guy Stewart Callendar**, a British engineer who dabbled in climate science, was the first to point to human-produced greenhouse gases as a possible cause of already observed warming. Callendar estimated that a doubling of CO_2 could produce roughly 2°C (3.6°F) of warming. Though he continued to study the problem into the 1960s, Callendar's work gained little notice at the time, perhaps in part because global temperatures were cooling slightly at that point.

In any case, many scientists discounted the message of Arrhenius and Callendar because of the fact that CO_2 and water vapor absorbed energy at overlapping wavelengths. Laboratory tests seemed to show that the two components of the atmosphere were already doing all the absorbing of infrared energy that they could: enlarging the atmospheric sponge, as it were, could have only a minuscule effect. Only after World War II did it become clear that the old lab tests were grievously flawed because they were carried out at sea level. In fact, carbon dioxide behaved differently in the cold, thin air miles above Earth, where it could absorb much more infrared radiation than previously thought. And CO_2's long lifetime meant that it could easily reach these altitudes. Thus, the earthbound absorption tests proved fatefully wrong, one of many dead ends that kept us from seeing the power of the greenhouse effect until industrialization was running at full tilt.

The tale told by a curve

If there's one set of data that bears out the inklings of Arrhenius and Callendar, it's the record of CO_2 collected on top of Hawaii's Mauna Loa Observatory since 1958 (see graph below). **Charles Keeling** convinced the Scripps Institution of Oceanography to fund the observing site as part of the International Geophysical Year. Because of CO_2's stability and longevity, Keeling knew that the gas should be well mixed throughout Earth's atmosphere, and thus data taken from the pristine air in the center of the Pacific could serve as an index of CO_2 valid for the entire globe. After only a few years, Keeling's saw-toothed curve began to speak for itself (see graph). It showed a steady long-term rise in CO_2, along with a sharp rise and fall produced each year by the wintering and greening of the Northern Hemisphere (whose land area and plant mass far outstrip those of the Southern Hemisphere).

With Keeling's curve as a backdrop, a growing number of scientists began to wonder whether human-induced warming might take shape much sooner than Arrhenius predicted, as population grew and industrialization proceeded. In the 1960s, climate scientists devised their first primitive renditions of Earth's atmosphere in **computer models**. Like the first crude

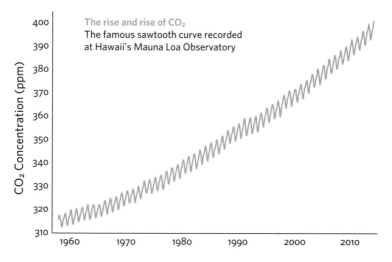

Carbon dioxide measurements taken on top of Hawaii's Mauna Loa since 1958 have produced this famous saw-toothed curve. The annual ups and downs reflect the cycle of plant growth and decay concentrated in the Northern Hemisphere, where land areas are more extensive. (Scripps Institution of Oceanography)

motion pictures, even these early models had a compelling quality. One model created by Syukuro Manabe and Richard Wetherald at the U.S. Geophysical Fluid Dynamics Laboratory confirmed in 1967 that a doubling of CO_2 could produce warming of 2.0°C (3.6°F).

The increasingly complex models developed since the 1960s have refined the picture, but they haven't changed it all that much (see chapter 12). The most widely accepted estimate of warming for a doubling of CO_2 is in the ballpark of 3.0°C (5.4°F). The Mauna Loa data continue to show carbon dioxide concentrations rising inexorably, and despite ratification of the Kyoto Protocol, the world remains perilously far from even beginning to stabilize CO_2 levels in the atmosphere. Although Arrhenius foresaw a century ago that humans could accentuate the natural greenhouse effect, he would doubtless be amazed at how quickly the process is unfolding.

CHAPTER THREE

Who's Responsible?
WHICH COUNTRIES, REGIONS, AND ACTIVITIES ARE WARMING THE WORLD?

It took an awful lot of coal, gas, and oil passing through engines and stoves to boost the atmosphere's level of carbon dioxide by close to 40% over the last century and a half. There's been able assistance from other heat-trapping greenhouse gases, such as methane, as well as from deforestation and other changes in land use. Still, most of the human-produced global warming to date appears to be due to CO_2.

Where did it all come from, and what are the main sources today? Asking these questions isn't merely an exercise in finger pointing. In order to get a handle on where global emissions are headed, it's critical to know what parts of the world and what economic sectors are contributing the most to the current situation.

When it comes to the greenhouse effect, one nation's emissions are everyone's problem. Ordinary pollution disperses in a few days and tends to focus its worst effects in the region where it was generated, plus some areas downwind. Carbon dioxide is much longer lived: once added to the air, it stays there for more than a century. That gives the natural mixing processes of the atmosphere time to shake and stir CO_2 evenly around the globe. Some of the countries producing minimal amounts of greenhouse gases, such as tiny island nations, are among the most vulnerable to the climatic fallout from emissions produced largely by a few big countries.

Scientists can't yet draw a straight line from emissions to a specific bout of extreme weather, but in some cases they can now characterize the extent to which greenhouse warming hiked the odds of a particular event, such as the 2003 European heat wave (see p. 69). It's not too much of a stretch to imagine courts in the future using such findings, along with the tally of who emits what and where, in order to assign partial responsibility for major climate shocks.

Most of the data below are drawn from analyses by the **Intergovernmental Panel on Climate Change (IPCC)**, the **U.S. Carbon Dioxide Information Analysis Center**, and the **Netherlands Environmental Assessment Agency**.

How much greenhouse gas is in the air right now?

The global atmosphere currently carries about **3100 metric gigatons** of carbon dioxide, which includes about 850 metric gigatons of carbon. That's by far the largest presence of any human-produced greenhouse gas. The atmosphere also holds about four metric gigatons of carbon in the form of methane, which is a much stronger but shorter-lived greenhouse gas. (A metric gigaton is a billion metric tons, or about 1.1 billion short tons, the measure used in the United States.)

How much are we adding each year?

As of 2012, according to the Netherlands Environmental Assessment Agency, humans were directly putting more than 34 metric gigatons of carbon dioxide into the atmosphere per year—a stout increase from around 15 metric gigatons in 1970 and about 26 metric gigatons as recently as 2002, according to a 2013 report from the Netherlands Environmental Assessment Agency. This is mainly from burning fossil fuels, though about 4% of the total comes from cement production. Almost half of this total yearly output remains in the atmosphere, while the rest is absorbed (more on this later). Deforestation adds still more carbon dioxide to the air—anywhere from 6% to 20% beyond the totals above, depending on the acreage and density of the forests felled in a given year.

The global **emission rate** of CO_2 doesn't go up every year. During economic slowdowns, the rate can actually decline slightly, as it did in 1992 and again in 2009. Even in those years, though, we're still adding lots of carbon dioxide to the atmosphere. Indeed, the **total amount** of CO_2 in the air—as seen in the graph on p. 39—has risen every year since measurements began in the late 1950s.

To put these figures in perspective, consider **per-person emissions**. On the basis of the figures above, each human on Earth is responsible for putting an average of 4.8 metric tons of carbon dioxide into the air each year. That's more than 4800 kg, or about 10,700 lb. In other words, someone who weighs 68 kg (150 lb) puts the equivalent of her weight in carbon dioxide into the atmosphere in less than a week. Of course, the real figures vary enormously on the basis of where and how you live. For most people in Europe and North America, the average is much higher, as shown by the chart later in this chapter. For more on the "carbon footprints" of individuals, see chapter 17.

What happens to the greenhouse gas we put into the air?

It's easy to assess how much CO_2 is burned each year, at least in principle. And because CO_2 is mixed throughout Earth's atmosphere over its long lifetime, it's also fairly straightforward to measure the global amount of CO_2 in the air by taking measurements at a few well-chosen observation points. From these two factors, we know that in a typical year, about 45% of the carbon that enters the atmosphere each year due to human activity stays there, adding to airborne CO_2 for a century or more. Individual molecules cycle through the system more quickly than that, however. Imagine water in a park fountain that's continually going down the drain and returning through the fountain pump. If you increase the inflowing water without changing the outgoing drainage, the water level in the fountain's pool will slowly rise, even though large amounts of water are coming and going all the time.

The other 55% of human-produced emissions is either absorbed by the ocean or taken up by land-based ecosystems—trees, crops, soils, and the like. Plants take up carbon dioxide when they photosynthesize and return it to the soil and the atmosphere when they die and decompose. Plants and soil-based microbes also add some CO_2 to the atmosphere through respiration, just as people do when they breathe. Of the carbon that a typical tree takes up, about half is quickly returned to the air by respiration; the rest goes into wood or leaves that store the carbon until they decompose.

This simplified picture gets a bit more complex when you consider that Earth's blanket of vegetation goes through changes from year to year and decade to decade. El Niño and other atmospheric cycles can spawn drought

Comparing greenhouse gases

Because greenhouse gases vary so widely in the power of their climatic effects, researchers often rely on a unit called **global warming potential**. The global warming potential of a gas is a measure of its contribution per unit mass to greenhouse warming in the atmosphere over a given time span as compared to that of carbon dioxide. Methane, for example, is shorter-lived than CO_2 but much more powerful in its ability to trap heat in the atmosphere. Thus, over a century's time, methane's global warming potential is estimated to be somewhere around 21 (a number that's not yet set in stone), compared to a value of 1 for carbon dioxide. These figures can be multiplied by the prevalence of each gas to produce a **carbon equivalent** that enables all emissions to be considered as a group. By using carbon equivalents, researchers get a better sense of the impact of the atmosphere's total greenhouse gas burden. Some studies use **carbon dioxide equivalent**, a number obtained by multiplying the carbon equivalent by 44/12 (the ratio of the molecular weight of carbon dioxide to that of carbon).

over huge areas. The warmer, drier conditions make for struggling plants that respire more CO_2 and absorb less of it. These conditions also allow for more deforestation and large-scale fires that add CO_2 to the atmosphere. Recent work hints that the absorption of CO_2 by plants and oceans could decrease in the long term as the atmospheric load increases. It's very hard to measure and separate out all these effects; for one thing, what we add is folded into a much larger exchange of carbon that takes place naturally. Over the last few years, the balance has played out as follows:

* **Land-based ecosystems** took up roughly 30% of the CO_2 emitted by fossil fuel use during the period 2003–2012. However, there are large variations from year to year due to climate cycles and changes in land use, and scientists aren't yet sure which regions account for most of the absorption (see below). Also, this doesn't factor in the CO_2 returned to the air by deforestation and other changes in land use, which was equal to about 8% of fossil fuel emissions.
* **Oceans** absorb between 25% and 30% of fossil fuel CO_2 emissions.
* **The atmosphere** retains the rest of the added CO_2. On average, that's about 45%, though in a given year the percentage can range anywhere from 30% to 80% depending on how much CO_2 is taken up on land.

Will Earth's land-based carbon sink get stopped up?

One of the critical issues in climate change science is why and how Earth's land areas are able to soak up anywhere from 20% to 40% of our carbon emissions from year to year, even as those emissions keep growing. One possibility is that the CO_2 we're adding to the atmosphere may be stimulating enough plants to grow more vigorously. After all, CO_2 is a fertilizer. However, this fertilization effect doesn't appear to fully explain the increasing strength of the carbon-based land sink, especially since there are other constraints on plant growth, such as the supply of nitrogen and other nutrients. Another part of the mystery is respiration by plants and soil microbes, which puts a great deal of CO_2 into the air. In fact, before human emissions skewed the picture, this natural process was long balanced with plant uptake to keep global carbon dioxide levels remarkably steady over the long run.

Reforestation is also in the mix. Many parts of eastern North America that were deforested and then farmed for decades are now tree-covered once more. That growing forest is pulling additional CO_2 from the air, though this effect will slow down as the trees reach maturity. On the other side of the equation, the massive deforestation of Earth's tropics has served as a huge carbon source, although it now appears the tropics might be taking up more CO_2 than previously thought—perhaps even enough to compensate for deforestation. On top of these regional differences, Earth's plant life goes through large multiyear changes related to climate patterns. A strong El Niño, for example, can cause so much ecosystem stress in so many parts of the world that Earth's land areas can briefly serve as a net source of carbon (see chapter 3). Researchers are still working on ways to measure the year-to-year and decade-to-decade variability. One key area of inquiry is the evolving balance between the increased absorption of CO_2 due to the fertilization effect above and the potential for decreased CO_2 absorption, especially in the tropics, as changes in temperature and precipitation put stress on plants.

For now, the bottom line is that Earth's land areas as a whole continue to serve as a sink for carbon. The land sink may have even strengthened over the last couple of decades, according to a 2010 analysis by Jorge Sarmiento (Princeton University). Happily, this takes a substantial edge off the impact of our greenhouse emissions, at least for the time being.

Because most of Earth's plant life is north of the equator, and because land-based uptake of CO_2 is more seasonally driven than oceanic uptake, the northern spring produces a "breathing in" effect (a dip in CO_2 levels), with a "breathing out" (a rise in CO_2) each northern autumn. These ups and downs show up as the sawtooths atop the steady multiyear rise in CO_2 visible on the graph on p. 39.

How much do different activities contribute?

The chart opposite shows two different ways to analyze the sources of human-produced greenhouse gas: where the emissions come from (left) and the activities that generate them (right). As our emissions climb at roughly 1%–2% per year, the balance among various sources is slowly shifting.

By and large, the world's most technologically advanced nations have become more energy efficient in recent years, with **industrial emissions** actually declining a few percent in some developed countries since 1990. Counterbalancing this is the explosion of industry in the world's up-and-coming economies, such as China's and India's. Another area of progress is deforestation, which accounted for more than 25% of emissions in the 1990s but closer to 10%–15% in recent years.

The emissions produced by heating, cooling, and powering **homes** have been increasing by more than 1% per year. Even though energy use per square meter has been dropping, today's homes have gotten bigger, and they're stuffed with more energy-hungry devices. Consumer electronics, including computers, now make up around 15% of global household electricity use, according to the International Energy Agency. In many parts of the world, there are ambitious plans to reverse the growth in household emissions, but it remains to be seen whether such targets will be met. Meanwhile, **commercial buildings** are tending to become more efficient more quickly than homes, with companies motivated by the potential for long-term savings.

Emissions in the **transportation** sector are climbing at well over 2% per year, stoked by the global growth in motor vehicle use and the ever-longer distances many people are driving and flying. Transport-related emissions are growing most quickly in the developing world, as millions of people in newly prospering nations take to the roads for the first time ever in their own cars.

For more on how personal choices in home life and transportation feed into the global picture, see chapters 18 and 19.

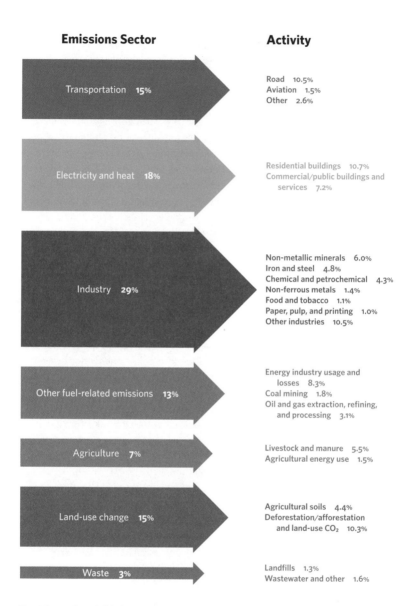

Emissions Sector

Transportation **15%**

Electricity and heat **18%**

Industry **29%**

Other fuel-related emissions **13%**

Agriculture **7%**

Land-use change **15%**

Waste **3%**

Activity

Road 10.5%
Aviation 1.5%
Other 2.6%

Residential buildings 10.7%
Commercial/public buildings and
services 7.2%

Non-metallic minerals 6.0%
Iron and steel 4.8%
Chemical and petrochemical 4.3%
Non-ferrous metals 1.4%
Food and tobacco 1.1%
Paper, pulp, and printing 1.0%
Other industries 10.5%

Energy industry usage and
losses 8.3%
Coal mining 1.8%
Oil and gas extraction, refining,
and processing 3.1%

Livestock and manure 5.5%
Agricultural energy use 1.5%

Agricultural soils 4.4%
Deforestation/afforestation
and land-use CO_2 10.3%

Landfills 1.3%
Wastewater and other 1.6%

Breakdown of total global greenhouse gas emissions (CO_2 equivalent) by sector (left) and end use (right). Because of rounding, columns do not total 100%, and some emissions sectors at left contribute to more than one end use at right. (Adapted from World GHG Emissions Flow Chart 2010, World Resources Institute, wri.org.)

Aviation: Taking emissions to new heights

The advent of low-cost flying has sent millions more people hopping from country to country over the last decade, especially across Europe. In many developed countries, aviation is the fastest-growing transportation segment, and it threatens to counteract progress in reducing greenhouse gas emissions on other fronts. The Netherlands Environmental Assessment Agency estimates that CO_2 emitted by international transport now totals more than 1 billion metric tons per year. That's less than 3% of all global emissions, but in the developed countries where air travel is most common, the percentage is rising quickly. In fact, though huge homes and hulking SUVs are familiar symbols of emissions excess, frequent flyers are among the people with the very biggest carbon footprints.

If you consider only the fuel needed per passenger for a given trip, air travel doesn't look so bad. You could easily burn more gas driving a big, lumbering car on your own than by covering the same distance in a newer, more fuel-efficient plane that's flying at full capacity. The problem is that convenient, inexpensive air travel makes it easy to go vast distances quickly, so the net effect tends to be more people traveling more miles on more planes. In countries such as the United Kingdom, emissions from aviation have more than doubled since 1990. To make matters worse, it appears that the contrail clouds generated by aircraft also contribute to warming, perhaps doubling aviation's greenhouse impact (though the numbers are still being researched—see chapter 10).

International airplane and ship traffic aren't considered in the Kyoto Protocol, so their emissions don't play a role in whether the United Kingdom or other nations meet their Kyoto requirements. However, they're likely to enter into any post-Kyoto agreement, and the European Union (EU) has already taken action to control emissions in the sky. As of 2012, all international flights to or from member nations—including flights operated by U.S. carriers—fell under the EU's emissions trading system. Airlines were mandated to keep their greenhouse emissions 3% below 2004–2006 levels, with a 5% decrease that began in 2013, or else acquire permits for the carbon burned beyond those amounts. The advent of the new rules triggered a wave of protest from more than two dozen nations, including China, Russia, and the United States, which forced the EU to back off on enforcement until at least 2014. Even so, some activists consider the emission targets weak ones, as they allow aircraft to continue sending far more greenhouse gas into the atmosphere than they did in 1990, the benchmark year for the Kyoto Protocol.

It's a good thing that aviation is a fairly small piece of the greenhouse emissions picture for the time being, because it may be one of the more troublesome

ones to address in the long term. Electric cars are already taking to the streets, with some of them drawing from solar power, but there's no obvious way to wean aircraft off fossil fuels in the next several decades. Innovations such as improved wingtips and composite materials are paring down how much fuel an airplane needs, and some prototype designs promise to improve efficiency by 50% or more. However, it will take decades for fuel-sipping aircraft to replace the airlines' vast existing fleets, and the physics of flight demand a certain amount of energy consumption. Biofuels have been tested, but these are unlikely to make up more than a sliver of airlines' carbon footprint. Some progress, however, is being made in how airplanes fly. Scandinavian Airlines began experimenting with a go-slower strategy in 2008. Test flights between Oslo and Bergen showed that cutting in-flight speeds by about 10% reduced emissions for the entire flight by a few percent per passenger. Landings can become more than 30% more efficient (and quieter to boot) when planes descend in a smooth manner, rather than the traditional stair-step technique. Even though "green landings" (referred to as continuous descent arrival in Europe and optimized profile descent in the United States) require more airspace, they're growing in popularity in Europe, with tests now under way at many U.S. airports.

Carbon intensity: An easy way out?

George W. Bush's administration in the United States found itself under in-tense pressure in 2002 to ratify the Kyoto Protocol, which was then making its way through the world's legislatures. Instead, Bush steered the United States away from Kyoto and toward a different way of assessing progress on climate change. His plan emphasized greenhouse gas intensity, aka **carbon intensity**. This is a measure of how much fossil fuel it takes to produce a certain amount of economic output. Thus, carbon intensity is not the actual amount of carbon emitted, but a number prorated by the gross domestic product (GDP) or pur-chasing power parity (PPP). For example, if GDP and emissions both climbed 3% in a given year, the carbon intensity would remain unchanged even though the actual emissions had risen.

The Bush administration called for an "ambitious but achievable" reduction in carbon intensity of 18% by the year 2012. Environmentalists pointed out that the U.S. carbon intensity dropped 17.4% from 1990 to 2000 without any spe-cial attempt to reduce it. Thus, they claimed, the plan offered little more than business as usual. The picture is similar elsewhere. Globally, carbon intensity (looking only at CO_2) dropped by 13% from 1990 to 2000, even as total emissions grew. In China, a booming economy helped reduce carbon intensity by 47% at the same time that CO_2 emissions climbed by 39%. And the carbon intensity of the world's energy sector—the amount of CO_2 emitted per unit of energy produced—has changed little over the last 40 years, with jumps in renewable energy running neck and neck with ever-increasing fossil fuel use.

Despite its shortcomings, the concept of carbon intensity is still widely used. As part of their 2007 Sydney Declaration, the 21-nation Asia–Pacific Economic Cooperation group held back from any direct targets for emission reduction, but agreed on an "aspirational" goal of reducing energy intensity by at least 25% by 2030, as compared to 2005 values. Similarly, China put forth carbon-intensity

Which countries are most responsible?

Establishing which countries are most responsible for climate change is in some ways an arbitrary exercise, since the atmosphere isn't particularly concerned about national boundaries. In any case, it's more complicated than simply adding up the amount of fossil fuels that each nation burns each year. First, there's the fact that a populous country's total greenhouse gas emissions may be relatively large even if its per capita emissions are relatively small, as is the case with China. (Consider this: if China were

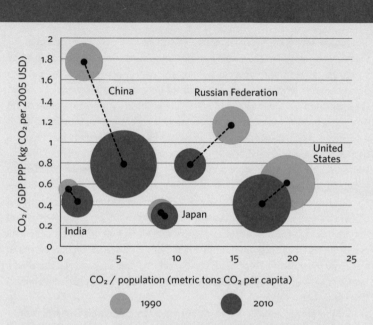

Trends in CO_2 emission intensities for the top five emitting countries. The size of each circle represents total CO_2 emissions from the country in that year. The location of each circle reflects energy use per capita (increasing from left to right) and energy use per dollar GDP/PPP (increasing from bottom to top). (International Energy Agency)

goals at the 2009 Copenhagen summit rather than pledging any specific level of emission cuts.

In the long run, then, intensity is a useful way of gauging the impact of greenhouse gas reductions on the economy. But when it comes down to effects on the physical world, a molecule of gas is still a molecule of gas.

suddenly divided into four nations, each with roughly the population of the United States, then U.S. emissions would far outstrip those from each of the four new countries.) Then there's the issue of greenhouse gases emitted in the past (some of them before global warming was a known problem), many of which still remain in the atmosphere.

The question of how countries are using—or abusing—their landscapes also needs to be considered. Countries undergoing major deforestation, such as Brazil and Indonesia, would rank significantly higher in the list

Outsourcing emissions

The nation-by-nation figures cited in this section look only at the direct emissions of greenhouse gases within those countries, but there's an important side effect of globalization to be considered: the shift it produces in the balance of greenhouse emissions. When a country imports consumer goods, should the emissions produced by the manufacture of those goods be assigned to the destination country rather than the supplier? In recent years, as much as half of the increase in China's greenhouse emissions has arisen from the manufacture of exports. On the receiving end, one preliminary study led by Shui Bin (U.S. Pacific Northwest National Laboratory) and Robert Harriss (Houston Advanced Research Center) estimated that U.S. emissions would have been 6% higher in 2003 if the products imported by Americans had been made in the United States. Extending this concept worldwide, a 2010 study by Steven Davis and Ken Caldeira (Carnegie Institution for Science) found that U.S. imports from all countries together accounted for nearly 11% of the nation's carbon footprint in 2004, the most recent year for which international trade figures were available (see graphic below). The trends behind these statistics have only intensified in the years since.

Shifting emissions isn't the main purpose of U.S. trade with China, of course, because the United States is not a party to the Kyoto Protocol. But what if a country whose emissions are limited by Kyoto did decide to transfer its greenhouse-intensive industry to a nation unfettered by the protocol? The risk of this so-called **carbon leakage** has been studied in some depth. Economists are still tussling over how big of a concern it is. A 2005 study by Mustafa Babiker of the Massachusetts Institute of Technology argues that carbon leakage could be as much as 30% in some countries—more than enough to counteract any domestic

shown on p. 55 of nations with the highest total emissions if the greenhouse-boosting effects of forest destruction were taken into account. To add another level of complexity, there's the claim that emissions figures are currently rigged in favor of nations that tend to import, rather than export, goods (see p. 52).

When it comes to overall greenhouse gas output, however, two countries stand above the rest. The United States is responsible for around 25% of the **cumulative** CO_2 emissions to date (that is, all emissions since the industrial era began) and continues to generate about 15% of annual greenhouse gas emissions—more than three times its share, considering that the

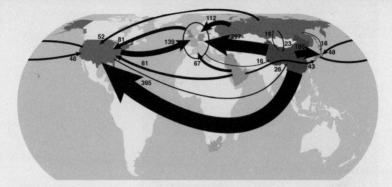

This schematic from Ken Caldeira and colleagues shows the largest international flows of emissions embodied in trade between major partners as of 2004, with nine western European nations considered as a group. (Steven J. Davis and Ken Caldeira/PNAS)

emissions reductions made in those nations under Kyoto. Confirming the point, the Davis-Caldeira study found that some European nations were already outsourcing more than 30% of emissions by 2004. According to Caldeira, "Policies aimed at achieving reductions in greenhouse gas emissions must account for international trade, so that these policies do not simply offshore carbon-intensive industries." Outsourcing can occur even within nations. A 2013 study led by Kuishuang Feng (University of Maryland, College Park) found that up to 80% of the goods consumed in China's thriving coastal provinces, where emission controls tend to be most strict, were imported from less developed provinces farther west.

U.S. represents less than 5% of the world's population. Some of the nation's outsized emission rate is clearly due to a lack of emphasis on energy efficiency and a focus on economic growth as opposed to environmental virtue. There are also historical factors in the mix that are difficult to change: a car-loving and car-dependent culture, an economy built on vast reserves of fossil fuels, and the simple fact that the United States is large in both population and land area. The country now has more than 60 years' worth of suburban development that virtually forces millions of Americans to drive to work, school, and just about anywhere else outside the home. Despite all these factors, and to the surprise of many, U.S. emissions actually dropped

Suburban Chicago's grid, ablaze at dusk, stretches to the horizon.

by roughly 12% from their 2005 peak to 2012. Among the factors at play: the major recession from 2008 onward, increased use of mass transit, higher auto efficiency standards, a boom in solar and wind energy, and—perhaps most influential—a major shift from coal to natural gas, which produced about 30% of U.S. power in 2012 compared to 19% in 2005.

Quickly outpacing the United States in terms of current emissions is China. With its billion-plus citizens, the country's emissions per person are still comparatively low. But as the nation continues to industrialize, its total emissions have been climbing at a staggering pace—close to 10% annually for the decade 2002–2011. At the turn of this century, it was thought that China might surpass the United States in total emissions by the year 2025. That point was actually reached in 2006, according to the Netherlands Environmental Assessment Agency. In recent years the annual *increase* in China's emissions has run almost twice as high as Britain's *total* emissions.

The **United Kingdom**, for its part, has produced about 5% of the world's cumulative emissions, thanks to its head start as the birthplace of the In-

Percentage of global CO_2 emissions (fossil fuels, cement, and gas flaring only)	Emissions per capita (metric tons of CO_2 from fossil fuel use)	Carbon intensity (metric tons of CO_2 emitted per thousands of U.S. dollars in GDP/PPP, using 2000 prices)
China 24.7	Qatar 40.1	Zimbabwe 2.71
United States 16.2	Trinidad and Tobago 37.7	Netherlands Antilles 1.59
India 6.0	Kuwait 34.2	Turkmenistan 1.41
Russia 5.2	Netherlands Antilles 23.5	Trinidad & Tobago 1.38
Japan 3.5	Brunei 22.9	Kazakhstan 1.30
Germany 2.2	United Arab Emirates 22.3	Uzbekistan 1.27
Iran 1.7	Aruba 21.6	Mongolia 1.19
South Korea 1.7	Luxembourg 21.3	Iraq 1.02
Canada 1.5	Oman 20.6	Ukraine 0.96
United Kingdom 1.5	Bahrain 19.2	Bahrain 0.88
Saudi Arabia 1.4	Falkland Islands 19.1	Estonia 0.83
South Africa 1.4	United States 17.3	Saudi Arabia 0.80
Indonesia 1.3	Saudi Arabia 16.9	Russia 0.79
Mexico 1.3	Australia 16.8	China 0.77
Brazil 1.2	Gibraltar 15.7	South Africa 0.73
Italy 1.2	New Caledonia 15.6	Bosnia & Herzegovina 0.72
Australia 1.1	Kazakhstan 15.5	Kuwait 0.71
France 1.1	Canada 14.7	Kosovo 0.70
Poland 0.9	Faeroe Islands 14.6	Serbia 0.66
Ukraine 0.9	Estonia 13.7	Kyrgyzstan 0.64

Note: The first two columns show 2010 data from the U.S. Carbon Dioxide Information Analysis Center (cdiac.ornl.gov), with the data converted from carbon to CO_2 for the second column. The third column shows 2010 data from the International Energy Agency (www.iea.org/statistics/topics/co2emissions).

dustrial Revolution. In the course of a major industrial transformation, the United Kingdom has cleaned up its act significantly. It now generates less than 2% of the yearly global total of greenhouse gases. Still, that's about twice its rightful share, considering that the United Kingdom houses less than 1% of Earth's population.

Above is a list of the world's top 20 greenhouse emitters, measured in three different ways: percentage of all global emissions, emissions per capita, and carbon intensity (see p. 50). For **total emissions**, the list is a mixed bag, with the contributions slightly weighed toward developing versus developed nations. (If considered as a unit, the European Union would rank third in total emissions behind the United States and China.)

The **emissions per capita** column tells a different story, with the list topped by the tiny oil-producing nations of **Qatar**, **Trinidad and Tobago**, **Kuwait**, **Brunei**, and the **Netherlands Antilles**. These countries have so few residents that their contributions to the global greenhouse effect remain small, but because they are heavy producers and consumers of oil and gas, they have a high per capita emissions rate. Otherwise, industrialized nations lead the way, with the sprawling, car-dependent trio of the **United States**, **Australia**, and **Canada** not far behind the oil producers. For more about **carbon intensity** (the data presented in the third column), see p. 50.

The Symptoms

WHAT'S HAPPENING NOW, AND WHAT MIGHT HAPPEN IN THE FUTURE

The Shadow Lake fire burns near Santiam Junction in central Oregon, September 2011. (Jeff McLaughlin, Oregon Department of Transportation)

CHAPTER FOUR
Extreme Heat
TOO HOT TO HANDLE

To those who think of Russia as a perpetual icebox, the summer of 2010 came as a sizzling shock. The nation roasted in unprecedented heat from July into early August. Moscow, whose summer afternoons are typically close to room temperature, soared above 38°C (100°F) for the first time on record and hit 30°C (86°F) every day for more than four weeks. Smoke from peat and forest fires blocked out skyscrapers and clogged lungs. Even before the heat wave subsided, it was clear that thousands of lives had been lost, and Russian leaders—long skeptical on global warming—openly expressed concern about the nation's climatic future.

The horror in Russia triggered memories of the super-hot European summer of 2003—a relentless barrage of heat that produced an even greater slow-motion catastrophe, killing more than 50,000 people. In at least one study, a group of climate scientists confirmed the suspicions held by many: the odds of getting the great heat wave of 2003 were boosted severalfold by the helping hand of fossil fuels.

Heat waves themselves are nothing new, of course. As of 2013, many central U.S. states had yet to top the all-time records they set during the 1930s Dust Bowl, including a reading of 49.5°C (121°F) as far north as North Dakota. (In this context, an all-time record is the warmest reading observed at a given location on any day of any year.) However, climate-modeling studies and our understanding of the greenhouse effect both indicate that the next few decades could bring hot spells that topple many long-standing records across midlatitude locations. In 2010 alone, at least 19 nations—an unprecedented number, most of them in Eurasia and Africa—set all-time temperature peaks (their highest single readings on record). The roll call included 37.2°C (99.0°F) in Finland, 48.2°C (118.8°F) in Niger, and a blistering 52.0°C (125.6°F) in Saudi Arabia. More nations set all-time highs in 2010 than set all-time lows in the entire preceding decade. In 2011, another seven countries notched all-time highs, including 53.5°C (130.3°F) in Pakistan.

Striking as these statistics are, a heat wave doesn't have to bring the warmest temperatures ever observed to have catastrophic effects. All you need is a long string of hot days combined with unusually warm nights. Stir in a complacent government, cities that were built for cooler times, the lung-clogging effects of air pollution, and a population that can't or won't respond to the urgency, and you have a recipe for the kind of disaster that 2003 brought to Europe and 2010 delivered to Russia.

Off the charts

Extremes in summertime heat are one of the long-anticipated outgrowths of a warming planet. Part of this is due to simple mathematics. If the average temperature goes up, then the most intense spikes in a hot summer ought to climb in parallel. Computer models and recent data bear out this intuitive concept. In the contiguous United States, daily record highs have been outnumbering daily record lows by an increasing margin in each decade from the 1980s onward (see graphic). The ratios are even more skewed for monthly and all-time records. Looking ahead on the global scale, the 2013 IPCC report considered it virtually certain that most of Earth's land areas will see more frequent warm extremes and less frequent cold ones, on both a daily and seasonal basis. Of course, it'll still be possible to have a record-setting blast of cold every now and then, just as a baseball player on steroids will occasionally strike out. Over time, though, warm records should increasingly outpace the cold ones.

Could heat spikes themselves become larger? That's a different and more complex question. Any rise in average summer temperature—the base on which any heat wave builds—is itself enough to cause plenty of trouble. But if heat waves also gained amplitude, thus straying even further from a rising average, then the resulting double whammy would have huge implications for people and ecosystems.

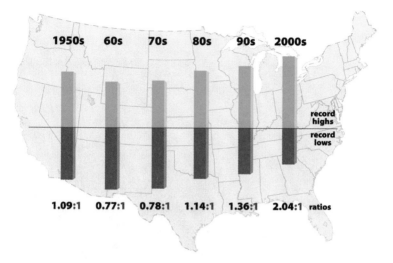

For the contiguous United States, the last 40 years have seen an increasing proportion of daily record highs to record lows, as reflected in data from about 1800 weather stations. (The ratio shown for the 2000s extends from January 2000 to September 2009.) (UCAR)

Only in the last few years have scientists looked more closely at temperature variability and whether it might change as the planet warms. Globally, several recent studies point to a surge in three-sigma heat, or warm spells and heat waves that go at least three standard deviations beyond the norm. Periods this extreme were virtually absent in the mid-twentieth century. However, when current summer conditions are compared to the climate of 1951–1980, three-sigma months now occur over about 5% of Earth's land area each year. The heat was so intense across parts of western Russia in 2010 and Texas in 2011 that the entire summer reached the three-sigma point, according to analyses led by NASA's James Hansen. If the climate weren't changing, you'd expect that to happen in any given spot only about once every 770 years. Hansen and colleagues interpret their results as a sign of increasingly variable temperatures.

Drawing on the most recent IPCC-related modeling, and using the 1951–1980 benchmark, Dim Coumou (Potsdam Institute for Climate Impact Research) found that the global coverage of three-sigma hot months should at least quadruple by midcentury, regardless of how much emissions are trimmed. Even five-sigma months, virtually unheard of today, could cover 3% of the globe by the year 2040. Yet these future scorchers may not

Many Muscovites wore masks in August 2010 when a heat wave—accompanied by smog and smoke from nearby wildfires—gripped the Russian capital. The city's mortality rate doubled as the concentration of atmospheric carbon monoxide soared to more than double acceptable safety norms.

need enhanced variability to do their dirty work. Coumou and coauthor Alexander Robinson found that most of the model-based trends could be explained simply by a rising baseline temperature. Several other modeling studies do point toward a future with increased variability on the hot end in some areas, including parts of Europe and North America, but this could end up being location dependent rather than a consistent worldwide trend.

In the long run, hot is hot, whether it manifests because of intensified spikes, a rising baseline, or both. "While we may want to sort out the 'variability issue' for scientific reasons, it does not alter the fact that the future will be bleak with regard to heat extremes," note Lisa Alexander and Sarah Perkins (University of New South Wales) in a 2013 essay for *Environmental Research Letters*.

However high it might get, the peak temperature on any given day isn't the ideal measure of a heat wave's ferocity. When it comes to the toll on people, animals, and plants, the **duration** of a heat wave and the warmth of the **nighttime lows** are the real killers. A weeklong stretch of severe heat can be far more deadly than just two or three days of it. If a heat wave is defined at a certain location by a certain number of days with readings

Pollution: Heat's hidden partner in crime

Some victims of heat waves die not because the air is so warm, but because it's so dirty. The sunny, stagnant conditions prevalent during heat waves make an ideal platform for the sunlight-driven processes that create **ozone**, which is a blessing in the stratosphere (see p. 36) but a dangerous pollutant at ground level. Ozone irritates the lungs and makes them more vulnerable to other chemicals. Moreover, the relative calm of a heat wave allows tiny bits of heavy metals, sulfates, nitrates, and other substances to accumulate in the air. These are often grouped into the classes **PM10** (particulate matter smaller than 10 microns or 0.0004 in.) and **PM2.5** (particles smaller than 2.5 microns). The smallest of these particulates can sneak past the body's natural respiratory filters, causing a variety of lung problems and raising the risk of heart attacks.

After focusing on other pollutants for decades, scientists more recently found how deadly ozone and fine particulates can be. The World Health Organization estimates that mortality goes up by 0.3% during low-level ozone episodes. The comprehensive Global Burden of Disease project found that particulate air pollution contributed to roughly 2.3 million deaths globally in 2010, with ozone pollution linked to another 150,000 deaths. (Indoor pollution, largely from home cookstoves in developing countries, was tied to nearly 3.5 million fatalities.)

Several studies connected a substantial fraction of the 2003 heat wave's deaths in Europe—anywhere from around 20% to 40%, depending on the nation—to ozone and particulates. Looking ahead, the growing megacities of the developing world are at particular risk. If there's an upside, it's that a concerted effort to reduce ozone and particulate pollution might help many of the people who die in the worst heat waves. One analysis led by U.S. scientists Mario and Luisa Molina showed that a 10% reduction in fine particulates in Mexico City's air could save roughly 1000 lives per year.

above, say, 35°C (95°F), then a warming climate would be expected to bump more days above such a threshold over time.

There's a different connection between global warming and steamy nights. As noted in chapter 5 (Floods and droughts; see p. 77), higher temperatures have boosted **atmospheric water vapor** on a global scale. When there's more water vapor in the air, nights tend to stay warmer. Moreover, cool nights act as a critical safety valve, giving people a physiological and psychological break from the intense daytime stress of a heat wave. If the nights warm up too much, that safety valve goes away.

When a fever hits

Europe's 2003 heat onslaught is a textbook example of how the variables above can add up to disaster. During the first two weeks of August, large swaths of the continent topped 35°C (95°F) and found themselves hard-pressed to cool below 20°C (68°F). Such temperatures might not sound like a big deal to an American from the Sunbelt, but most of Europe's urban areas lack the ubiquitous air conditioning of the hottest U.S. cities, and certainly most Europeans aren't accustomed to dealing with weather like this. Among the history-making statistics from 2003:

* August 10 was the hottest day on record in **London**, with highs of 37.9°C (100.2°F) at **Heathrow Airport** and 38.1°C (100.6°F) at the Royal Botanic Gardens. Not to be outdone, Faversham, Kent, climbed to 38.5°C (101.3°F) that same afternoon. It was the first day in nearly 300 years of recordkeeping that any place in Britain had topped the 100°F mark.
* All-time national records were also broken with 40.4°C (104.7°F) in **Roth, Germany**; 41.5°C (106.7°F) in **Grono, Switzerland**; and a scorching 47.3°C (117.1°F) in **Amareleja, Portugal**.
* Germany notched its warmest night on record, with the weather station at **Weinbiet**, a mountain near the Rhine, recording nothing lower than 27.6°C (81.7°F) on August 13.

Numbers can't do full justice to the impact of this atmospheric broadside on society. In **England**, railway tracks buckled, tube stops resembled ovens, and the London Eye and its glassed-in Ferris wheel pods had to shut down. Schools and offices closed for "heat days" in parts of Germany, but perhaps no place suffered so acutely as **Paris**. With little air conditioning and block after block of tall, closely packed stone buildings, the French capital—ill prepared for weather that was more suited to Paris, Texas—became a heat-trapping machine. The city ended up falling a whisker short of its all-time high of 40.4°C (104.7°F), which was set in 1947, but it did endure nine consecutive days that topped 35°C (95°F). Even worse for many were the nights of surreal heat. On August 12, the temperature never dropped below 25.5°C (77.9°F), making it Paris's warmest night on record. The early August heat was so sustained that it kept the temperature within the intensive care unit of one suburban Paris hospital, which lacked air conditioning, at or above a stifling 29°C (84°F) for more than a week.

Russia's heat wave of 2010 brought similar misery. A vast dome of high pressure encased Moscow and surrounding areas from mid-July to early

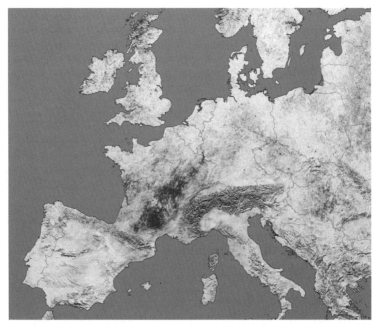

The areas hit hardest by the 2003 heat wave (red), relative to July averages for 2000, 2001, 2002, and 2004. (NASA)

August, setting the stage for a vicious brew of heat, smog, and smoke from nearby fires. During the worst bouts of pollution, visibility dropped to a few city blocks. On these days, Muscovites faced the no-win choice of opening windows for a smoky breeze or keeping their homes sealed and sweltering. Wind shifts pushed the smoke away from Moscow at times, but the heat was virtually constant, with high temperatures running more than 10°C (18°F) above average for weeks on end. Alexander Frolov, the head of Russia's weather service, dubbed the heat wave the worst in the nation's 1000-year history. Relief wasn't easy to come by, as smoke and heat also plagued the thousands of summer homes (dachas) where Muscovites traditionally flee from urban swelter. Many people took to the beach after drinking, which—combined with the disorientation that extreme heat can produce—helped lead to more than 1200 drownings in June alone. The heat also decimated wheat crops across a huge swath of Russia's breadbasket, which sent world grain prices soaring. All told, the months of July and August brought 56,000 more fatalities across central Russia in 2010 than in the year before, with most of those deaths presumably related to the heat and pollution.

The disaster galvanized Russian leaders, who'd long viewed global warming science with suspicion. Early in 2010, President Dmitry Medvedev had dubbed climate change "some kind of tricky campaign made up by some commercial structures to promote their business projects." Only a few months later, the heat had produced a remarkable about-face. In late July, Medvedev said, "What's happening with the planet's climate right now needs to be a wake-up call to all of us."

In 2012, it was North America's turn to experience unprecedented heat on a grand scale. The first salvo was pleasant rather than painful: an astoundingly warm March that sent temperatures rocketing to summerlike levels—in many cases, soaring above 27°C (81°F) for days on end—across much of the midwestern and northeastern United States and southeastern Canada. At least three U.S. locations experienced evening lows that broke daily record *highs*. The largest cities in the Canadian provinces of New Brunswick, Nova Scotia, and Prince Edward Island each saw readings balmier than anything they'd previously observed in March *or* April. Perhaps because it was so tempting to simply enjoy the warmth, the event got little serious scrutiny at the time, but it was actually one of the most extreme temperature departures in North American history. Millions of dollars in Midwestern fruit crops were lost when premature blossoms froze in April.

Things got considerably worse that summer, as drought and heat punished much of the central and eastern United States. July 2012 ended up as the hottest month ever observed in cities from Denver to St. Louis to Indianapolis, as well as for the United States as a whole. That entire year was also the nation's warmest, by a full 0.6°C (1.0°F)—a huge margin for such a record. The massive drought, which lingered into 2013, became one of the nation's costliest natural disasters on record, with crop losses totaling tens of billions of dollars. Huge wildfires also devastated large parts of the U.S. West. Surprisingly, less than 100 deaths were initially confirmed from the summer heat of 2012. Though air conditioning is far more common in the States than in Europe, the low death toll also suggests that heat wave safety practices are gaining ground (see below).

The human cost of heat

It can take a surprisingly long time to gauge how many people die in a heat wave, but the broad strokes of the disaster often show up quickly. Moscow's death rate in July 2010 doubled that of the preceding July, and the number of funerals soared. At the height of the heat, one Moscow crematorium operated around the clock for days. (In a gruesome but telling bit of irony, its furnaces overheated.)

Likewise, it was clear in early August 2003 that people were dying in large numbers across Europe, but it took several weeks to start getting a handle on the heat's human cost. By September more than 20,000 Europeans were listed as casualties of the summer of 2003, and the summer's toll continued to mount over time. It hardly made the news when, in 2005, authorities in Italy abruptly raised their nation's toll from 8000 to 20,000. Simply adding the various national counts would imply that more than 50,000 people died as a result of the heat wave. Later estimates brought the toll as high as 70,000, though the exact number is difficult to discern (see below). No heat wave in global history has produced so many documented deaths.

How could heat victims go untallied for a year or more? The answer lies partly in the way mortality statistics are compiled and partly in the way heat waves kill. Many people die indirectly in heat waves—from **pollution**, for instance (see sidebar "Pollution: heat's hidden partner in crime," p. 63), or from preexisting conditions that are exacerbated by the heat—so it's not always apparent at first that weather is the culprit. And unlike a tornado or hurricane, a hot spell doesn't leave a trail of photogenic carnage in its wake. People tend to die alone, often in urban areas behind locked doors and closed windows. The piecemeal nature of heat deaths over days or weeks makes it hard to grasp the scope of an unfolding disaster until the morgue suddenly fills up. Furthermore, it often takes months or even years for countries to collect and finalize data from the most far-flung rural areas.

On top of this, heat is a selective killer. It targets the very young and especially the very old, people whose metabolisms can't adjust readily to temperature extremes. In Rome, more than half of the 700 deaths attributed to 2003's heat were among people older than 85. Because the elderly die so disproportionately in heat waves, it's tempting to assume that weather is simply claiming those fated to die shortly anyhow. If that were the case, however, then you'd expect mortality to dip below average in the months following a heat wave. Careful studies of major heat disasters have, in fact, shown that such dips typically account for only about 20%–30% of the spikes above average, or what epidemiologists call the **excess deaths**, that were observed in the preceding heat. Thus, it seems that most victims of heat waves, even the elderly, die well before their time.

In France, the situation was exacerbated by local culture. The worst of the heat struck during the August vacations, when much of professional Paris heads out of town. Left behind to fend for themselves were many thou-

The nights Chicago fried

A cool breeze might have been appreciated in the Windy City in the second week of July 1995. During four days of searing conditions in Chicago, more than 700 people died from heat-related causes. The days were hot indeed—ranging from 37° to 41°C (98°–106°F) at the in-town Midway Airport—but in this case, it truly was the humidity as much as the heat. Tropical amounts of moisture in the air, coupled with the heat-island effect (see p. 238), helped produce overnight lows of 27° and 29°C (81° and 84°F) on two consecutive nights. As in Paris, many of the city's older multistory buildings lacked air conditioning. When children and teens sought relief by opening fire hydrants, police put a stop to it.

Chicago's heat-emergency plan sat on the shelf, and the city's mayor and other key officials remained on vacation until the disaster became dire. Then things really heated up—at least politically, according to New York University sociologist Eric Klinenberg, author of *Heat Wave: A Social Autopsy of Disaster in Chicago*. Mayor Richard Daley acknowledged the heat but asked people not to blow it out of proportion. The city's commissioner of human services blamed victims for not taking care of themselves. "They were often interpreted as individual failure," says Klinenberg of the Chicago heat deaths. "In Europe, the heat wave was immediately framed as a political event."

Chicago did learn from its calamity. In 1999, when another heat wave struck, the city's action plan included not only the usual warnings but also free bus rides to "cooling centers." Crews of city workers phoned and checked in person on elderly people living alone. Those steps helped keep the death toll to 110, a number that Klinenberg still finds too high. As global warming unfolds, he says, "we know that more heat waves are coming." And, he adds, dying from heat is not a nice way to go. "If you look closely at the police reports, or the medical autopsies, they're just horrific. These are isolated, lonely, painful deaths."

In Chicago, densely settled neighborhoods with busy streets and public spaces, such as the heavily Latino Little Village, fared much better than areas where people were more disconnected from neighbors and had few places to gather, such as the African-American neighborhood of North Lawndale. As Klinenberg sees it, heat waves are "invisible disasters that kill largely invisible people. Perhaps that's the reason we don't care enough about them."

sands of elderly Parisians, the ones most susceptible to heat-related health problems and least able to seek refuge in a cool location. As many as half of France's fatalities occurred at rest homes, few of which had air conditioning.

Could a better **warning system** have saved lives in Europe? That's been the case in the United States, where most parts of the country can easily top 35°C (95°F) in a typical summer. After years of perfunctory heat advisories, issued by the U.S. National Weather Service (NWS) but seldom heeded, the country was shocked into awareness by a 1995 heat wave in Chicago that killed more than 700 people (see box, previous page). That same year, Laurence Kalkstein of the University of Delaware teamed up with the NWS to launch a new watch–warning system for extreme heat. The scheme also included intensive **neighborhood interventions**, such as block captains assigned to check on neighbors. It's been estimated that Philadelphia's heat plan saved more than 100 lives in four years. U.S. heat can still be deadly, though, especially when it strikes where people aren't accustomed to it. Hundreds of Californians died in an intense 2006 heat wave, many of them living alone in homes where interior temperatures topped 40°C (104°F).

As it happened, Kalkstein and Italian colleagues had just launched a U.S.–style warning system in Rome when the 2003 heat wave struck. The city ended up declaring heat emergencies on 18 days that summer. Even with its new plan in place, hundreds more Romans died than the scientists expected. In a subsequent analysis, the team found that the heat was worse than forecast and, indeed, transcended anything they'd considered in planning their warning system.

Attributing the heat

How much of today's heat can be pinned on greenhouse gases? The European disaster of 2003 brought that question to the fore. The ever-growing heat islands of modern European cities no doubt played a role, and these are a form of climate change in themselves (see p. 238). Yet, the continent-wide, rural–urban scope of the heat pointed to something more going on. Many activists trumpeted the heat wave as a classic example of global warming at work, but it was a year later before scientists had marshaled a statistical case for this accusation. The resulting study, published in *Nature* in 2004 by Peter Stott (University of Reading) and colleagues at the University of Oxford, was a landmark—one of the first to lay odds on whether a specific weather or climate event was rendered more likely to occur by global warming. Using a climate model from the Met Office Hadley Centre, Stott's team simulated European summers since 1900 a number of times, sometimes including the century's ramp-up in greenhouse gases and sometimes omitting it. The authors found that human-induced greenhouse emissions made it roughly four times more likely that a summer like 2003 might occur.

Such efforts—often dubbed attribution studies—are a crucial way that scientists and policymakers get perspective on the effect of climate change on extreme weather events. However, the results can be open to varying interpretations. For example, Randall Dole and Martin Hoerling of the National Oceanic and Atmospheric Administration (NOAA) asked whether the Russian heat wave of 2010 could have been foreseen on the basis of twentieth-century temperature trends. They concluded that there was enough variability inherent in the region's July climate to allow for such an intense event to occur regardless of climate change. However, using a different technique, Stefan Rahmstorf and Dim Coumou (Potsdam Institute for Climate Impact Research) found that the odds of setting a July heat record in Moscow have been boosted fivefold by longer-term warming. Such studies may not be as contradictory as they seem. Even where a record-smashing heat wave can occur without a boost from climate change, long-term warming may still make a particular record value far more likely to be achieved.

The future of summer sizzle

On a broader scale, the European disaster of 2003 and Russia's 2010 meltdown were merely blips—albeit spectacular ones—in a long-term warming trend that spans most of the world, particularly large parts of North America, Eurasia, and Australia. While the planet as a whole has warmed nearly 0.8°C (1.4°F) since 1900, the rise has generally been greater over the Northern Hemisphere continents (especially at higher latitudes) than over the tropics. That simplified picture doesn't hold quite as well, though, when you go beyond the average temperature and look at various types of heat extremes, some of which have increased in the tropics as well as midlatitudes.

It's taken a long while for scientists to verify how temperature extremes have changed globally over the last few decades, not to mention how they might evolve in the future. Why is this so? After all, the global average has been tracked for decades (see p. 235), and individual stations keep close tabs on daily weather. Between the local and the global, though, it's surprisingly difficult to assess regional trends in temperature and precipitation extremes.

Nations have traditionally shared more general data, such as the average high and low temperatures for each month at major reporting sites. But knowing that the average August high in a given city rose, say, 1°C in 50 years doesn't tell you whether stretches of the most intense heat are becoming more frequent, intense, or longer lived. To find that out, you need a consistent way to measure heat extremes, and you need day-by-day data extending over decades. That's been hard for researchers to obtain. Some nations balk at releasing such data, considering it valuable intellectual property, and others aren't set up to provide the data easily.

Climate scientists made big headway on this problem in the 2000s through a series of regional workshops. These meetings, accompanied by the creation of an Expert Team on Climate Change Detection and Indices, provided a way in which countries could freely share some parts of their daily-scale data for research without jeopardizing the revenue they earn by selling those data. The expert team also forged a set of 27 core indices of heat, cold, and precipitation that were used in preparing the 2007 and 2013 IPCC reports. Emerging from this and subsequent work is a more detailed picture of how temperature-related extremes are evolving across the world and what we can expect going forward.

Drawing on these core indices, a 2013 analysis of climate trends from 1901 to 2010, published in the *Journal of Geophysical Research: Atmospheres* and led by Markus Donat (University of New South Wales), found the following:

* Stations around the globe are seeing an average of roughly 16 more "warm days" per year than they did in 1900. (Warm days are defined as those when the high temperature falls in the top 10% of the spectrum for a given calendar date; you'd expect 36 such days in a typical year.) Cool days are down by about 14. Changes after dark are more dramatic: warm nights have surged by about 30 per year, and cool nights are down by about 26.
* Averaged across the planet, the hottest day recorded at each spot each year rose on the order of 1°C (1.8°F) since 1900, while the coldest night jumped by a full 6°C (10.8°F). Russia, in particular, is seeing nighttime winter temperatures often holding well above where they used to bottom out.

Not all geographic areas have seen temperatures climb. One region where the twentieth century's overall warming barely made a dent is the southeastern United States. Donat and colleagues found a small but significant decline in the number of warm days across the U.S. Southeast since 1950, although warm nights have increased there. Parts of this "warming hole" saw average temperatures drop by as much as 1°C (1.8°F) between 1930 and 1990. Why the hole? It's still a topic of active research. Some scientists have found possible links to sea surface temperatures in the Atlantic and Pacific, while a study led by Eric Leibensperger (Harvard University) points to sunlight-blocking particulate pollution across the region, which grew during the postwar years but is now on the wane.

The climate modeling featured in the latest IPCC assessment shows many of the major observed twentieth-century trends increasing over the next few decades. How much they accelerate depends largely on how much and how soon greenhouse emissions can be controlled. Jana Sillmann (Canadian Centre for Climate Modelling and Analysis) led a 2013 paper in the *Journal of Geophysical Research: Atmospheres* detailing the globally averaged picture painted by these models for the period 2081–2100, as compared to the period 1981–2000. Some results include the following:

* The total percentage of nights that fall in the "warm nights" category, as defined above, ranges from 30% if greenhouse gas concentrations begin to drop by midcentury (the pathway dubbed RCP2.6) to almost 70% if emissions rise unchecked (RCP8.5). Warm days will range from around 25% to 60%, depending on RCP. In contrast, only about 1%–4% of all days would qualify as cool days. Cool nights would virtually disappear under RCP8.5 and would drop to less than 4% even for RCP2.6.
* The hottest day of each year will warm by about 1.5°C (2.7°F) for RCP2.6 and by more than 5°C (9°F) for RCP8.5. The coldest night will be about 2°C (3.8°F) warmer for RCP2.6 and more than 6°C (10.8°F) warmer for RCP8.5.

Again, as one might expect, there is liable to be a good deal of variety in how and where these temperature changes play out. Sillmann and colleagues found that large parts of Brazil and Sub-Saharan Africa could see as many as 100 or more additional "tropical nights" per year, when temperatures fail to drop below 20°C (68°F). Major boosts in the number of such balmy nights may also occur as far poleward as the southeastern United States and southern Europe. The number of nights that dip below freezing should see a drop worldwide, but the decrease could be as much as 60 nights or more per year in a coastal belt stretching from the Pacific Northwest states to Alaska.

As for heat waves themselves, the 2013 IPCC report concluded that by the 2080s and 2090s, it's very likely that heat waves will occur with a higher frequency and duration, with somewhat more uncertainty on how much their intensity might be boosted. Although there wasn't a specific outlook

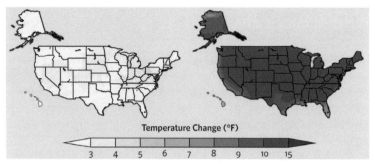

Temperature Change (°F)

3 4 5 6 7 8 9 10 15

2014 U.S. National Climate Assessment maps show projected increases (in degrees Fahrenheit) for the average temperature on the hottest days by late this century (2081–2100) relative to 1986–2005, under scenarios RCP2.6 (left) and RCP8.5 (right). "Hottest days" are defined here as those so hot they occur only once in 20 years, on average. Such days would be about 10–15°F hotter for most U.S. residents under RCP8.5. (U.S. National Climate Assessment/NOAA National Climatic Data Center/Cooperative Institute for Climate and Satellites, North Carolina)

provided for earlier in the century, the summary did note that "models project near-term increases in the duration, intensity, and spatial extent of heat waves and warm spells."

A wild card in the Atlantic

There's one character waiting in the wings that could change this picture, at least in theory: a slowdown or shutdown of the **Atlantic thermohaline circulation**, which helps keep Europe unusually warm for its latitude. This is the scenario made famous in the 2004 film *The Day After Tomorrow* (see p. 345). Although evidence shows that the thermohaline circulation has ground to a halt more than once in climate history, it's believed that this process takes at least a few years to play out, and sometimes many decades, rather than the few days portrayed in the movie.

In any case, a complete shutdown of the thermohaline circulation isn't expected any time soon. However, the 2013 IPCC report deems a slowdown "very likely" this century. According to the IPCC, that weakening could be 20%–30% if emissions begin dropping by midcentury (the RCP4.5 pathway) or 36%–44% if they continue rising through the century (RCP8.5). A slowdown of this magnitude could largely erase long-term warming across the United Kingdom and diminish it across other parts of Europe. However, summers could still be warmer and more drought-prone across the United Kingdom and Europe than they are now. Rather than

What counts as a heat wave?

Half the battle in assessing climate extremes is simply deciding what to measure. What defines an extreme spell of heat, aside from a sticky shirt or a wilting garden? Experts have tried out a variety of indices to capture climate extremes. In the realm of heat, these include the following, but there are many other possibilities, including measurements that incorporate both heat and humidity.

* **Absolute thresholds** are the number of days that exceed a given temperature. While this is nicely concrete, one threshold doesn't always fit all. A week of July afternoons at 35°C (95°F) might feel miserable in London but normal in air-conditioned Houston. Thus, each location might need a different threshold, which makes it hard to compare heat intensity among locations.
* **Monthly maximums and minimums** are changes over time in the highest and lowest single temperature observed during a given month of the year. These can provide useful, easy-to-grasp illustrations of a shifting climate, like the all-time highs recorded during the 2003 European heat wave. However, a single day or night of record warmth doesn't necessarily correspond to the kind of sustained multiday heat that causes major problems.
* **Threshold departures** are the number of days when temperatures climb above average by a fixed amount, such as 5°C (9°F). This gives a more location-appropriate sense of how unusual a hot stretch might be. However, it doesn't acknowledge that one city might normally have more variability than the next. For example, a 5°C jump in Denver's dry climate, where temperatures can gyrate wildly from day to night or across a few days, would be less obvious than the same leap in Miami's sultry summers, where high and low temperatures often change little from day to day.
* **Percentile departures** are the number of days that land among the hottest of all days in that month's long-term record, based on percentage (the hottest 10%, 5%, 1%, etc.). This index provides a tailored-to-fit measure of a heat wave's intensity on the basis of each city's unique characteristics. No matter where you live, a day that is among the warmest 1% observed in the past decade or century means something (especially if such heat were to start occurring 5% or 10% of the time, rather than 1%). Some researchers combine this with a measure of duration—for instance, the number of consecutive days on which the temperature reaches a given percentile ranking of heat.

fully counteracting greenhouse warming, then, a thermohaline slowdown would only make the picture even more complex. For a full discussion of the thermohaline shutdown hypothesis, see p. 161.

Handling the heat

In the long run, even if the world tackles climate change wholeheartedly over the next few years, Europe will clearly need to adapt to the risk of heat waves like the one it endured in 2003. Better warning systems will help; more air conditioning, and the associated cost, seems inevitable. The latter may put a dent in Europe's goal of reducing greenhouse emissions, though it's possible that some of the energy spent to cool the continent will be counterbalanced by a drop in the need for wintertime heating fuel. However, nobody knows how many of the poorest and most vulnerable of Europeans will simply be left to suffer through future hot spells in un-air-conditioned misery.

Poverty is certainly a major co-factor in heat deaths across the **developing world**. A blistering two-month-long heat wave in 2013 brought major suffering to hundreds of millions of people across eastern China, including rural residents as well as the urbanites of Shanghai. That city beat its all-time temperature record, set in 1942, at least five times during the summer, finally topping out at 40.8°C (105.4°F) on August 7.

In India, people are well accustomed to spells of intense heat during the late spring, just before the monsoon arrives, when temperatures can soar well above 40°C (104°F) across wide areas. As is the case elsewhere, it's the extremes on top of that already scorching norm that cause the most suffering. Several premonsoonal heat waves in recent years have each killed more than 1000 people, many of them landless workers forced by circumstance to toil in the elements. The heat was especially fierce in 2010, as India endured its hottest premonsoon weather on record. On May 26, Mohenju Daro, Pakistan, reported an astounding 53.5°C (128.3°F)—the highest temperature reliably observed in Asia up to that point.

As for the United States, much of its Sunbelt, including the vast majority of homes and businesses in places like Atlanta and Dallas, is already equipped with air conditioning. Even as temperatures soar further, these regions may prove fairly resilient—as long as the power stays on. Older cities in the Midwest and Northeast appear to be more at risk of occasional heat crises, even though their average summer readings fall short of those in the South, where air conditioning is the accepted standard. Indeed, fatalities appear to be more common in places where intense heat is only an occasional visitor. Another factor is diabetes, a global epidemic hitting the United States especially hard. Because diabetes inhibits sweating and destabilizes blood sugar levels, it could put millions of Americans at an increased risk of heat-related illness, according to a 2010 study by the Mayo Clinic and NOAA.

CHAPTER FIVE

Floods and Droughts
TWO SIDES OF A CATASTROPHIC COIN

Although it's natural to think of temperature first when we think of global warming, the impact of climate change on precipitation may be even more important in the long run for many places and many people. Too much rain at once can cause disastrous floods, while too little can make an area unproductive or even uninhabitable.

Anyone who was caught in the epic 24-hour rainfall of 944 mm (37.2 in.) in Mumbai, India, in July 2005, or who experienced Britain's destructive summer rains in 2007, might wonder if the planet's waterworks are starting to go into overdrive. That thought also must have crossed the minds of people in Pakistan during the summer of 2010, as torrential monsoon rains inundated an area the size of England. At least four million people were left homeless—one of the worst such displacements in modern times.

Full-globe observations of rainfall are difficult to obtain, in no small part because 70% of the planet is covered by oceans. For the world's land areas as a whole, the available data do not point to any clear trend in total precipitation over the last 50–100 years. However, some regional patterns have emerged, as noted in the 2013 IPCC assessment. There's generally more rain and snow falling across middle and higher latitudes of the Northern

A more potent monsoon for India?

There's no better example of seasonality in rainfall than the Asian monsoon, which delivers life-giving water—and flooding that is sometimes deadly—to more than two billion people from Pakistan to China. The monsoon normally kicks into gear as the Indian subcontinent cooks in the intense heat of late spring. Updrafts rising from the sunbaked ground help pull in moist air from the nearby ocean, just as a hot day by the seashore stimulates a sea breeze. While the basic process is the same from year to year, the strength of a given year's monsoon can vary greatly. For instance, heavy winter snows over the Tibetan Plateau can slow the process by which land heats up in the spring and pulls in summer monsoon moisture. Also, El Niño events tend to dampen the Indian monsoon's strength, while most La Niña years are wetter than usual. Seasonal forecasters have gained skill at incorporating these and other factors into outlooks for each year's monsoon, although most Indians still wait on tenterhooks until the first raindrops arrive.

Since the last ice age, the monsoon has apparently strengthened and weakened in tandem with Earth's temperature. About 10,000 years ago, summer sunlight across the Northern Hemisphere was a few percent stronger than today, due to a cyclic variation in Earth's orbit (see p. 259). The extra sunlight not only helped melt the ice sheets but also intensified the Asian monsoon. After that effect started to wane about 6000 years ago, the trend has been toward weaker monsoons during cooler periods and stronger monsoons when Earth is relatively warm. Oddly enough, the link is especially strong with the North Atlantic: when those surface waters are warm, it's a sign that the global conveyor belt of ocean circulation is vigorous (see p. 161), and this appears to help keep the Indian monsoon humming as well.

Consistent with this picture, most of the global models analyzed in the 2013 IPCC report show an increase in average summer monsoonal rains across South and East Asia over the next century, including India. However, the models also suggest that year-to-year variability in the Indian monsoon's strength could increase as well.

Since the 1950s, India's average monsoon rainfall has actually dropped by 5%–10% even as global temperatures have warmed. Rainfall totals fell below the long-term average during 9 of the 14 years from 2000 to 2013. There's still plenty of variation from year to year, however. The monsoon of 2009 was the third weakest in a century, and the tepid 2012 season reduced power from India's hydroelectric dams enough to help trigger one of the world's biggest-ever electrical outages. Yet the 2013 monsoon was quite healthy, racing inland at the fastest pace in at least 50 years and retreating in the autumn weeks later than usual.

At least some of India's recent drying may be due to the rapid growth of sulfate pollution, according to a 2006 study by Chul Eddy Chung and V. Ramanathan of the Scripps Institution of Oceanography. The soot and other pollutants blowing offshore appear to be blocking enough sunlight to shift heating patterns across the Indian Ocean and in turn weaken the circulations that drive India's monsoon. Another possible culprit behind weaker Indian monsoons is the very irrigation that helped fuel the much-touted Green Revolution. A 2010 study led by Purdue's Dev Niyogi used satellite data to reveal that the average amount of soil moisture during the hot weeks before the monsoon's arrival jumped threefold during the period 1988–2002. However, rainfall over the same region has plummeted as much as 40% since the 1950s. It's possible that, even as parched conditions prod farmers to irrigate, the moistened soil might be making things worse by weakening the heat-driven updrafts that help pull monsoon flow inland.

No matter how the average monsoon evolves, it's the extremes that cause havoc. Monsoon failures of the past have led to massive famines, and excessive rains—which may become more intense in a warmer, moister atmosphere—can put enormous stretches of land underwater. A classic case in point is 2010's monsoon, which came on the heels of some of the hottest weather in Asia's history (see previous chapter). Instead of India's usual wet-to-the-east/dry-to-the-west pattern, which has generally intensified in recent years, 2010 saw the tables turned. Some of India's lush eastern states received as little as half of their usual rainfall, while desert areas in the far west got close to double their normal quotient. But 2010 also brought far too much of a good thing to some areas. In India's far northwest, the normally bone-dry town of Leh was raked by a vicious flash flood on August 6 that killed more than 150 people and injured thousands more. Farther south, along the Indus River valley in Pakistan, monsoonal torrents led to a mind-boggling round of flooding that took more than a thousand lives and displaced millions of residents across a huge area. (Far to the northwest, the same locked-in-place weather pattern helped lead to that summer's deadly heat wave and fires in Russia.)

Hemisphere. (In the United States, average precipitation has increased by about 5% over the last century, according to the National Oceanic and Atmospheric Administration.) Meanwhile, the distinction between the wet tropics and dry subtropics has intensified over the last 30–60 years, with the wet areas tending to moisten and the dry areas trending drier.

Over the next century, researchers expect a further increase in rain and snow, especially in winter, poleward of about 50°N (including most of Canada, northern Europe, and Russia), with rainfall also increasing in tropical areas that benefit from monsoons. Meanwhile, precipitation may decrease—particularly in the summer—across the subtropics and in drier parts of the midlatitudes, including much of Australia, Southern Africa, the Mediterranean and Caribbean, and the southwestern United States and Mexico. While confidence in this general picture has become stronger, there's still important uncertainty about just where the transition zones between increasing and decreasing precipitation will lie.

When it comes to floods, though, it's the character of the precipitation that really counts. How often is rain concentrated in short, intense bursts that produce flash floods, or in multiday torrents that can cause near-biblical floods across entire regions? And how often is it falling in more gentle, nondestructive ways? Here, the global-change signal is worrisome. Data show a clear ramp-up in **precipitation intensity** for the United States, Europe, and several other areas over the last century, especially since the 1970s. When it rains or snows in these places, it now tends to rain or snow harder, over periods ranging from a few hours to several days. Projections from computer models suggest that this trend will intensify in the decades to come, although the details will vary from place to place and not every location will follow the trend. Modeling carried out in support of the 2013 IPCC assessment indicates that the heaviest five-day precipitation events one might expect per year at a given spot could increase by anywhere from 5% to 20% this century, depending on the pace of greenhouse emissions.

There's a cruel twist to this story. You might expect droughts to diminish on a global basis as rainfall intensity goes up, but higher temperatures not only allow more rain-producing moisture to enter the atmosphere, they also suck more water out of the parched terrain where it hasn't been raining. Atmospheric circulations associated with a warming planet also tend to reduce the number of lighter precipitation events. Thus, in addition to triggering more intense rainfall, global warming could also increase the occurrence of drought, a seeming paradox that already appears to be taking shape in some areas. Where and how the worst droughts manifest hinges on a number of factors, including a tangled web of influence through which the world's tropical oceans can help trigger multiyear drought in places as far flung as the southwestern United States and southern Africa. Even in a single location, both drought and downpours can become more common if the timing of rainfall shifts in not-so-fortunate ways.

A wetter world (but not everywhere)

The stakes are high when it comes to assessing the future of floods and droughts, because both are notoriously deadly and destructive. According to the World Meteorological Organization, flooding affected 1.5 billion people between 1990 and 2001. China's Yangtze River flows over its banks so reliably during monsoon season (its floods and related crop failures have claimed more than a million lives since 1900) that it's prompted the world's largest flood-control project, the colossal and controversial Three Gorges Dam. Many other flood victims die on a smaller scale, unlucky enough to be caught solo or in a small group along a rampaging creek during a lightning-quick flash flood.

Though drought is a less dramatic killer than flooding, its toll is even more staggering. Monsoon failures across India reportedly killed more than a million people each in 1900 and 1965–1967. More than 30 million Chinese died in the first half of the twentieth century as a result of drought and related famine, according to the U.S. National Climatic Data Center. The World Meteorological Organization estimates that more than a million people died throughout Africa's Sahel during 25 years of poor rainfall that peaked in the devastating droughts of 1972–1975 and 1984–1985 (see box on p. 87). Southern and eastern Africa have also been prone to recurring drought over the last several decades. Looking further back, the history of civilization is checkered with cultures believed to have met their downfall partly or mainly because of drought (see chapter 11).

The roots of both flooding and drought lie in the physical process known as evaporation. As global warming heats the world's oceans, the water molecules near the sea surface become more energetic and tend to evaporate into the atmosphere more readily. Thus, the air gains water vapor—or, as it's often put, the air "holds" more water vapor. For every degree Celsius of warming, the water vapor in the atmosphere is able to increase by up to about 7%.

Because the connection between warmth and evaporation is so well established, scientists have considered it a fairly safe bet that a warmer atmosphere would carry more water vapor around the globe. That's been confirmed by observations since the 1970s, which point to an increase in

England and the rains of autumn

You'd have to have been a contemporary of King Charles I in order to have seen Yorkshire's River Ouse higher (in 1625) than it was in autumn 2000. A series of punishing rains and floods swept much of the United Kingdom during October and November 2000, flowing into over 10,000 homes and prompting more than 1400 flood warnings. The total cost topped US$1 billion. It was the United Kingdom's most widespread flooding since March 1947, when many areas were inundated by the melting of a six-week snowpack. The floods in 2000 came during the wettest autumn in the 230-plus years of record keeping for precipitation in England and Wales. Moreover, the November damage could have been far worse, according to the consulting firm Risk Management Services: "It is clear that a flood event of even a slightly greater magnitude would have overtopped and/or breached many defenses, causing a significantly greater loss." Only a few years later, in 2013–14, England and Wales slogged through their wettest winter on record, with the London area getting more than twice its typical rainfall.

By themselves, the floods of autumn 2000—or summer 2007, or winter 2013–14—can't be laid at the doorstep of climate change, but the character of rainfall in the United Kingdom is clearly changing, says Hayley Fowler of Newcastle University. As part of a European Commission–funded project, she and Chris Kilsby joined colleagues in analyzing rainfall extremes at 204 U.K. stations in nine regions for the period 1961 to 2000. They focused on "return periods," the amount of time one would expect to go between experiencing extreme rainfall or flooding of a given magnitude.

The project produced a cloudburst of attention-grabbing statistics. Across eastern Scotland, for instance, a 10-day rainfall of 150 mm (5.9 in.) could be expected every 50 years during the mid-twentieth century. But since 1990, it's now moved closer to every eight years. Other parts of the United Kingdom have also seen substantial changes, with many return periods in the north less than half of what they were before 1990. Such dramatic shifts have come unexpectedly soon, says Fowler: "This pattern of change is the same as that projected by climate models under global warming for the end of the twenty-first century." Subsequent work by Fowler and colleagues, extending through 2009, confirmed the trend toward more frequent bouts of intense rain in most parts of the United Kingdom across most of the year, although research continues on how summer thunderstorms might evolve.

In the past, the northern and western United Kingdom would tend to experience its worst floods in autumn, while the southern United Kingdom would get

them in wintertime. But now, the worst floods in the south—including London and environs—seem to be occurring in the September to November time period. That is likely a reflection of the recent shift in extreme rainfall events toward autumn over the last century, also documented by Mari Jones (NCAR). "This has huge economic and social implications, especially if these trends continue under global warming, as climate models suggest," says Fowler.

If the trends do continue, they may dovetail with projections from the U.K. Climate Impacts Programme, whose twenty-first-century outlook (UKCP09), released in 2009, projects wetter winters and drier springs, summers, and autumns across Britain. Fowler's work using climateprediction.net simulations (see sidebar, chapter 12) also supports a shift toward increasingly soggy British winters. However, UKCP09 also indicates that autumn-to-autumn variability in rainfall may increase markedly across southeast England. Thus, it's possible that tranquility could alternate with torrents from one October to another.

atmospheric moisture near ground level of roughly 0.1% per decade. (The increase largely stalled after 2000, in tandem with the recent flattening of global temperatures, but this may be only temporary.)

Rain and snow tend to develop where air is converging and rising. If the air is warmer and has a bit more water vapor, it ought to rain or snow more intensely, all else being equal. This appears to be exactly what's happening on a global average, as noted above. However, rainfall and snowfall often vary greatly over small distances. Thus, a modest increase in global precipitation intensity could mask regional and local trends that are more dramatic, downward as well as upward. To complicate the picture further, rain or snow totals at some locations can rise or fall sharply for a year or more because of the climate cycles such as El Niño discussed below.

As with temperature, it's the extremes that matter most in rainfall and the lack of it. The United Kingdom's soggy summer of 2007 vaulted into the record books largely on the strength of two extremely wet days: June 25, which produced 103.1 mm (4.06 in.) of rain in Fylingdales, North Yorkshire, and July 20, when 120.8 mm (4.76 in.) fell at Pershore College in Worcestershire. In both cases, the amount of rain that fell over 24 hours was roughly half of what one would expect during an entire summer. The resulting floods put large sections of northern and western England under water. Hundreds of thousands of residents lost power and access to drinking water, and damages were estimated at more than £2 billion.

To help quantify the links between downpours and climate change, the scientists who came up with a set of criteria for measuring temperature extremes (see chapter 4) did the same for precipitation. Those indices are a key part of research that fed into the IPCC's 2007 and 2013 assessments. In 2002, Povl Frich (Met Office Hadley Centre) and colleagues confirmed that precipitation extremes seem to be increasing in general, especially across the planet's higher latitudes. Highlights from Frich's study, which looked at the second half of the twentieth century, included the following:

* **The number of days that see at least 10 mm (0.4 in.) of rain or melted snow** rose by 5%–15% across large stretches of the United States, Europe, Russia, South Africa, and Australia. These locations also tended to see upward trends in the peak five-day totals of rain or snow in a given year.
* **Much of North America and Europe** showed jumps in the fraction of rainfall and snowfall that fell on the soggiest 5% of all days with precipitation. In other words, the wettest days became wetter still.

Subsequent research, including modeling in support of the 2013 IPCC assessment, backs up the general picture of global trends toward more intense rain and snow events. A 2013 overview led by Markus Donat (University of New South Wales) found a particularly strong jump since the 1980s in the ratio of total annual rainfall to the number of wet days—in other words, a jump in the average amount of rain per wet day. However, there's still a lot of smaller-scale variability in the global patchwork of wet and dry. Even a region where precipitation is generally on the increase can have some locations that buck the overall trend.

Are floods increasing?

Among the lineup of atmospheric villains often associated with climate change, river and stream flooding is one of the toughest characters to nail down. There's no doubt that rising seas will enhance the flood threat from hurricanes and coastal storms along low-lying shorelines (see chapters 7 and 8), but inland flooding is another matter.

Although there's a good deal of confidence that the most intense bouts of rain and snow are getting heavier, it's surprisingly hard to compile a global picture of whether river floods are becoming more frequent or intense. In part, that's because the chain of events that leads from an unusually heavy rain to a flood involves many factors other than immediate weather: for example, how wet the region's soils already are, how high rivers and reservoirs are running, what kind of flood-control devices are in place and—perhaps most critically—how much the landscape has been altered by development. Putting aside any potential contribution from climate change, development and population trends alone will boost the world's financial exposure to flood risk severalfold from 2010 to 2050, according to a study led by Brenden Jongman (University of Amsterdam) that considered both river and coastal flooding.

One of the few attempts at a global flood census was published in the journal *Nature* in 2002. Led by Christopher Milly of the U.S. Geological Survey, the study examined 29 of the world's largest river basins. It found that 16 of their 21 biggest floods of the last century occurred in the century's second half (after 1953). However, as noted above, a warming climate makes both downpour and drought more intense, so an increase in floods doesn't necessarily mean rivers are running consistently higher around the globe. A 2009 study from the National Center for Atmospheric Research estimated that about a quarter of the world's largest rivers saw a reduction in streamflow between 1948 and 2004, while only about 10% of the rivers showed an increase.

Land-use changes play a huge role in flooding potential. Deforestation appears to exacerbate the risk of flooding and landslides in most cases, as the water that falls is less likely to enter the soil and more likely to move unimpeded toward the nearest stream or low area. Water flows particularly easily across the acres of pavement that are laid down as cities expand into the countryside. Faster-flowing water is especially likely to feed into small-scale flash floods, which are even harder to monitor and analyze than larger river floods.

Timing is another important factor when looking at how global precipitation and flood risk have evolved. For instance, are the heaviest one-hour deluges getting heavier, which would presumably help make flash flooding more likely? This is virtually impossible to calculate for many parts of the globe, because many national meteorological centers do not compile such fine-scale data or won't release them without collecting a hefty fee. Also, the heaviest of heavy downpours tend to be so localized that they can be measured only with the help of the dense observing networks found mainly in highly industrialized countries. This leaves us in the dark as to changes in flash floods across poorer nations.

It'll likely remain challenging for scientists to untangle any greenhouse-related trends in river flooding from the myriad other factors involved. In its special 2012 report on managing risks from extreme events, the IPCC noted that "[m]any river systems are not in their natural state anymore, making it difficult to separate changes in the streamflow data that are caused by the changes in climate from those caused by human regulation of the river systems." The report expressed low confidence in analyses of recent change in the global magnitude and frequency of floods, as well as in projections of future change, albeit with a few regional exceptions. This lack of confidence does not necessarily mean that changes related to greenhouse gases haven't happened or that they won't occur in the future, only that they're difficult to confirm, as noted above.

Defining drought

Anybody can see drought at work: wilting crops, bleached grass, trickling streams, a receding lakeshore. Yet scholars have debated for many years how best to define this strange condition—the result of the absence, rather than the presence, of something.

Which way will the Sahel go?

Only a few years after the promising 1960s, when many African nations were shaking off their colonial past, drought stepped into the role of oppressor. A prolonged period of low rainfall—which hit hardest in 1972-1975 and 1984-1985—caused despair across much of the continent, with the north-central Sahel region at the epicenter. The region's name is derived from the Arabic *sahil*, or shore, and it's an apt label. Each summer, like the tide coming in, monsoon moisture sweeps north from the Gulf of Guinea. It brings the only substantial rain of the year to the semiarid belt of land that extends from Senegal to Sudan across the great breadth of Africa. Stretching more than 7000 km (4400 mi), the Sahel separates the wet tropics to the south from the parched Sahara to the north. There's a lot of climatic contrast packed into this thin ribbon, which spans only about 800 km (500 mi) from north to south. Across that span, the average yearly rainfall drops from about 750 mm (30 in.) toward the south—more than in London—to about 250 mm (10 in.) toward the north, less than in Los Angeles.

It doesn't take much of a shift in the monsoon to put the region in jeopardy. Major droughts struck the Sahel in the 1910s and 1940s, but the dry spell of the early 1970s was particularly intense, and the world responded with a mass outpouring of relief. Another widespread drought struck in the mid-1980s, this one extending east to Ethiopia. This time the world's response—though unconscionably delayed until the BBC captured the misery on camera—was even more resounding. Bob Geldof organized the supergroup charity single "Do They Know It's Christmas?," followed by the U.S. counterpart, "We Are the World," and the first Live Aid concert.

Despite the loss of dozens of rain-measuring stations as the tragedy unfolded, the drought's signature is clear enough. Analyzing the years 1920-2003, a team of U.S and British climate scientists found that rainfall during the worst of the drought, in 1984-1985, fell nearly 30% short of the long-term average of about 500 mm (19.7 in.). In more recent years, the Sahel's rainfall figures have been both unsettling and encouraging. Though the region has moistened gradually, some years, such as 1994, 1999, and 2003, have produced more rain than the pre-1970 average while others have been notably dry (occasionally due to strong El Niños, as in 1997-1998). Wet years now appear to be driven more by short, intense downpours rather than the sheer number of rainy days, according to Alessandra Giannini (International Research Institute for Climate and Society).

It appears that the North Atlantic holds the key to trends in Sahelian rainfall over the last century. James Hurrell (NCAR) and Martin Hoerling (NOAA) analyzed a major round of climate simulations in 2005 that helped clarify this connection. During the worst of the Sahel droughts, the North Atlantic surface waters ran cool relative to the South Atlantic's, which appears to have fostered rain-inhibiting atmospheric circulations. That state of affairs reversed in the 1990s, as the North Atlantic began to warm more strongly and monsoon moisture returned more reliably to the Sahel, bringing back the long-sought rains.

Subsequent work by Giannini and colleagues showed that tropical oceans as a whole can be juxtaposed against the North Atlantic to provide an even stronger way to diagnose Sahel rainfall. When the globe's tropical oceans are warming more quickly than the North Atlantic, then drought-producing circulations become more likely across the Sahel. Likewise, if the North Atlantic is warming relative to tropical oceans, then the odds of drought-producing patterns in the Sahel go down. As for why the 1970s and 1980s droughts were so fierce, it's possible that widespread industrial air pollution in North America and Europe blocked enough sunlight to contribute to the noteworthy cooling that occurred in the North Atlantic.

The outlook for Sahelian rain in the twenty-first century offers mixed signals. In part, geography remains the Sahel's destiny: the region will stay poised between desert and wet tropics, leading to fluctuations in year-to-year rainfall. But there's also been persistent disagreement among climate models in what might happen over the longer haul. Michela Biasutti (Lamont-Doherty Earth Observatory) examined output from 20 models used in the IPCC's 2013 assessment. About half of the models project the average summer rainfall to increase this century—perhaps by 10% or more—while the other models are split between little change and significant drying. As is the case elsewhere around the world, warmer temperatures should cause whatever rains do fall to evaporate more quickly.

Other challenges are sure to tax the ability of Sahelians to adapt to future climate trends. Right now, the region is one of the fastest growing on Earth: its population has risen from 30 million to more than 100 million since 1950, and it's projected to top 300 million by 2050, according to a 2012 report led by the University of California, Berkeley. The resulting stress on wood-based fuel sources and on food stocks will make the region even more vulnerable to climatic swings. The UC Berkeley report warns that without immediate large-scale action, "death

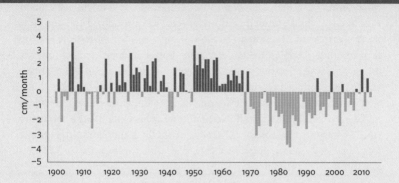

Sahel precipitation was above the long-term mean from 1915 through the late 1930s and 1950s–1960s, after which it was persistently below the long-term mean, with the largest negative anomalies in the early 1980s. (University of Washington/Joint Institute for the Study of the Atmosphere and Ocean)

rates from food shortages will rise as crops wither and livestock die, and the largest involuntary migration in history could occur."

The story of Lake Chad shows what can happen when climate and land use conspire. Once the sixth-largest freshwater lake in the world, Lake Chad shrank by more than 90% from the 1960s to the 1980s, a victim of the droughts discussed above as well as intensive use for irrigation. Recent improvements in rainfall may have halted the decline, although the lake appears unlikely to regain its mid-twentieth-century size without sustained wetness and major changes in water use.

Despite these grim signs, there's some ground for optimism. Though many farmers and pastoralists have pushed into ever more marginal lands across the Sahel, there's also growing diversity in the ways land is used—a sign of resilience in a place where adaptation is essential—and more local involvement in development decisions. David Gressley, the United Nations' regional humanitarian coordinator for the Sahel, has called for a strengthened multidecadal partnership to boost the region's resiliency to drought, one that would involve public, private, and nonprofit parties from the local to global scale. Such work may be critical, as paleoclimate data shows that even greater shifts in rainfall have occurred across this climatically precarious land in the distant past. From that angle, says NCAR's Hurrell, "the recent African dryings appear to be neither unusual or extreme."

Meteorological drought is the most obvious type: the rains stop, or at least diminish, for a given length of time. In places where it normally rains throughout the year, such as London or Fiji, you might see regular rain and still find yourself in a meteorological drought if the amounts don't measure up. One way to gauge a drought in such places is to add up the cumulative rainfall deficit over some period of time (for example 30, 60, or 90 days). Another would be to calculate how long it's been since a single day produced a given amount of rain. In semiarid or desert locations, such as Melbourne or Phoenix, you could measure how many days it's been since it's rained at all. No matter where you are, it's the departure from average that makes the difference, since flora and fauna adapt to the long-term norms at a given locale. A year with 500 mm (20 in.) of rain would be disastrously dry for Tokyo but a deluge in Cairo.

Since drought affects us by drying up water supplies and damaging crops, experts now classify drought not only by rainfall deficits but also in terms of **hydrological** and **agricultural drought**. How low is the water level in the reservoir that your city relies on? Are the lands upstream getting enough rain or snow to help recharge the system? These factors, and others, help determine if a hydrological drought is under way. Especially when a dry spell first takes hold, it may take a while for rivers and lakes to show the effects. Likewise, long-parched ground may soak up enough of a drought-breaking rain to keep the hydrologic system from responding right away. Thus, hydrological drought often lags behind meteorological drought, a process that may unfold over a few weeks in a fairly moist climate or across months or even years in a semiarid regime.

Crops can be affected in even more complex ways by drought, making a declaration of agricultural drought highly dependent on what's normally planted in a given location. If a dry spell is just starting, the topsoil may be parched but the deeper subsoil may still moist enough to keep an established crop going. Conversely, a long-standing drought may break just in time to help germinating plants get started, even if the subsoil remains dry.

What do oceans have to do with drought?

Oceans might seem to be the last thing that would control the onset or departure of a drought, but the more scientists learn about the 70% of Earth's surface that is covered by water, the more they learn about how oceans can dictate the amount of water falling on the other 30%. And as greenhouse gases heat them up, the oceans may start influencing drought in hard-to-predict ways.

The drying of southern Australia

Water—and the lack of it—are key elements in the psyche of Australia. The unofficial national poem, Dorothea Mackellar's "My Country," extols a land "[o]f droughts and flooding rains." As a severe dry spell raged in 1888, Australian Henry Lawson bemoaned his fate in verse: "Beaten back in sad dejection, / after years of weary toil / On that burning hot selection / where the drought has gorged his spoil."

With much of the continent at the mercy of wild swings in precipitation, many Australians have viewed the southwestern coastal belt, roughly from Perth to Cape Leeuwin, as a corner of relative climatic sanity. One of the world's most biodiverse areas, it features a Mediterranean climate with hot, dry summers and mild, dependably damp winters. A 1920 book called *Australian Meteorology* noted the region's reliable moisture: "Here the rains rarely vary 10% from their average amount, and the lot of the farmer should be a happy one."

The farmers and the urban water managers haven't been so happy lately. Rainfall across far southwest and eastern Australia has declined notably over the last half century. Since the mid-1970s, wet-season rainfall around Perth has consistently run some 10%–15% lower than before, most noticeably in the late autumn and early winter. These days, a wet year in Perth is one that merely reaches the long-term average of 848.3 mm (33.4 in.). As of 2013, the city had gone more than four decades without mustering a yearly total of 1000 mm (39 in.), a mark once hit every few years. For a booming city with more than 1.7 million thirsty people, those are scary statistics.

The story is little better across the southeast, Australia's most populous corner. Places like Sydney, unlike Perth, do get a substantial amount of summer rain in the form of showers and thunderstorms. However, new analyses for the last 50 years show that even warm-season rains have slackened slightly across far southeastern Australia. Intense drought dogged the region during the first decade of the twenty-first century, and relentless heat helped deplete water for cities and agriculture even more than the rainfall deficit might suggest. The pattern fluctuated dramatically in 2011, as atmospheric cycles teamed up to drench Australia with so much rain that sea levels measurably dropped as a result. By 2013, precipitation had returned to normal levels and Australians watched warily to see if and when drought would make a comeback.

How much of Australia's drying is related to greenhouse gases, and will it continue? Long-term computer simulations hint that natural variability can push the region into and out of multi-decade dry spells. Like much of Australia, the southwest corner tends toward dryness during El Niño years, so the increase in El Niño's frequency and strength since the 1970s may be a factor. Furthermore, much of the countryside was deforested after European settlement, and some models show that the resulting landscape evaporates less moisture into the air, perhaps exacerbating drought. Even with all that in mind, however, the last few decades of drying do bear some of the fingerprints of global warming, especially for southwestern Australia. Scientists have verified a poleward shift in the storm track that girdles the Southern Ocean, encircling Antarctica. The low-pressure centers that race eastward along this track are the source of nearly all of Perth's rainfall, and they also provide important cool-season rains in Melbourne and Sydney. These days, the storm tracks still bring rain, but increasingly it falls over the Southern Ocean rather than over the parched land.

Global computer models tend to agree that the storm track's winter position is likely to shift even farther south as the century unfolds (see the Southern Annular Mode discussion, chapter 7). This could mean real trouble. Australia's Bureau of Meteorology and the nation's Commonwealth Scientific and Industrial Research Organisation employed an ensemble of 23 computer models for a major set of climate projections released in 2007. The predictions showed an increased risk of further warming and drying across all but northernmost Australia in the coming century. Some models indicated winter and spring drying of up to 10%–15% across southern Australia as soon as 2030. Rising temperatures will only add to the water woes, helping evaporate what rain does fall before it can recharge water supplies.

Desalination, the conversion of ocean water to drinkable water by removing salt, has quickly become a major tool in Australia's hydrologic arsenal. Between 1997 and 2005, streamflows in the Perth region dropped 30% below

Some parts of the world are dry by virtue of their location. Most of the planet's major deserts are clustered around 30°N and 30°S, where sinking air predominates at many longitudes, while rising air keeps most of the deep tropics between 15°N and 15°S quite rainy on a yearly average. The fuel for those showers and thunderstorms is the warm tropical water below. Should a large patch of ocean warm or cool substantially—because of a transfer of heat by ocean currents, for example—it can take months or years for that region to return to more typical conditions. While they last, though, the unusual water temperatures can foster atmospheric circula-

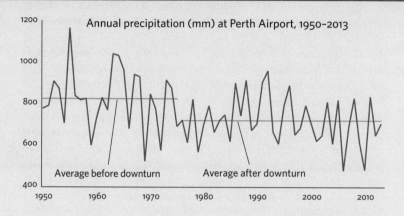

Annual precipitation (mm) at Perth Airport, 1950–2013

Average before downturn   Average after downturn

those observed in the previous 23 years, which were themselves on the dry side. Faced with a deepening crisis, the city responded by launching Australia's first major desalination plant in the industrial town of Kwinana, about 25 km (16 mi) south of Perth. The larger Southern Seawater plant opened in 2011, with the two now capable of meeting more than 30% of Perth's water needs. Other major desalination facilities have come online this decade near Sydney (Kurnell), Adelaide (Port Stanvac), and Melbourne (Wonthaggi). The massive Wonthaggi plant—now Australia's biggest—is expected to provide as much as a third of Melbourne's water supply. Although these projects are generally offsetting most or all of their greenhouse emissions through dedicated wind farms, they've often been lambasted by activists and the nation's Green Party, with fossil fuel use as a key sticking point.

tions that bring persistent wetness or dryness to various parts of the globe.

The supreme examples of such an ocean-driven shift are El Niño and La Niña (see chapter 7), which bring warmer- or cooler-than-normal surface waters, respectively, to the eastern tropical Pacific. This region of warming or cooling can be as large as the continental United States, and the resulting shifts in rainfall patterns can affect much of the globe. Typically, El Niño hikes the odds of drought across Indonesia, Australia, India, southeast Africa, and northern South America, and it can cause winter-long dryness across parts of Canada and the northern United States. Most El Niño droughts only last a year or so, but that's enough to cause major suffering where societies can't easily adapt (by changing crops or shifting their planting schedules, for instance). Some of the worst North American droughts have occurred during multiyear periods when La Niña predominated.

Although El Niño and La Niña are the undisputed champions in drought-making, other oceanic cycles (see chapter 7) also have an influence. Two of the most important are the Atlantic Multidecadal Oscillation (AMO) and the Pacific Decadal Oscillation (also called the Interdecadal Pacific Oscillation when both the North and South Pacific are analyzed). Despite the contrast in their names, both play out over more than a decade: the PDO tends to switch phases about every 20–30 years (though with a good deal of irregularity in the mix) while the AMO takes closer to 30–40 years. There's some evidence that the warm phase of the AMO and the cool phase of the PDO are linked to enhanced odds of drought in the U.S. Southwest. Both of these phases predominated from around 2000 through at least 2013.

On top of these natural waxings and wanings, there's human-induced climate change. Warmer temperatures help promote more evaporation and thus create a tendency for droughts to be more severe where they do occur. Global-scale warming also fosters a poleward shift of arid subtropical zones. These two factors are expected to hike the drought risk substantially along a belt from the southwestern United States to the northern Mediterranean. Already, years of on-and-off drought have left Arizona's Lake Powell and Nevada's Lake Mead—both critical for water supply across the populous U.S. Southwest—struggling to remain even half full. Scientists such as Richard Seager (Lamont-Doherty Earth Observatory) and Tim Barnett (Scripps Institution of Oceanography) have voiced concern that the southwestern United States and northwestern Mexico may enter a state of semipermanent drought later this century. Factors such as El Niño will still lead to periodic wet years, and substantial water may still flow to the region via the Colorado River from the high-elevation Rocky Mountains, where precipitation appears less likely to decrease over time. Still, the long-term trends suggest major trouble for agriculture and water use in the U.S. Southwest.

How will these broad shifts intersect with oceanic cycles and changes? That picture remains fuzzy, because it's exceedingly difficult for climate models to replicate ocean cycles and how they might evolve and interact in a warmer climate. This, in turn, makes it challenging to know when and where oceanic influences on precipitation may shift. Some researchers have argued that the Pacific might move toward a long-term state that resembles

El Niño, while others see a greater risk of an ongoing La Niña–like state. All in all, it seems there is much yet to learn about how ocean-modulated droughts will unfold as the world warms—and the stakes are high for many millions of people.

The plow and its followers: Farming and rainfall

Perhaps no other part of weather and climate humbles human ambition as does rainfall and the lack of it. For centuries we've flattered ourselves by thinking we can control the atmospheric tap, whether it be through rainmaking rituals or cloud seeding. Indeed, one of the first ways in which Western thinkers conceptualized human-induced climate change was the notion that "rains follow the plow"—an interesting spin-off from the manifest destiny mindset that led to U.S. and Australian expansion in the 1800s.

As Europeans moved into progressively drier parts of North America and Australia, they believed that cultivating the prairies and bushland—watering the crops and thus moistening the air above them—would help bring rainfall. Occasional setbacks, like Australia's vicious drought of 1888, didn't quash the belief at first. It wasn't until the 1930s, when severe multiyear drought created the Dust Bowl across the U.S. heartland, that farmers and land managers began to rethink their relationship to climate.

The flip side of this faith that humans could produce a rainy regime via agriculture was the belief that improper use could dry out the land permanently. The notion of **desertification** got its start in Africa's Sahel (see p. 87), where it was postulated by colonial explorers as far back as the 1920s and long supported by actual events. During moist years, government policy in some African nations had encouraged farmers to cultivate northward toward the desert's edge. In turn, that pushed nomadic farmers and their herds into even more marginal territory farther north. When the drought hit, both grazing and farming were too intensive for their land's now-fragile state. Wind and erosion took hold, and the zone of infertile land grew southward. At the time, it was accepted by many that the disaster was being produced, or at least perpetuated, by its victims.

Now it seems that researchers have come around to the idea that neither drought nor rain may follow the plow in a predictable way. It stands to reason that both dry and wet conditions have a certain self-sustaining quality. All else being equal, when the ground is soaked, the air above it is more likely to produce further rain than when the ground is dusty. But the big question is, how powerful is this process against the vast energy reserves of the oceans and the natural variations of weather?

A dust storm approaches the Texas Panhandle town of Stratford in April 1935. (NOAA/ George E. Marsh Album)

Thus far, there's no consensus on exactly how much of a difference land use can make. One possibility is that human-produced changes to the landscape can sometimes modulate a drought that's driven by broader forces. For example, some computer models fed with ocean conditions from the 1930s produce multiyear drying across the southern U.S. Great Plains. However, the actual Dust Bowl was larger and farther north than these model-generated versions. One reason, according to Lamont-Doherty's Richard Seager, might be the influence of dust from overplowed, desiccated farms. As the dust blew across a vast area, it likely fostered dry, subsident conditions that suppressed rainfall over a larger swath than usual.

No matter what might cause their rains to stop, drought-prone areas can take simple, useful steps to reduce their vulnerability. Some of the techniques that proved helpful in the U.S. Great Plains, such as wind-breaking belts of trees, were successfully adapted to the Sahel. Prudent land management can also help people get the most benefit out of the rain that does fall. At the same time, a bit of modesty is in order: oceanic shifts and other natural processes appear to be able to produce drought in some cases no matter how much we do to keep it away.

CHAPTER SIX

The Big Melt

CLIMATE CHANGE IN OVERDRIVE

If there's any place where global warming leaps from the abstract to the instantly tangible, it's in the frigid vastness of the Arctic. Here, the ice is melting, the ground is heaving, plants and animals are moving, and people are finding themselves bewildered by the changes unfolding year by year. Because of a set of mutually reinforcing processes, climate change appears to be progressing in the Arctic more quickly than in any other region on Earth.

Not that some Arctic residents haven't had to adapt to climate change before. Signs of Stone Age life dating back nearly 40,000 years—well before the last ice age ended—have been found near the Arctic Circle in northwest Russia. What appears to be different this time is the pace of the change and the sheer warmth now manifesting. Long-time residents of western Canada and Alaska have seen wintertime temperatures climb as much as 4°C (7°F) since the 1950s. That's several times beyond the average global pace. Some of the permafrost that's undergirded Alaska for centuries appears to be thawing. Sea levels are slowly rising across Arctic coastlines, as they are elsewhere around the globe, and longer ice-free stretches are increasing the risk of damaging, coast-pounding waves. Looking ahead, experts believe the twenty-first century will likely bring a summer when the Arctic Ocean is virtually free of ice for the first time in at least five millennia (and perhaps more than 100,000 years). As we will see, these changes threaten to reorder

Arctic ecologies in dangerous and unpredictable ways, with potentially vast long-term consequences for other parts of the planet.

The big melt isn't limited to the poles, however. Many glaciers in the highest mountains of the tropics and midlatitudes are receding at a startling pace. It's tempting to chalk this up to a warmer atmosphere, but drying may in fact be a bigger culprit than warming in the erosion of glaciers on some tropical peaks, including the one best known to many: Africa's Kilimanjaro (see pp. 126). Meanwhile, the biggest ice repositories of all—the vast sheets that cover most of Antarctica and Greenland—are sending frozen and melted water to the ocean at accelerating rates, an ominous development that will touch every corner of the world's oceans and coastlines.

On thin ice

The North Pole sits at the heart of an ocean ringed by the north fringes of eight countries: **Canada**, **Greenland**, **Finland**, **Iceland**, **Norway**, **Russia**, **Sweden**, and the **United States**. All this adjoining land provides plenty of toeholds for sea ice, which overspreads the entire Arctic Ocean and portions of nearby waters each winter. This ice takes on many roles: a home base for polar bears, a roadbed for people, and a shield for coastal towns and underwater creatures.

In the summer, as warm continental air sweeps poleward from North America and Eurasia, the ice releases its grip on land and contracts into the Arctic Ocean, with only the oldest, most resistant sections clinging to Greenland and the islands of the Canadian Arctic. Much of the rest disintegrates into a messy patchwork that allows icebreaker cruise ships to penetrate all the way to the North Pole (see box on p. 101). While some patches of ice erode in place, winds and waves push other fragments south to meet their doom, primarily through the Fram Strait between Greenland and Spitsbergen, Norway. Eventually, winter returns, and by February the Arctic Ocean is once again encased in ice.

Nobody expects the Arctic to completely lose its wintertime ice cover anytime soon, but summer is another matter. As of 2013, the average coverage in September, when the ice usually reaches its minimum, was only about **65%** of what it was in the late 1970s, according to the U.S. National Snow and Ice Data Center (see diagram on p. 99). Even the wintertime ice extent is shrinking by 2%–3% per decade. Satellite images paint a stark portrait of these trends, which have reached truly dramatic proportions (see illustration on p. 99).

Aug 26, 2012

The yellow contour shows the average minimum extent for Arctic sea ice achieved each year from 1979 to 2010. Though the sea ice normally doesn't reach its minimum until September, it had already shrunk to the lowest value on record by the time this satellite image was taken on August 26, 2012. (Scientific Visualization Studio)

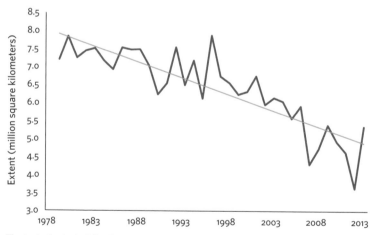

The typical extent of Arctic sea ice—averaged across September of each year, when the ice normally bottoms out—has plummeted during the last several decades, though not every year has shown a drop. (U.S. National Snow and Ice Data Center)

The summer of 2007 smashed all prior records, with even the most experienced ice watchers surprised by the pace of melt. By that September, the ice extent was barely over 4,000,000 km², more than 1,000,000 km² less than its previous low, set in 2005. Only two-thirds of the usual late-summer pack remained, and ice-free waters stretched from Alaska and Siberia nearly to the North Pole. For the first time in memory, the fabled Northwest Passage north of Canada was clear enough of ice so that a typical ocean-going vessel could have sailed directly from Europe to Asia unimpeded.

Scientists later connected the melt of 2007 to unusually sunny skies in early summer, along with an ice-threatening pattern of atmospheric pressure called the Arctic Dipole. Summer winds normally tend to encircle the Arctic, which helps preserve ice locked at high latitudes. But when the dipole sets up, warm air shuttles across the Arctic from North America to Eurasia, and ocean currents push ice toward the North Atlantic, where it's more likely to melt. A team led by Xiangdong Zhang (University of Alaska Fairbanks) has targeted the dipole pattern—only recently recognized, but increasingly common—as evidence of a drastic change in Arctic climate.

The sea ice didn't retreat quite so dramatically the next summer, a point that was trumpeted by climate change contrarians. However, it's now clear that 2007 marked a turning point for the Arctic. In every September since then (up through at least 2013), the minimum ice extent has dipped below anything observed prior to 2007. Moreover, in each of these years, the Northwest Passage opened again, as did the Northern Sea Route above Siberia. Dozens of adventurous sailors now trek to the region each year, and in 2010 two yachts—*Northern Passage* and *Peter 1*—became the first to complete both routes in a single summer, circumnavigating the Arctic.

At first, it looked as if a brutally cold winter and spring across the western Arctic in 2012 might lead to less summer melt, but once again, the ice deteriorated at a shocking pace. High winds from an unusually powerful summer cyclone helped to speed the ice's breakup across much of the central Arctic. By September, the record-low extent from 2007 had been broken by a wide margin, falling well below the consensus outlook issued just weeks earlier by a panel of experts. A noteworthy rebound in summer ice extent did occur from 2012 to 2013, as was the case from 2007 to 2008 (see graphic on p. 99). Again, this may simply be another pause in what resembles an inexorable, two-steps-forward-and-one-step-back process.

The Arctic's ice isn't just contracting horizontally; it's also getting shallower. Widespread summer melt and shifting ocean currents are eliminating much of the ocean's thick multiyear ice, which can extend 2 m (6.6 ft) or more below the surface. The thinner coat that forms in place of multiyear ice is easier to erode when summer returns. Data on ice thickness collected by U.S. nuclear submarines from the late 1950s to 2000 show that sea ice thinned by up to 40% over the period. The submarines traversed only a

Where did that North Pole go?

A group of Arctic tourists got less than they bargained for in the summer of 2000 when they headed for the North Pole on board an icebreaker ship. Planning to set foot on the ice at 90°N, they made it to the pole—and found only a few scattered floes strewn among the cold water. The group's unnerving experience soon made it onto the front page of *The New York Times*. A paleontologist and a zoologist on board the ship were quoted beneath the headline "Ages-Old Icecap at North Pole is Now Liquid, Scientists Find." Climate experts hastened to point out that winds and summer warmth can easily push open sections of ice as far north as the pole in a typical summer, but the image of an unnaturally soggy North Pole persisted, a symbol of the very real Arctic melting that's been gaining momentum over the last few years. Extreme swimmer Lewis Gordon Pugh called on the archetype in 2007 when he swam a kilometer amid chunks of ice at the pole to draw attention to climate change.

few select routes below the ice, though. Estimating ice thickness over a broader area from space isn't easy, but NASA started doing so in 2003 using a laser-based sensor on board a satellite called ICESat. (Similar data are being collected by a European satellite launched in 2010, Cryosat-2.)

Computer models are stitching together the limited thickness observations with depictions of ocean, ice, and atmospheric behavior to produce a comprehensive portrait of ice decline in three dimensions. At the University of Washington, a model called the Pan-Arctic Ice Ocean Modeling and Assimilation System (PIOMAS) indicated that by 2013 the April maximum in ice volume was down about 20% from its 1979–2012 average, and the September minimum had plunged by roughly 60%—far more than the surface data shown on p. 105 might lead one to believe.

Even the stubborn remnants of multiyear ice hanging on for dear life aren't what they used to be. A team led by David Barber (University of Manitoba) trekked deep into the Arctic Ocean on board an icebreaker in September 2009. They expected to grind their way through multiyear or thick first-year ice along most of their route.

> **The rules are starting to change, and what's changing the rules is the input of greenhouse gases."** —*Mark Serreze, U.S. National Snow and Ice Data Center*

What makes the Arctic so vulnerable?

Trouble seems to beget trouble when it comes to far-northern climate. As a team of scientists noted in the American Geophysical Union's weekly newspaper, *Eos*, "The Arctic system balances on the freezing point of water. Each summer, the system swings toward the liquid phase; each winter, it returns to the solid phase." While a tropical city might warm by a degree or two without obvious effects, the same amount of warming could transform parts of the Arctic if it were to bring the system from just below to just above the freezing point. If the tropics are the lumbering long-haul trucks of the climate world, responding slowly but powerfully to greenhouse gases, then the Arctic is a finely tuned sports car, reacting sensitively to even small changes.

Much of this responsiveness is because of positive feedbacks: self-reinforcing processes that tend to amplify change. In its definitive 2005 study, the Arctic Climate Impacts Assessment (ACIA) noted several **positive feedbacks** that allow the Arctic to warm more quickly than other parts of the globe in response to a given increase in greenhouse gases. These include the following:

* **A darkening surface.** Several kinds of change in the Arctic all help to produce a darker surface that absorbs more sunlight and, in turn, warms the air above it more strongly. Open ocean typically absorbs more than 90% of the solar energy reaching it, while snow and ice absorb as little as 10%, reflecting the rest of the sunlight off their bright white surface. Thus, a patch of ocean that remains unfrozen can soak up as much as nine times more solar energy. Likewise, when snow cover over land disappears for longer periods, the exposed tundra absorbs more of the sunlight than the snow did. Shrubs and forests, in turn, absorb more solar energy than the lighter tundra does, so as these larger plants migrate northward (see p. 113), they act to further warm the region. (This is in contrast to the tropics, where trees tend to cool the climate as they catch rainfall and evaporate it back into the air.) Pollution wafting into the Arctic from more populated areas, including fast-industrializing Asia, helps darken some of the snow and ice, increasing its absorptivity. A 2007 study led by Mark Flanner (University of California, Irvine) found that the effects of regional soot may explain a third or more of the Arctic's warming since preindustrial times. Mark Jacobson (Stanford University) found that reducing soot and related gases could be the quickest way to limit Arctic ice loss, and a 2013 project called Dark Snow zeroed in on the question of how pollution might be hastening Greenland's ice melt.

* **A more direct route to warming.** As the sun passes over the tropics, much of its energy goes into evaporating water but the evaporation rate in the frigid Arctic is far less. Thus, of the sunlight that reaches the Arctic, a higher percentage goes directly into warming the air.
* **A thinner atmosphere.** The Arctic's troposphere, or "weather layer," typically extends about 8–10 km (5–6 mi) up, compared to 16–18 km (10–11 mi) in the tropics. In the shallow troposphere of the poles, it takes less energy to produce a given amount of warming. There's another wrinkle as well. Especially during clear, calm, cold spells, Arctic landscapes often experience a thin, ground-hugging layer of air, called an inversion, that's much colder than the air just a few hundred meters higher up. When the inversion breaks up, the surface air can warm dramatically. Thus, anything that punctures inversions or makes them form less often, such as a change in atmospheric circulation or a loss of snow cover, could produce major ground-level warming.

Potential changes in **heat-trapping clouds** are a huge uncertainty in polar regions and elsewhere across the globe (see chapter 10). Over the Arctic, summer cloudiness appears to be increasing. This could help tamp down snow and ice loss, since low clouds and fog in summer screen out the midnight sun, whereas clear skies in winter allow heat to easily escape to space. However, the variegated, complex behavior of clouds can lead to unexpected results. One analysis found that distinctive low-lying clouds over central Greenland in July 2012 allowed sunlight in while blocking outgoing heat, which helped lead to record-breaking ice melt.

With all this in mind, we might hope to find a few **negative feedbacks**—processes that tend to dampen warming over time, such as the apparent cloud changes noted above. One possibility is the vast flow of heat from the tropics toward the poles. If the Arctic continues to warm at a faster pace than lower latitudes, as expected, then the pole-to-equator contrast would be weakened and the flow of heat ought to slacken. However, those weakened winds will carry more moisture than before—and that's a warming influence, all else being equal.

All of these feedbacks, plus the tendency of glacial melting to accelerate, point toward a potential cascade of effects—a so-called "tipping point," as it's been characterized by some scientists and journalists. In fact, there may be a number of tipping points, each with its own setting. If so, each point we pass makes it that much harder to avoid the others. Conversely, the sooner that greenhouse warming is stabilized, the less likely the Arctic will enter a period of runaway change.

Instead, they encountered chunks of older, thicker ice beneath a coating of fresh ice only about 5 cm (2 in.) thick. When scanned by NASA's satellite-based sensor, this new icescape resembled a more typical, multiyear ice field—a masquerade hiding a far more serious decline than scientists had realized. The tenuous state of today's Arctic ice was also highlighted by the Great Arctic Cyclone of August 2012. This unusually large and intense storm delivered several days of strong wind that fragmented large swaths of ice across the central Arctic. Before-and-after satellite images showed a huge impact on the ice pack, but it looks as if the storm mainly hastened the destruction of ice that was already doomed, according to an analysis led by Jinlun Zhang (University of Washington).

How long will the Arctic Ocean hang onto its summertime ice? Until the last few years, the best consensus among top climate modelers was that the Arctic would retain some ice throughout each summer until at least 2080, and perhaps into the next century. But with more recent data and model results, that consensus has fragmented even more quickly than the ice itself. The 2013 IPCC report asserts that the Arctic is likely to see a September virtually free of sea ice before midcentury, assuming that emissions follow a middle-of-the-road path. That's based largely on global climate models that show improved skill in depicting Arctic ice. However, even the best models still tend to deplete sea ice at a pace slower than that actually seen over the last few years.

As pointed out by James Overland and Muyin Wang in a 2013 review for *Geophysical Research Letters*, there are three common approaches to predicting how quickly ice loss will proceed.

* If you simply extrapolate from recent trends in sea ice volume, which is diminishing more rapidly than sea ice extent, you might conclude that a nearly ice-free Arctic could arrive by 2020, or even sooner (see graphic, next page).
* However, weather and climate don't usually follow straight lines or smooth curves. Natural variability in the Arctic could easily lead to plateaus in ice loss. With this in mind, you might expect it to take a while longer to reach the nearly ice-free point, perhaps around 2030.
* If you go strictly by climate models, you'd call for the ice-free benchmark to occur still later—perhaps around midcentury, with some models pegging a date as early as 2040 and others around 2060 or later.

Predicting when we'll see the first summer days with a wide-open Arctic is one of the great forecasting endeavors of our time, especially given the

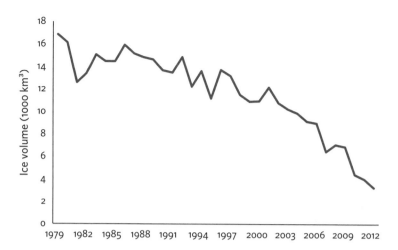

Until 2004, the volume of ice remaining at the last day of each melt season, as computed by the University of Washington PIOMAS model, topped 10,000 km³ each year. In 2012, that volume was a mere 3261 km³. If this pace of decline were to continue, the Arctic Ocean could be virtually ice-free at some point before 2020. However, many experts believe that the rate of ice volume loss will slow down before then. (PIOMAS/ University of Washington)

rollercoaster developments of the last decade. But this is much more than an intellectual challenge. As Overland and Wang put it, "Whether a nearly sea ice-free Arctic occurs in the first or second half of the twenty-first century is of great economic, social, and wildlife management interest."

People, animals, and ice

So what's wrong with a few extra weeks of open ocean in a place that's usually locked tight in ice? Indeed, the prospect of ice-free summers has many industries and entrepreneurs salivating, especially those hoping to cash in on trans-Arctic shipping (see p. 392). Along the Arctic coast, however, the increasingly tenuous state of the ice during autumn and spring complicates life for indigenous residents. Subsistence whalers and seal hunters use sea ice to venture beyond the coastline. The timing of those seasons, and the quality of the ice, has now become less dependable. Also, the nearshore ice normally helps keep wind-driven waves well offshore during the violent weather of autumn and spring, but storms now smash into coastal towns unimpeded by ice well into autumn. The resulting erosion threatens the very survival of some Arctic settlements (see p. 108).

The loss of sea ice has especially grave implications for **polar bears**, which have become a familiar symbol of climate change for good reason. In some areas, these white giants remain on the sea ice throughout the year, retreating to land only when forced there by melting ice. Land areas are also used by pregnant polar bears, who then bring their hungry young onto the sea ice to find seals who are bearing and raising pups in the spring. In areas where the ice melts completely in summer, the bears go without food while on land, relying on fat reserves built up during the spring. Now that ice-free spells are lengthening, the reduced duration of sea ice duration (earlier breakup and later freeze-up) is a stressful affair for both mothers and nursing cubs. At Hudson Bay, near the southern end of the polar bears' range, the summer fast is now a month longer than it was decades ago. The body condition of bears arriving on land has declined and has resulted in both lower survival (especially for young and old bears) and lower reproductive rates.

Greater extremes in precipitation (see p. 77) are also a concern. Fires that rage through drought-stricken boreal forests destroy permafrost and the dens lodged within it, and more frequent spring rains have collapsed some of the bears' dens. Rain in place of snow can also affect the availability of prey for polar bears, because ringed seals rely on snow lairs to give birth to and raise their pups. When sea ice forms later than usual, it can reduce the amount of winter snowpack on top, thus posing another risk for the seals' lairs. Then there's the threat of interbreeding between polar bears and grizzlies now taking shape in Canada, as a warming climate shifts the grizzlies' range toward coastal lands where polar bears roam in between shorter sea-ice seasons. One bear shot in 2006 in northern Canada was later found through DNA testing to be a grizzly/polar hybrid: a "pizzly."

Despite all of these concerns, the polar bear is hardly disappearing from Earth just yet. Because the bears' territory is so remote, it's tough to obtain a firm count, but the most common estimate is that there are now between 20,000 and 25,000 of them. That's more than twice the number tallied in the 1960s, but the increase has been attributed to careful management of hunting that allowed depleted populations to recover. Going forward, the challenges facing polar bears are very different. If in fact the bears become less of a presence in the Arctic over time, there's another predator waiting in the wings. Killer whales, which normally avoid sea ice because of their long dorsal fin, are now expanding into polar bear habitat during the ice-free season. Killer whales would likely become the top predator in the Arctic marine ecosystem if sea ice were to disappear.

Recent genetic studies indicate that polar bears as a species have some 600,000 years of history behind them, which implies that they've survived interglacial periods when temperatures were slightly warmer than today. However, the pace of warming is much faster this time around (with still higher temperatures on the way), and there's added stress from human civilization that didn't exist back then. With all this in mind, it's unclear how well the bears will be able to adapt. Only a few months after the Arctic's dramatic melt of 2007, the U.S. government listed polar bears as a threatened species. In a study released that year, the U.S. Geological Survey (USGS) estimated that two-thirds of Earth's polar bears will be gone by 2050.

The safest haven for the species may be in the waters adjoining northern Greenland and the archipelago of far northern Canada, where year-round sea ice is expected to linger the longest. The World Wildlife Fund is working with nearby residents on designating a potential "last ice area" that could serve as a refuge for the bears as well as for the ways of life practiced by indigenous residents.

Several species of **seal** are also at serious risk from ice loss. **Ringed seals**, the Arctic's most common type, are a common food source for many Inuit communities as well as for polar bears. As noted above, the seals nurse their pups in lairs on top of the snow-coated ice sheet, so the stability of springtime ice is critical for them. Fortunately for the seals, Arctic sea ice hasn't diminished as rapidly in spring as it has in summer. One survey of ringed seal populations near the northwest coast of Hudson Bay showed a seesawing pattern, with the numbers dropping in the 1970s and 1990s but rising in the 1980s, in sync with the snowfall that the seals use to build their winter dens. As the ice retreats, ringed seals may get a short-term boost from richer marine life, but those benefits will vary from place to place. Pups may find themselves exposed to the cold ocean—and to predators—at earlier points in the spring. In 2012 the National Oceanic and Atmospheric Administration proposed listing several species of ringed seals as threatened, noting the same climatic pressures that drove the polar bear listing. There's no evidence that ringed seals will take to land, which goes for several other seal species as well, so in the long run the loss of summertime ice may prove devastating. In contrast, gray seals and other temperate species already accustomed to land have a better shot at migrating northward and/or dealing with reduced Arctic ice successfully.

Birds—made for mobility—appear to have a head start in adapting

> "You don't have to be a polar scientist to see that if you take away all the sea ice, you don't have polar bears anymore." —*Andrew Derocher, University of Alberta*

Move or drown: One town's tough choice

Poised at a geographic crossroads, Shishmaref, Alaska, also lies in the cross-hairs of a warming climate, earning the village of 600 people a modest measure of fame. Shishmaref is perched on a slender island the length of New York's Central Park but barely half as wide. It lies off the northern coast of the Seward Peninsula, not far from Russia (in fact, Shishmaref is closer to Moscow than to Washington, D.C.). Inupiat peoples, with close ethnic ties to their counterparts across the Bering Strait, have lived on the island for centuries. Today, like many indigenous communities, their town is a hybrid of the modern (a school, con-temporary homes) and the frontier (no television, no running water).

Each winter the Chukchi Sea surrounds Shishmaref with a thick layer of ice that's integral to the subsistence-based, hunting-and-fishing culture. The ice also keeps out seawater that would otherwise smash into the island's cliffs during violent winter storms. Residents have long counted on the freeze-up to occur in October, but lately there have been patches of open water near shore as late as December, leaving the town at the mercy of autumn elements. Recent October storms have flung winds of 144 kph (90 mph) and waves of nearly 4 m (13 ft) at the tiny island, whose top elevation is only 6.7 m (22 ft).

In an earlier time, the people of Shishmaref might have simply packed up and moved on, but abandoning a modern township with ancient roots isn't so easy. Though it goes against their cultural grain, locals pleaded for state and federal officials to recognize their plight. An Erosion and Relocation Commission was

to changes across the Arctic. The task won't be easy for all species, though. Ivory gulls are scavengers that feed mainly on the carcasses of ringed seals left by polar bears, so a decline in those species will at least force the gulls to find other meal tickets. Fish and other marine creatures may also find their diets shaken up, as schools shift with the warming oceans and as industrial fishing moves northward into formerly ice-encased areas. One of the major challenges facing birds in the far north is mismatched ecosystem timing, which is especially critical here given the Arctic's short window of warmth each summer. Some species, such as Arctic terns, travel all the way from the Antarctic to exploit the brief seasonal abundance of food in the Arctic. If the timing of the productive season shifts, many species are expected to be out of sync with their food.

It's hard enough for some forms of Arctic wildlife to deal with climate change, but **pollution** may be making matters worse. A witch's brew of

formed in 2001, and in 2002 the townspeople voted by a count of 161 to 20 to move. But as of 2013 it still wasn't certain where, or even whether, Shishmaref would relocate. The most likely place for a move is Tin Creek, a mainland site about 8 km (5 mi) away—close enough that many homes could be moved there and townspeople could maintain connections to the land and sea. At one point the U.S. Congress had considered moving the residents to the larger towns of Kotzebue or Nome, but the idea of their community's identity being swallowed up didn't sit well with many residents.

Wherever they end up, the move will be a costly process for Shishmaref, with the tab estimated at up to $200 million and no ready source of funding available. Other settlements along the Arctic coastlines of Alaska, Canada, and Russia may also be forced to move, a prospect that serves as eloquent—and expensive—testimony to a climate in flux. In 2008, one of those Alaskan villages, Kivalina, filed a lawsuit (thus far unsuccessful) against 22 energy companies for their role in boosting greenhouse gases.

> " It's not our fault that the permafrost is melting, or that there's global warming that's causing us to go farther away from our home in Shishmaref. But we'll survive." —*Luci Eningowuk, Shishmaref resident*

contaminants, including mercury and other heavy metals, has drifted into the Arctic over the years from both developed and recently industrializing nations. Many of these contaminants can work their way up the food web, and they've been found at dangerously high levels in the fat reserves of some creatures near the top of the chain, most notably polar bears. Wildlife stressed by climate change may be less able to deal with the negative consequences of these imported toxins.

A softening landscape

As the Arctic Ocean slowly sheds its mantle of year-round ice, terrestrial life nearby is taking on new forms. Perhaps the most emblematic is the "drunken forest"—larch and spruce trees that lean at crazy, random angles as the soil beneath them thaws out.

Permafrost—land that's been frozen for at least two years—holds the key to this and many other transitions across the Arctic landscape. Permafrost covers about a quarter of the Northern Hemisphere's land area, including half of Canada, most of Alaska, and much of northern Russia, as well as a small part of Scandinavia and several mountainous regions closer to the equator. On the fringes of the Arctic, the permafrost is **sporadic** or **discontinuous**, forming in a mosaic of favored locations. In the coldest areas, **continuous permafrost** extends throughout the landscape.

Permafrost is topped by an **active layer** that typically thaws and re-freezes each year. The permafrost itself, which starts anywhere from a few centimeters to a few tens of meters below ground, remains below freezing year round. It can extend hundreds of meters farther down, until the soil temperature again rises above freezing en route to Earth's molten center. The thicknesses of the active layer and the permafrost depend on both recent and past climate, making them convenient diaries of global change. As part of the International Polar Year (which ran, despite its name, from 2007 to 2009), scientists drilled more than 500 boreholes across the frozen north to gauge the health of its permafrost. Most areas showed a continuation of the rising soil temperatures documented over the last few decades, with the coldest permafrost warming the most quickly—a finding reinforced in other recent studies.

It's the deepening of the active layer, and the thawing of permafrost just below it, that's causing problems in certain parts of the Arctic. Tucked within the permafrost is water, some of it in crystals bonded to the soil structure and some in much larger ice beds or wedges. As thawing gets deeper, it can liquefy ice deep within the ground, leading to slumping of the surface and the formation of marshy hollows and small hummocks. When such features come to dominate a landscape, it's referred to as **thermokarst**. Such is the process that has put many trees and buildings in a slow-motion fall, leaning at ever-sharper angles over time, especially across the fast-warming lower Arctic between about 55° and 65°N. Because of the variability of year-to-year warming and the random locations of ice pockets, it's almost impossible to predict which house or tree in a melt zone will start leaning next.

In **Yakutsk**, the major city of eastern Siberia, many buildings are on stilts, which can not only stabilize a structure but also prevent it from warming the ground further. More than 300 structures and an airport run-way have been compromised by unstable permafrost. In this case, however, media and activists may have given climate change a bit too much credit for Yakutsk's woes. Less-than-optimal building practices are likely behind some of the structural failures. Moreover, as noted in NOAA's 2012 Arctic Report Card, the observed thawing of permafrost around Yakutsk can be directly related to soil disturbances triggered by forest fires, construction, and agriculture. (Of course, as we'll see in chapter 9, the massive fires across Siberia in recent years have gone hand in hand with climate warming.)

Across large stretches of the Arctic, it's evident that rising temperatures are indeed having an impact on permafrost. Much of northern Russia has seen permafrost warming on the order of 1°–2°C (1.8°–3.6°F) since the 1970s. In 2012, most locations across northern Alaska saw record-warm permafrost temperatures as deep as 20 m (66 ft). Active layers are more mercurial, thinning and deepening mainly in response to the pace of ice melting within them, the warmth of a given summer, and the depth of the precedent winter's snow. However, even active layers have shown longer-term deepening in some regions, including northeast Siberia and interior Alaska, where **Fairbanks** is a poster child for permafrost trouble.

Fairbanks lies on top of a fairly thin permafrost layer—only a few meters deep in many locations. Record warmth in the past 20 years has percolated downward, heating permafrost across Alaska by 1°–2°C (1.8°–3.6°F). Around Fairbanks, some sections of permafrost have warmed by more than 3°C (5.4°F), bringing its temperature above –1°C (30°F) in some locations. The results are startling: drunken spruce trees, heaving bike trails, and sinkholes pockmarking the landscape. Thus far, the sinking has been patchy, focused where digging and construction have opened up layers of landscape to warming and melting. More widespread problems could occur if and when the bulk of Fairbanks' permafrost gets to the dangerous side of the freezing point.

As one might expect, experts and climate models agree that the Arctic's permafrost will continue to degrade through this century and beyond. The overall pace will depend largely on how quickly the atmosphere warms. By 2100, the area covered by permafrost within 3 m (10 ft) of the surface could shrink by anywhere from 37% to 81%, depending on greenhouse emissions, according to the 2013 IPCC report. While this would be a major ecosystem change in its own right, the repercussions would be much larger if thawing permafrost were to release huge volumes of carbon dioxide and methane. Some 1700 metric gigatons of carbon—about twice as much as the entire global atmosphere now holds—are estimated to be trapped in the top 3 m (10 ft) of northern permafrost. Could this storehouse trigger a massive positive feedback that makes climate change even worse? This is a critical long-term question that's yet to be comprehensively answered. The amount of carbon that might be released depends greatly on the extent and evolution of features such as thermokarsts, which can generate new methane-producing wetlands, and waterways, which can carry organic matter out to sea before it has a chance to enter the atmosphere. The main player in the mix, however, is most likely soil-based microbes that release carbon dioxide. Right now, the icy, saturated conditions in and near permafrost inhibit the growth of these microbes, but as temperatures rise, the soil should tend to dry out (despite the increased precipitation projected for polar areas). Warmer, less-soggy soil, in turn, could allow CO_2-spewing microbes to proliferate.

One of the first major analyses of permafrost–climate feedback, from scientists from NOAA and the University of Colorado Boulder, was published in 2011 in the journal *Tellus*. The study estimated that enough carbon could be released through permafrost melt by 2200 to boost atmospheric CO_2 concentrations by some 60–120 ppm. The high end of that range is on par with the full amount of CO_2 added to date by human activity. Other studies have pointed to the potentially dire consequences if a large pulse of methane, on the order of 50 metric gigatons, were to escape to the atmosphere over the next several decades from permafrost or seabeds. However, as yet there's no widely accepted mechanism by which rising temperatures would trigger such a vast and geologically quick release of methane. Most global climate models don't yet simulate permafrost, much less the potential effects of permafrost thaw on the carbon cycle, so the IPCC didn't include possible effects from permafrost–climate feedback in its 2013 projections. The panel did note that such effects could be significant. (For more on this topic, see p. 115).

Northern fire

Trees aren't just toppling across the Arctic; they're also burning. In 2004 and 2005, **fires** swept across more than 10% of the tundra and forest in Alaska's interior. Across Siberia, which holds half of the world's evergreen forest, losses have reportedly risen more than tenfold over the last few decades. In 2012, Russia experienced its worst fire season on record, with flames consuming an area larger than Colorado (300,000 km², or 115,000 sq mi). Sharply warmer summers in recent years have helped dry out the landscape much sooner than usual, lengthening the fire season and making the fires that start that much more intense. In July and August 2013, Fairbanks saw a record 23 consecutive days reaching 70°F (21°C), with 11 of those days hitting 80°F (27°C). That same summer, temperatures topped 30°C (86°F) in far northern Russia, and wildland fire reached well poleward of its usual territory—extending within 200 km (125 mi) of the Arctic Circle.

Insects are also playing a part. Epic infestations of pine and spruce bark beetles have swarmed across western North America over the last 15 years. The bugs have killed off millions of trees and increased the forests' vulnerability to intense fires for several years afterward. These infestations have been driven in part by warmer temperatures, especially in winter, which reduce the likelihood of insect die-off and help foster multiple breeding cycles in a single year (see chapter 9).

To be sure, a milder Arctic paves the way for the northward migration of many flora and fauna. The problem is that many parts of the ecosystem can't change with the fluidity of the climate itself. For instance, the weather in 30 or 40 years may be conducive to boreal forests across some regions now covered with nutrient-poor tundra. However, it might take decades longer for decaying grasses to prime the soil so it can support tree growth. Shrubs might fill the gap more quickly: their coverage across northern Alaska has increased by more than 1% per decade since 1950, and satellites indicate that Siberia might also be seeing more shrub-covered areas. Still, models suggest a northward migration of high-latitude forest later this century, especially across western Siberia and north-central Canada. This would pull an increasing amount of carbon dioxide out of the atmosphere, though probably not enough to counteract the warming effect caused by the forests' propensity to absorb sunlight (see sidebar, p. 102). Increased shrubbery may also have its own mix of impacts, shading the ground and insulating snowfall but also absorbing sunlight—again, perhaps enough to provide a net warming where the shrubs thrive.

The Arctic's signature land beasts, **reindeer** (known as **caribou** in North America), face their own challenges. Heavier spring runoff and rains across the North American Arctic could widen rivers enough to jeopardize the seasonal trek of the continent's vast herds of caribou, some of which hold more than 100,000 animals. In some parts of the Arctic, autumn and early winter are already proving treacherous. Caribou and reindeer normally browse below the surface of early-season snow to reach the tender lichen that spread at ground level. In the past few years, however, layers of ice within the snow—fostered by increasingly prevalent warm spells and bouts of freezing rain—have encased the lichen, blocking the animals from feeding. This process has been implicated in the decline of Peary caribou, whose numbers across far north Canada and Greenland have plummeted by more than 80% over the last 50 years, according to the Committee on the Status of Endangered Wildlife in Canada. Reindeer in Lapland have

Hundreds of buildings in Yakutsk, Russia, have been compromised by permafrost instability, although construction practices are likely a factor as well. (Will Rose/futureworldproject.org)

experienced similar problems in accessing winter food, raising concerns among ecologists and local herders.

The Arctic's patchwork of **freshwater lakes** and **rivers** is also being rewoven as the climate warms. Across a Texas-sized patch of Siberia, researchers used satellite photos to identify more than a hundred small lakes that disappeared between the 1970s and 1990s. These were victims of thermokarsts that allowed the waters to drain out while new lakes opened to the north.

How will Greenland's fate affect ours?

The biggest chunk of land in the far north, **Greenland** has languished in frozen obscurity for many centuries. Two areas on its chilly southwest coast were populated by Erik the Red and a few hundred other Danes and Icelanders around AD 1000. The Western Settlement disappeared around AD 1350 during a marked cooldown, while the Eastern Settlement hung on for some 200 years more before giving way. Since then, the island's brutal

Methane lurking in the muck

Among all the by-products of the melting Arctic, one stands out in its sheer horror-movie potential. Trapped within permafrost—especially below Arctic ice—are billions of tons of **methane clathrates** (also known as **methane hydrates**). These are molecules in which water and the potent greenhouse gas methane are bonded under high pressure and/or low temperature. Besides their presence in permafrost, methane clathrates are even more extensive in seafloor sediments around the margins of continents across the globe. In their supercompacted form, methane clathrates are more than 150 times more concentrated than gaseous methane. It's not yet clear what will happen to methane clathrates as the permafrost continues thawing and the ocean warms up. Scientists and activists have pointed to the risk that vast quantities of methane could be released as the planet warms—perhaps enough to dwarf the greenhouse gases emitted by human activity. Should that happen, global warming could go well beyond current projections. Dramatic images of methane bubbling out of Arctic lakes and seabeds have only added to the anxiety. However, since methane only lasts about a decade in the atmosphere, any surge would have to be truly colossal to kick off a positive-feedback cycle of enhanced warming. The 2013 IPCC report expresses high confidence that a catastrophic release is very unlikely to occur this century. Over the longer term (thousands of years), the ocean could heat up enough to make methane release a major player in keeping a warmer planet warm. Looking back, it's believed that methane clathrates may have been involved in some of the most intense warming episodes in Earth's history (see chapter 11).

Gaseous methane burns off a lump of seabed hydrate, releasing liquid water below. (Gary Klinkhammer)

Two ways to look at glacier motion

Can a glacier speed up and retreat at the same time? Interestingly, the answer is yes—depending on how you define the terms. Glaciologists measure the speed of glacial ice as it moves toward the sea, but they also evaluate the location of the glacier's leading edge, or terminus. When a glacier is described as speeding or slowing, that's a reference to the first variable—the ice motion (typically in meters or kilometers per year) as measured at a given point along the glacier's length. However, the leading edge, or terminus—the second variable—may not behave in line with the ice's overall motion. A glacier that extends into the sea (i.e., an outlet glacier) may be a floating ice tongue, or it may be grounded on the seafloor. Other glaciers may terminate well inland. Even as a glacier accelerates, its terminus can advance (grow farther toward the sea) or retreat (melt back). The latter becomes especially likely if the waters surrounding an outlet glacier's front edge are warming. Think of dipping an icicle very slowly into a glass of hot water: even as the whole chunk goes downward, the bottom edge of the ice is melting its way upward, perhaps even more quickly than the downward motion of the cube. A third glacial variable is mass balance, the change in total water content over time. This hinges not only on motion, but also on how much ice is accumulating and how much is melting and/or evaporating across the entire breadth of the glacier.

cold has kept human occupation limited to a scattering of Inuit towns and, in the last few decades, a sprinkling of seasonal research camps. Several of these are parked on top of the colossal **ice sheet** that covers about 85% of the island. Greenland's vast coating of ice could end up shaping where millions of people live a few generations from now. Actually, "sheet" may not be the best word for this icy monolith. It extends over 1,300,000 km² (500,000 sq mi)—the size of France and Spain combined—and the bulk of it extends more than 2 km (1.2 mi) high. That makes the ice itself a topographic feature on a par with the Appalachian Mountains. Although some of the sheet presumably melted in between each pair of ice ages, scientists believe it's been at least three million years since Greenland has been ice-free. Under the sheer weight of all its ice, Greenland's center has sunk roughly 300 m (1000 ft) below sea level.

Until recently, scientists were debating whether the amount of water locked up in Greenland ice had increased or decreased over the last several decades, but recent data show that the continent is indeed losing ice mass (see below). If and when the entire Greenland ice sheet was to melt into the sea, it would trigger a truly catastrophic sea level rise of close to 6 m (19 ft). Such an inundation would swamp coastal cities around the world and, in some low-lying areas, close the book on millennia of human history.

Thankfully, most glaciologists agree it would take many centuries for the Greenland ice sheet to fully melt. A major ice-core analysis released in 2012 bolstered that view. During the Eemian, the period between the last ice age and the one before that (about 130,000 to 115,000 years ago), temperatures averaged as much as 8°C (14°F) warmer than today at the location in north Greenland where the ice core was drawn (a much greater rise than occurred globally). But it appears that only about a quarter of the ice sheet's volume was lost during the interglacial period, adding perhaps 1–2 m (3–6 ft) to global sea level, and it took roughly 6000 years for that to happen.

Although this finding offers at least a bit of short-term reassurance, it's worth keeping in mind that human-produced climate change is amplified strongly in the Arctic, so any global rise in temperature could be much larger in Greenland. Moreover, human-caused climate change probably won't follow the Eemian script to the letter. Greenhouse gases are accumulating far more rapidly now than they did during the Eemian, and there are small but important differences in Earth's orbit around the sun now versus then (see chapter 11).

The 2013 IPCC report notes that if global temperatures were sustained above a certain threshold—which isn't yet certain, but perhaps between 1° and 4°C (1.8°–7.2°F) above preindustrial values—the Greenland ice sheet could almost completely melt over the course of a millennium or more. Right now, with rapid cuts in global emissions looking unlikely, we're on a trajectory that could bring global temperature well within that temperature range this century. For Greenland's ice, the point of no return wouldn't be obvious at first. Once under way, though, the melting would become virtually unstoppable through a set of positive feedbacks. For example, as meltwater runs off from the top of the ice sheet, the surface would gradually lose elevation and thus encounter progressively warmer air temperatures. This would allow for still more melting, and so on, putting the ice in a cascade toward oblivion.

Even if that scenario isn't expected to culminate for at least a thousand years, there's still cause for concern in the shorter term, because Greenland's peripheral glaciers appear to be losing ice at a faster clip than either scientists or their computer models had reckoned. Some of the ice processes now at work aren't represented well—or at all—in computer models because of their complicated nature.

Moreover, some dramatic thaws have been occurring across Greenland as a whole. In the summer of 2007, melting occurred across fully half of the ice sheet's surface. The thaw was even more jaw-dropping in 2012, when locations that normally see only a few days of melting experienced two months of it. On July 12, satellites detected wet snow across 97% of the entire Greenland ice sheet (see graphic). Ice-core analysis shows that such widespread melts have happened only eight times in the last 1500 years. Looking ahead, a study led by Daniel McGrath (University of Colorado Boulder) gave 50/50 odds that Greenland-wide melt will be a yearly occurrence by 2025. It's quite possible that such melt is already being hastened by soot from forest fires and fossil fuels landing on the ice sheet and boosting its solar absorption (see p. 121). Jason Box (Geological Survey of Denmark and Greenland) led an 2013 expedition called Dark Snow—funded largely through crowdsourcing—to dig into this hypothesis.

Another tool for diagnosing the health of Greenland's ice sheet has been the Gravity Recovery and Climate Experiment (**GRACE**), twin ice-monitoring satellites launched by the United States with German instrumentation. Analyses of GRACE data by Isabella Velicogna (University of California, Irvine) show that Greenland as a whole is losing ice mass each year, and there are signs that the loss is accelerating. During a four-year span from late 2006 to the summer of 2010, Velicogna estimates that Greenland lost an eye-popping one trillion metric tons of frozen water. That amount was nearly matched over the following two years alone. According to an exhaustive 2012 analysis published in *Science* that drew on GRACE and other data, meltwater from Greenland accounted for 10%–15% of sea level rise from 1992 to 2011. Data from 2012 suggest an even higher percentage. The IPCC's 2013 report expresses high confidence that surface melt from Greenland will continue to increase at a faster pace than snowfall, keeping the continent's water budget in the red.

The main routes for this water loss appear to be outlet glaciers, the frozen superhighways along which ice flows to the sea. Several of Greenland's largest glaciers have been moving seaward at a snappier pace than expected over the last 10–15 years. At the same time, so much melt is occurring at the forward edge of these glaciers that the net effect can be a retreat of that edge (see sidebar, p. 116). The bottom surface of the immersed ice tongue is especially vulnerable to melting because the added pressure of water lowers the melting-point temperature.

In some cases, a glacier's rush to the sea can be bolstered through a process called **dynamic thinning**. The idea is that, as meltwater pours into a flowing glacier through moulins (deep, river-like channels), it can lubricate

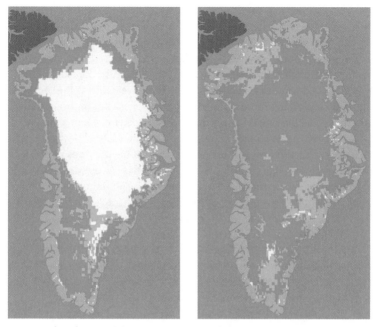

In 2012, surface thawing of the Greenland ice sheet (pink colors) exploded in extent from about 40% of the sheet on July 8 (left) to roughly 97% on July 12 (right). Areas in light pink (probable melt) show where at least one satellite detected surface melting, while the dark pink shows areas where at least two satellites detected melt. (Nicolo DiGirolamo, SSAI/NASA GSFC, and Jesse Allen, NASA Earth Observatory)

the base and hasten the glacier's seaward flow. Because it has been difficult to quantify this process and incorporate it into ice-sheet models, it has remained something of a wild card. While most research has emphasized the potential of dynamic thinning to push sea level rise well beyond standard projections, a dissenting view appeared in a modeling study in 2013 led by Sarah Shannon (University of Bristol). It suggests that the changes in ice flow triggered by dynamic thinning, though widespread, might only add a small fraction to Greenland's total contribution to sea level rise by 2100. This supports the idea that Greenland's contribution to sea level this century could end up being more than minor but less than catastrophic. Because observations can't easily confirm everything going on at the base of ice sheets and glaciers, there will no doubt be much more research on this evolving topic.

The last few years have seen Greenland's largest, speediest glaciers flowing seaward more quickly, yet more erratically. It's now apparent that these highly dynamic systems can move seaward in pulses that strengthen and weaken from year to year on top of longer-term depletion induced by climate change. One of the world's fastest-moving glaciers is **Kangerdlugssuaq**, along Greenland's southeast coast. Some 4% of the island's ice flows through the Kangerdlugssuaq Glacier toward the Atlantic. This glacier moved seaward at about 6 km (3.7 mi) per year during the late 1990s and early 2000s, but more than twice as quickly—14 km (9 mi)—in 2005. Also that year, the glacier's front edge pulled back a full 5 km (3 mi), by far its largest retreat on record. Kangerdlugssuaq then settled down over the next six years, still flowing more quickly than in the pre-2005 period, but with its forward edge actually advancing slightly. On the west coast, the **Jakobshavn Glacier**, the largest of Greenland's ice-draining glaciers, more than doubled its seaward pace between 1996 and 2009, when it moved a total of nearly 30 km (19 mi). The glacier's motion slowed dramatically in 2011 but surged to a new record speed in 2013.

There's been action in far northwest Greenland as well. This region appears especially prone to multiyear spells of highly dynamic ice loss, with relatively quiet periods in between. In the summer of 2010, Greenland's north coast came into the media spotlight when a gigantic "ice island" four times the size of Manhattan broke off from the **Petermann Glacier**. It was the Arctic's largest calving event since 1962, which also made it the largest to be portrayed through vivid satellite imagery (see photo above). Though calving is a natural process, the sheer scope of this one served as visual shorthand for the larger issue of long-term ice loss.

It's fair to say that the future of Greenland's ice sheet is one of the biggest unknowns looming over our planet's coastlines. This uncertainty is boosted by the hundreds of less-studied Greenland glaciers for which there's little more than a decade of data on hand. According to Eric Steig (University of Washington), the short record thus far "shows a complex time-dependent glacier response, from which one cannot deduce how the ice sheet will react in the long run to a major climatic warming, say over the next 50 or 100 years."

What about Antarctica?

Strange as it may seem, Antarctica has stayed mainly on the sidelines in the saga of global warming—at least until recently. This vast, lonely land took center stage in the mid-1980s, when the infamous ozone hole was discovered a few kilometers up. The Montreal Protocol and resulting ac-

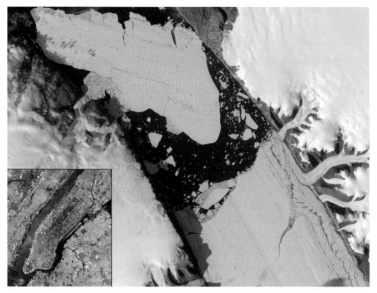

The cleaving of the Petermann "ice island"—satellite image of Manhattan to scale inset. (NASA Earth Observatory/U.S. Geological Survey)

Soot and dust make their presence known atop ice fields in southern Greenland. (Jason Box/darksnowproject.org)

tions by world governments are expected to eliminate the hole by later in this century (see p. 36).

Almost two-thirds of the planet's fresh water is locked in Antarctica's ice sheet, ready to raise sea levels by spectacular amounts if the ice were to melt. Fortunately, nobody expects the entire sheet to start melting any time soon. Temperatures in the interior of Antarctica are shockingly cold: Vostok Station, for example, has never recorded a temperature above freezing. Glaciers flowing from the interior off the shores of East Antarctica are sensitive to multiyear temperature swings, but thus far there's little sign of a long-term retreat. The main risk for now is at the continent's western end, where glaciers appear even more vulnerable than once thought and major collapses this century can't be ruled out.

It's been hard to determine exactly what's happening to temperatures across the great bulk of Antarctica, which has only a few reporting stations, but there are good reasons why parts of this frozen realm might run counter to Earth's long-term warming trend. For one thing, Antarctica is surrounded by the vast, cold Southern Ocean, which absorbs some of the heat that would otherwise boost air temperatures across the region. In addition, overall global warming has led to a strengthening of the Antarctic vortex, the ring of upper-level winds that encircles the continent. This helps keep the coldest air focused over the continent and inhibits warmer air at midlatitudes from working its way southward. Moreover, ozone depletion in the Antarctic stratosphere over the last several decades has helped to tighten the Antarctic vortex, shifting the strongest winds poleward. In East Antarctica, which spans the great bulk of the continent, satellite measurements of the ice sheet's temperature show cooling over some areas in recent decades, mainly during the summer and autumn. However, at the South Pole itself, surface readings that had been cooling have recently begun to warm, leading to little net change there since the 1950s.

As opposed to its eastern counterpart, West Antarctica is solidly in the warming camp—so strongly, in fact, that it more than compensates for any cooling to the east. The evidence mounted with a 2009 study in *Nature* that combined surface data with satellite-based temperature estimates gathered since the 1980s. The combined analysis shows a continent-wide average warming of more than 0.6°C (1.1°F) since 1950 (see graphic above). NASA's ongoing analysis of global temperature trends shows similar values of warming for the Antarctic since the late 1950s. The biggest changes are taking place on the **Antarctic Peninsula**, which juts northward toward South America. Average temperatures at weather stations along the peninsula's west coast—one of the few parts of the continent that sees major

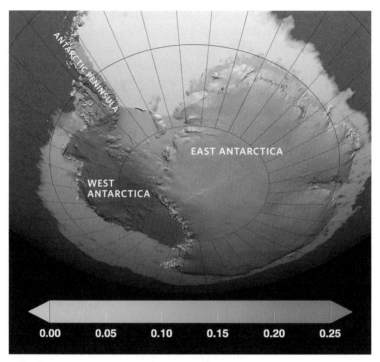

Red represents areas where temperatures have increased the most from 1957 to 2006, particularly in West Antarctica, where some points warmed even more than shown here. Temperature changes are measured in degrees Celsius. (Trent Schindler, NASA Goddard Space Flight Center Scientific Visualization Studio)

thawing and refreezing each year—have soared by as much as 2.5°C (4.5°F) since the 1950s, as rapid a warming as anywhere on Earth during that span. More recently, warming on this scale has been reported in the middle of West Antarctica, with at least one study suggesting significant rises there during the peak of the summer melt season.

It's also getting warmer above Antarctica, as revealed in a 2006 paper in *Science* by John Turner (British Antarctic Survey). Looking at the Amundsen–Scott South Pole Station plus eight coastal stations, Turner and colleagues found that the air at a height of about 5.5 km (3.4 mi) above sea level had warmed by 1.5°–2.1°C (2.7°–3.8°F) since the 1970s.

It's expected that warming will accelerate across most of Antarctica later this century, hastened by the globe-spanning effects of greenhouse gases as well as by a long-distance relationship with rising sea surface temperatures in the tropical Pacific (see below). The healing of the ozone hole later this century may also reduce the cooling effect of the polar vortex. Across the interior, the warm-up may do little at first other than slowly build the ice sheet, in much the same way as Greenland's ice sheet appears to be growing on top. Slightly warmer air (that is, air that's less bitterly cold) should bring more moisture with it, perhaps resulting in a boost to the tiny snowfall amounts across East Antarctica. However, West Antarctica and the Antarctic Peninsula are shedding ice and meltwater far more quickly than snow is piling up across the east. Ice cores at the tip of the peninsula show that the amount of summer melting there is well beyond anything observed in at least the past 1000 years. For the continent as a whole, the amount of ice lost was nearly five times greater in the decade 2002–2011 than in the preceding decade, according to the 2013 IPCC report.

The Antarctic Peninsula's huge Larsen B ice shelf made headlines in early 2002 when a Delaware-sized chunk—perhaps 12,000 years old—broke off. Compelling as they are, such calving events (as with the 2010 Petermann calving in Greenland noted above) have only minor overall effects on sea level, since most of the ice is already in the water. More concerning is whether such breakups might open the door to a faster seaward flow of ice from glaciers upstream. The hydrostatic force of the ocean pushes back on large ice shelves that are grounded on offshore bedrock; this process helps stanch the flow of ice from upstream. When the shelves break up, the adjacent glacier flows more easily into the sea. Six years after the Larsen B calving, the region's glaciers were moving roughly two to five times more quickly toward the sea.

West Antarctic glaciers are also speeding up, largely because of changes in ocean circulation that surround the glaciers' offshore tongues with relatively warm water. Interestingly, both oceanic and atmospheric shifts in West Antarctica are closely linked to the tropical Pacific, where recurrent clusters of thunderstorms can shape atmospheric circulation far afield in much the same way as El Niño and La Niña do. As the tropics are naturally variable, their influence on West Antarctica can wax and wane for periods of years. Thus, it's unclear how soon local changes might be enough to cause major ice shelf calving or collapse. That said, the fact that so many West Antarctic glaciers extend into a warming sea is major cause for concern. Glaciers on the Antarctic Peninsula are particularly vulnerable to future warming, given that the peninsula sits farther north and juts into the path of strengthening winds that bring warm air from lower latitudes.

It's a good thing that most climate models project the vast bulk of Antarctica's ice to stay onshore for at least the next century and probably much

Antarctica's sea ice holds its own

If you're interested in watching sea ice grow, head south. Each winter a huge ring of ice develops around the rim of Antarctica, spanning about 14,000,000–16,000,000 km² of ocean by its September peak. That's about 10% more area than the Arctic's sea ice encompasses at its maximum each March. In contrast to the dramatic, high-profile decline in northern sea ice, the southern ice extent has been quietly growing by about 1% per decade since the 1970s. In late 2012, Antarctic sea ice extent set a new high, topping 19,000,000 km² (even as Arctic sea ice extent fell to a new low), and late 2013 brought a higher value still. Some climate models didn't see the long-term increase coming, and many climate skeptics cite the growth as a natural compensation for the Arctic's ice loss. However, the more likely explanation is stronger winds ringing the continent due to both tropical variability and ozone loss; these winds tend to push sea ice outward and expand its range. Moreover, Antarctic coastal temperatures that are less frigid but still below freezing could actually boost the amount of moisture in the air above the sea-ice formation zone. This, in turn, would help more snow to fall on top of the newly formed ice, making it thicker and facilitating the ice's horizontal growth. Whatever its cause, the Antarctic's modest growth in sea ice pales next to the far more dramatic losses in the Arctic and the profound ecological impacts those could trigger. As Weather Underground blogger Jeff Masters points out, "Calling attention to Antarctic sea ice gain is like telling someone to ignore the fire smoldering in their attic, and instead go appreciate the coolness of the basement, because there is no fire there."

longer than that: a complete melt would raise global sea levels by a cataclysmic 53 m (174 ft). However, the IPCC's 2013 report notes that if a collapse of the West Antarctic Ice Sheet were to get under way later this century, it would raise the estimated upper limit of sea level rise at century's end by several tenths of a meter, causing significant impacts to coastal regions worldwide.

Tropics and midlatitudes: Goodbye glaciers?

Far above the lush forests and warm seas of the tropics, a few icy sentries keep watch on our planet's climate from a unique vantage point. These are Earth's tropical glaciers, perched atop the Andes and other peaks that extend from 4600 m to more than 6900 m (15,100–22,600 ft) above sea level.

Farther north, in the midlatitudes, thousands of small glaciers and a few larger ice caps are sprinkled across parts of the Rockies and Sierra Nevada, the Himalayas and other Central Asia ranges, and the Alps.

These scattered outposts of ice aren't very good at hiding the strain they're under. For several reasons, including their year-round exposure to warm air and their typically steep-sided structure, glaciers at lower and middle latitudes often respond especially quickly to climate change. A melting trend that kicked in during the mid-1800s has accelerated in recent decades across many of the tropical and midlatitude glaciers monitored by scientists. By later in this century, some of these glaciers could be little more than history.

Like rivers in slow motion, glaciers are fluid entities, with fresh water flowing through them. Snow accumulates on top and gets pushed downward and outward by the snowfall of succeeding years. For a healthy glacier, the change in its total ice mass per year—its **mass balance**—is either zero or positive. Today, however, most of the world's tropical and midlatitude glaciers have a mass balance seriously in the red. In short, they're shrinking and thinning.

It isn't just melting that spells doom for ice. Many glaciers erode largely through a process called **sublimation**. When dry air flows over ice, the moisture in the ice can pass directly from its frozen state to its vapor state without any intervening liquid form. Sublimation can occur even when the air is below freezing, and it's the dominant process of glacier loss in some areas. In general, though, melting is the main culprit. It takes eight times less energy for a molecule of ice to melt than it does for it to sublimate, so anything that pushes a glacier toward conditions favorable for melting can erode the glacier far more efficiently than sublimation. Under the right weather conditions, direct sunshine can melt glacial ice at temperatures as cold as −10°C (14°F), and, of course, a reduction in snowfall can take its toll on a glacier even if average temperatures stay constant.

It's the unique mix of conditions at each location that shapes the behavior of tropical and midlatitude glaciers. Those who study glaciers fear that this complex picture sometimes gets lost in the scramble to interpret glacial loss as a simple product of warmer temperatures. Still, they agree on the main point: with a very few cold-climate exceptions, climate change is eroding lower-latitude glaciers at a dramatic pace.

Trouble toward the equator

Thanks to a century of visits by adventurers, photographers, and scientists, the ice cap on top of Africa's **Mount Kilimanjaro** is the most widely

Lonnie Thompson: Putting tropical ice on the map

He may not be able to save them, but **Lonnie Thompson** has done as much as anyone to document and publicize the plight of tropical glaciers. Based at Ohio State University, Thompson paid his first field visit to Peru's Quelccaya ice cap in 1974 as a graduate student. He has since spent more than 840 days above 5486 m

Lonnie Thompson with Qori Kalis Glacier, the largest outlet glacier from Peru's Quelccaya ice cap. (Thomas Nash)

(18,000 ft) as well as 53 consecutive days at around 6096 m (20,000 ft). Early on in his career, Thompson toiled for years—often working on a shoestring budget with few collaborators—to collect data from the Andes and other regions of vulnerable ice. Few colleagues thought the tropics had much to offer glaciologists compared to the much vaster ice fields at the poles. Thompson proved them wrong in 1983 by drilling 164 m (538 ft) into the Quelccaya ice cap and retrieving a unique core that documents 1500 years of climate history near the equator. Thompson's subsequent expeditions have dug even deeper into climate's past. An ice core he retrieved in 1987 at Sajama, Bolivia, spans 25,000 years, while the Guliya cores from China's western Kunlun range extend back more than 700,000 years.

When Thompson re-cored Quelccaya in 2003, meltwater seeping downward had already compromised some of the ice layers from earlier years. As the glacier receded, it exposed plants that were covered by ice more than 6000 years ago. Discoveries like this help drive Thompson's growing urgency to capture the climate clues offered in tropical ice before they are gone. Only weeks after a heart transplant in 2012, Thompson led an expedition with Chinese colleagues to elevations topping 20,000 ft on Tibet's Zangser Glacier.

"The accelerating retreat of most tropical glaciers is very telling," says Thompson, "for it is occurring in a part of the world known for its temperature stability." Because glaciers are "great integrators of climate," as Thompson puts it, "the fact that most glaciers, from the poles to the tropics, are speaking in one voice and telling us the planet is warming should be a concern to us all."

The fast-disappearing ice of Kilimanjaro

> And then they were out and Compie turned his head and grinned and
> pointed and there, ahead, all he could see, as wide as all the world, great,
> high, and unbelievably white in the sun, was the square top of Kilimanjaro.
> "The Snows of Kilimanjaro," 1938

It's a safe bet that Ernest Hemingway wasn't thinking of the greenhouse effect
when he set his tale of a man dying of gangrene near a mountain whose ice cap
is itself dying. Yet the title of Hemingway's classic short story, "The Snows of
Kilimanjaro," is now a standard piece of climate change iconography. Snow still
falls on Africa's tallest mountain, but Kilimanjaro is indeed losing its year-round
ice, in a process that started decades before Hemingway penned his story. Of
the three peaks that make up this giant, extinct volcano, only one (Kibo, the
tallest, at 5893 m/19,334 ft) has anything left of the vast ice cap that apparently
covered all three peaks long ago. The most recent analysis, led by Nicolas Cullen (University of Otago), shows that the surface area of Kilimanjaro's ice has
decreased at a remarkably steady pace for a full century, shrinking from more
than 11 km² in 1912 to less than 2 km² by 2012. The ice can still hold its own in a
good snow year, but if these rates of melt continue, permanent ice fields will be
gone from Kilimanjaro by the 2020s.

Kilimanjaro is anything but a textbook case of "warmth equals melting,"
however. In fact, it's become something of a flash point, a topic of spirited debate among glaciologists as well as a favorite of climate change doubters who
extrapolate from a single mountain in an attempt to discount the message from
a world full of receding glaciers.

What makes the story of this ice so slippery? For one thing, the average air
temperature over Kibo's ice cap runs well below freezing, hovering close to –7°C
(21°F) year round. That in itself doesn't rule out climate change as a factor, but
it means that something more than sheer warming is probably at work. The ice
cap's shape offers more clues. Instead of being rounded, it features steep, sharp
sides about 20 m (66 ft) high. Georg Kaser of the University of Innsbruck argues
that if air temperature alone were shifting the balance toward melting, the ice
cap would quickly revert to a smoother-edged contour. This suggests that the
sun, rather than the air, is the prime melting agent.

More evidence for the sun's key role is the east–west orientation of the cliffs
along Kilimanjaro's ice fields. Since the mountain sits just 370 km (230 mi) south
of the equator, the north- and south-facing walls each face the midday sun at

different times of year. These happen to coincide with the start of the peak's two dry seasons (roughly December to February and June to September), which means few clouds sit between the sun and the ice. Moreover, the local winds generally stay light, which allows water from the sun-melted surface to escape rather than refreezing in strong, dry winds.

While all of this helps to explain the shape of Kilimanjaro's ice, it doesn't tell us what caused it to start disappearing. One possible solution to the mystery offered by Kaser and colleagues is a regional drying (confirmed by a drop in the level of nearby Lake Victoria) that began around 1880 and may have triggered the melting that continues today. What's unknown is how the ice cap behaved in other prolonged dry spells before 1880. Ice cores analyzed by glaciologist Lonnie Thompson indicate that some parts of the cap have survived for more than 11,000 years. However, Kaser believes the bulk of the cap may have built up as recently as a few hundred years ago. He doesn't rule out global warming as a trigger for changes in air circulation—perhaps driven by the nearby Indian Ocean—that could be helping to dry out the peak.

Apart from its symbolism, the loss of Kilimanjaro's snow may not harm the region all that much. Tourism could suffer, though snowstorms will still deliver fleeting images of grandeur. The mountain itself should certainly remain a draw, and it seems unlikely that locals will experience major water shortages if the glacier disappears. Much of the runoff appears to evaporate before it reaches springs and wells downstream, which get moisture from wet-season rains.

Although the decline of Kilimanjaro's ice has helped draw attention to climate change, some scientists would prefer that the spotlight be turned away from this celebrity victim and toward the thousands of other glaciers at risk. As Raymond Pierrehumbert (University of Chicago) noted on the RealClimate blog, "If Hemingway had written 'The Snows of Chacaltaya,' life would be much simpler."

Mt. Kilimanjaro's increasingly paltry cap of snow and ice, as viewed by satellite on February 21, 2000. (NASA Earth Observatory)

publicized of the tropics' disappearing glaciers (see box opposite), but others tell a similar tale. Not far away from Kilimanjaro, **Mount Kenya** lost 7 of its 18 glaciers over the last century; its long-studied Lewis Glacier spread over a mere 0.1 km² (0.04 sq mi) in 2010, only half of what it covered in 1993. On the Indonesian island of Irian Jaya, the glacial area atop Puncak Jaya shrank by 7.5% from 2000 to 2002. Its ice now covers less than 20% of the area it did in 1950.

All of these are mere slivers next to the ice atop the **Andes** of South America. With their string of peaks and high valleys, many extending above 5000 m (16,000 ft), the Andes hold more than 99% of the world's tropical ice. In Peru, north of Lima, more than 700 glaciers dot Peru's **Cordillera Blanca**, or White Mountains (a name deriving mainly from the range's granodiorite rock, rather than the ice it sports). Farther south along the Andes, not far from Bolivia, sits the vast, high **Quelccaya** ice cap, sprawling across a city-sized area that spanned 44 km² (17 sq mi) at last measurement.

Peru's tropical glaciers are far more than playgrounds for climbers and backdrops for photos: they carry both spiritual and practical significance. The ice melt provides water and hydroelectric power that's vital during the dry season for much of the country, including up to 80% of the water supply for the perennially parched Pacific coast. **Lima** spends months shrouded in clouds and drizzle, yet it averages less than 20 mm (0.8 in.) of rain a year, which makes it the world's second-largest desert city. Its eight million residents now depend on glacial ice that is measurably receding. So grave is Lima's risk of losing power that the city has begun expanding its portfolio of alternative energy sources—that is, power plants that burn natural gas. As for the potential water shortage, there's no solution in sight, though engineers have suggested a tunnel through the Andes.

What might be called "The Case of the Disappearing Tropical Glaciers" is the latest chapter in an ancient tale. Tropical ice has waned and waxed since the end of the last ice age, sometimes shaping the destiny of whole civilizations. The glaciers' most recent decline began in the mid-1800s, toward the end of the Little Ice Age but before human-produced greenhouse gases started to rise substantially. It's likely that a blend of forces, possibly including a slight rise in solar output, got the trend started. However, in recent decades greenhouse gases appear to have taken the reins, with assistance in the Andes from El Niño, the periodic warming of the eastern tropical Pacific. When an El Niño is in progress, temperatures generally rise across the Andean glaciers, and summer rainfall and snowfall decrease.

El Niños were unusually frequent from the 1970s through the 1990s, hastening melt across the Andes.

Today, even the most diehard skeptics are hard-pressed to deny that tropical glaciers are vanishing before our eyes. The Andes offer especially vivid examples of low-latitude ice loss, particularly in some of the most accessible parts of the ice (which often happen to be the lowest lying and the most tenuous to begin with). For half a century, a ski lift took adventurers onto Bolivia's Chacaltaya Glacier, billed as the world's highest developed ski area (the summit is at 5345 m/17,530 ft). But the lift wasn't used after 1998, and by 2007 Chacaltaya had lost some 80% of its surface area, with local experts predicting its demise within a year or two. As expected, the glacier vanished in early 2009.

Midlatitude melt

The signs of glacier decline are no less dramatic as you move poleward. Patagonia is home to South America's largest ice fields, as well as glaciers that cascade all the way to sea level. As with Antarctica and Greenland, the calving of glaciers that spill into the sea appears to be uncorking the flow of ice from higher up. NASA teamed up with Chile's Centro de Estudios Científicos (Center for Scientific Studies) to analyze the Patagonian glaciers in 2003. Combining space shuttle imagery with topographic data, they found that the rate of glacial thinning had more than doubled between 1995 and 2000, making it one of the fastest thinnings on the planet. The melting has proceeded into the 2000s, with even the highest glaciers of southern Patagonia now showing signs of thinning. So quickly is the Southern Patagonia Ice Field losing its ice that, from 2003 to 2006, the earth beneath it rose by nearly 12 cm (4.8 in.). It's the most rapid rebound from the weight of glacial ice measured by GPS sensors anywhere on the planet. Several factors may be involved in Patagonia's ice loss, including higher temperatures, reduced snowfall along the Andes' western flanks, and increasingly unstable glacier behavior, such as dynamic thinning (see p. 118) and enhanced calving.

Things are also changing quickly in the midlatitudes of the Northern Hemisphere. Glacier decline is obvious to millions in and near the **Alps**, whose ice has lost roughly half of its surface area and overall mass since 1850. (A few individual glaciers have bucked the trend due to local variations in climate.) Glaciers across the Alps took a big hit as industrialization of the late 1800s deposited sunlight-absorbing soot on the ice surface. This caused major retreat from 1850 to 1930 even as regional temperatures cooled. Over the last several decades, warming temperatures and reduced winter snow have joined to foster a second round of major melting. At heights of 1800 m (5900 ft), parts of the French Alps have experienced winters up to 3°C (5°F) warmer than those of the 1960s. During this period, glacier coverage across the French Alps has decreased by more than 25%. Some of the Alps' smaller glaciers lost as much as 10% of their ice mass in the warm summer of 2003 alone. A linear extrapolation from recent trends indicates that some 75% of Switzerland's glaciers could be gone by the year 2050. Any of these trends could accelerate or decelerate, of course, though there's no particular reason to expect the latter.

In Montana, **Glacier National Park**, one of the jewels in the U.S. park system, has lost nearly 80% of the 150 or so glaciers that it had in 1850. Most of the remainder are likely to vanish by 2030, according to a model developed by the U.S. Geological Survey. According to USGS ecologist Dan Fagre, recent temperature rises have outpaced that model's assumptions, which implies that many of the glaciers might actually disappear as soon as 2020. One observer has already suggested changing the park's name to Glacier Moraine National Park, a title less likely to become obsolete.

The loss of both ice and permafrost are major issues for the central Asia's highest mountains—the **Himalayas**, the vast **Tibetan Plateau**, and a number of nearby ranges. Together, these are home to the largest nonpolar set of ice caps and glaciers on Earth: more than 100,000 in all, covering an area the size of Pennsylvania (about half of that on top of the Tibetan Plateau). Across this region, sometimes dubbed the "third pole" of Earth, multiple threats to the ice are emerging.

Temperatures for the Himalayas as a whole rose by 1.5°C (2.7°F) from 1982 to 2006, more than twice the planet-wide rate of warming. Most glaciers across the region are shrinking and losing mass, in some cases at an increasing pace. Moreover, the amount of sun-absorbing black soot deposited on Tibetan glaciers has grown rapidly since the 1990s, a by-product of Asia's growing population and burgeoning industry. Because much of the Himalayan ice lies in thick swaths across broad valleys, it's likely that the glaciers will thin vertically more quickly than they shrink horizontally, at least for the time being. The possibility of increased glacial runoff poses problems for flood and water management downhill, especially toward the central and eastern Himalayas. While lowlands may be able to benefit from stepped-up hydroelectric power, flood risks could grow in the next few decades. The biggest threat is from what's known as glacier lake outburst floods, where natural moraine dams give way under the pressure of increased meltwater.

Landscape transformed: a lake at the foot of Alaska's Pedersen Glacier in 1917 had become grassland by 2005. (Louis Pederson [1917]/Bruce Molina [2005]/National Snow and Ice Data Center)

It's conceivable that the Himalayan glaciers might someday retreat enough to affect streamflows and related agriculture, but the annual monsoon is a more critical element in supplying water to much of South

The Athabasca Glacier in Jasper National Park, Canada. The glacier currently recedes at a rate of 2–3 m (6.5–10 ft) per year and has receded more than 1.5 km (0.93 mi) in the past 125 years, as well as lost over half of its volume. (Robert Shepard)

and East Asia. Certainly, there's no real risk of losing these glaciers by the 2030s, an erroneous conclusion that crept into the 2007 IPCC assessment (see p. 348). However, the timing of snow and glacier melt might change as the climate warms, and that could influence water supply during the dry premonsoonal months. Even without such complicating factors, the sheer growth in population across this part of Asia looks set to strain water resources.

The picture is a bit more nuanced farther north, in the sub-Arctic lands of North America and Europe. Certainly, **Alaska's** southern glaciers are eroding quickly, especially those that extend into the Pacific Ocean. The vast **Columbia Glacier**, at the end of Prince William Sound, has lost roughly half its thickness and volume since 1980, and its front edge has moved more than 20 km (12 mi) upstream. In 2007, the glacier's end point lost its grounding on the seafloor and became a floating ice tongue, the first such transformation observed in detail by scientists. The glacier now sends huge icebergs into the sound (which, in turn, are convincing cruise ships to stay much farther from the glacier's edge). The rapid three-decade retreat

This image from NASA's *Terra* satellite shows the termini of the glaciers in the Bhutan Himalayas. Glacial lakes have been rapidly forming on the surface of the debris-covered glaciers in this region during the last few decades. (NASA Earth Observatory)

of the Columbia's leading edge—now a matter of simple physics as much as climate change—may slow down within a few years, at least temporarily, once the front edge emerges onto the shoreline.

In contrast, the sub-Arctic reaches of Europe hold some of the most publicized exceptions to global glacier retreat. Several of **Norway's** coastal glaciers grew during the 1990s: for example, the leading edge of **Briksdalsbreen** advanced by more than 300 m (1000 ft). However, these advances were easily explained by the strong westerlies and enhanced snowfall that prevailed during that decade along Norway's west coast, thanks largely to the positive mode of the North Atlantic Oscillation (NAO; see chapter 7). With negative NAO indices and drier conditions becoming more common in the 2000s, coastal glaciers are again on the decline. Briksdalsbreen retreated some 500 m (1600 ft) between 1999 and 2011. Farther inland, virtually all of the glaciers in interior Norway have been shrinking for decades.

Many climate change skeptics have pointed to the growth of specific glaciers as evidence against global warming. However, since climate change isn't a globally uniform process, it's easy to see why a few individual glaciers scattered across the globe have grown or held their own in recent decades, especially in places where warming hasn't yet pushed increased precipitation from snow toward rain. The bigger picture becomes clear in global analyses published annually by the World Glacier Monitoring Service, which keeps track of 37 reference glaciers across 10 mountain ranges

A rocky future for skiers

Thanks to a warming climate, skiing and snowboarding conditions are heading downhill at some of the most popular resorts around the world. Skiing depends critically on each winter's snowfall, so there will remain both good and bad years. Overall, though, the good years are becoming less frequent and the bad ones are coming more often. One of the worst was 2006–2007: many slopes in the Alps and the northeastern United States remained snow-free from autumn into January, which ruined countless holiday plans and scotched more than a half-dozen World Cup events.

The future looks bleakest at lower latitudes and/or lower elevations, where winter temperatures often run close to the freezing point. In the Vancouver area, the 2010 Winter Olympics came close to disaster after a record-warm January turned many slopes to slush and forced truckloads of snow to be shipped in. The games struggled through multiple days of borderline conditions, and thousands of spectators missed out when their standing-room areas were declared unsafe by heavy rains. Russia pushed the envelope even further by hosting the 2014 Winter Olympics in Sochi, a palm-studded resort city on the Black Sea with a subtropical climate. A full year before the games, organizers carried out warm-day preparations by collecting hundreds of thousands of cubic meters of snow and storing them in a series of giant structures covered by sunlight-reflecting tarps. It's a good thing so much snow was saved, because above-average temperatures plagued Sochi during the 2014 games. With many spectators in shirtsleeves, and skiers dodging puddles, Sochi topped out at 20°F (68°F). It was the first time on record that a Winter Olympics host city failed to dip below freezing during the entire event.

In Europe, despite the bountiful snowfalls of 2012–2013, the long-term outlook for skiers and boarders is ominous. Temperatures in parts of the Alps have risen by more than 1.5°C (2.7°F) in the last few decades, more than twice the global pace. A 2003 report from the United Nations Environment Programme (UNEP) estimates that the Alps' average snow line will rise 200–300 m (660–990 ft) by the 2040s and predicts that "many mountain villages, above all in the central and eastern parts of Austria, will lose their winter industry because of climate change." Half of Italy's ski resorts lie below 1300 m (4300 ft), and many German resorts also sit at relatively low altitudes. Locations above 2000 m (6600 ft) appear less climate-vulnerable for the time being, but winter conditions have gotten more variable at all elevations.

Resorts are already being forced to adapt. Switzerland's Tortin Glacier, which supports the **Verbier** ski area, has receded so much that in 2005 the resort placed an insulating sheet the size of a city block atop the glacier's edge to slow summer melting. Other ski areas are thinking of moving uphill. The small French resort of Drouzin-le-Mont, located at 1400 m (4600 ft), floated the idea in 2012 of dismantling its ski area and promoting other forms of mountain recreation.

There's lots of income at stake: roughly 8% of Switzerland's workforce and 5% of Austria's gross national product are ski-based. Though the highest resorts in the Alps and Colorado may hang on, a 2007 report produced for Halifax Travel Insurance foresees quota systems as demand skyrockets for the few resorts where snow remains reliable. The firm ATMOS Research projects that ski seasons across many resorts in the western United States could be 10–20 days shorter on average by 2025.

Snowpacks across the **U.S. West** are projected to become smaller and more variable over the coming century. Many ski resorts in the **U.S. Northeast**—already vulnerable to winter warm spells—could become marginal. In response to all this, more than 70 of the nation's ski areas joined forces with the Natural Resources Defense Council to promote energy efficiency and renewables and push for federal emission controls. Their motto is "Keep Winter Cool." More recently, a group of hard-core skiers and snowboarders, bolstered by individual and corporate sponsors, has launched the activist group Protect our Winters, which is working with resorts, communities, and legislatures to raise awareness and to promote clean energy.

Skiers who flock to the modest mountains of southeastern Australia savor every flake of the white stuff. But at least one part of the **Snowy Mountains** has seen a 40% drop in the depth of spring snow since the 1960s, according to Australia's Bureau of Meteorology. "Although there is much variability from year to year, the overall downward trend in snow depth in the Snowy Mountains is clear," warns a 2013 report from the Australian Climate Commission. The future looks even gloomier for snow lovers, according to the nation's Commonwealth Scientific and Industrial Research Organisation (CSIRO). CSIRO predicted in a 2007 report that the area covered by snow for at least 60 days each year could shrink by 18%–60% by 2020 and by 38%–96% by 2050, depending on global emissions levels.

with data extending back to 1980. In a report for the period 2010–2011 that included data on 33 of these 37 reference glaciers, all but three lost mass during the period.

More than melting?

It's impossible to quantify just how much of the blame for the overall retreat in low- and midlatitude glaciers can be pinned on human activity, as glacier behavior today is partly a result of conditions present many years ago.

That said, tropical ice appears to be one of the most reliable of all witnesses to human-induced change. For starters, the wild temperature swings often found at higher latitudes are muted near the equator, especially as you go up in height. There's also little variation as you go from west to east around the globe. As a result, the average readings vary only about 1.5°C (2.7°F) across the entire equatorial belt at glacial heights of about 6 km (3.7 mi), and there's little change across seasons as well. This means that even a small trend should stand out more noticeably than it would in midlatitudes.

On the other hand, the tropics aren't the best-sampled part of the world, meteorologically as well as glaciologically. A few weather stations are cropping up at the highest tropical altitudes, but observations remain sparse, both on the ground and aloft. As a result, scientists are debating exactly how much temperatures across the higher altitudes of the tropics have risen. Thus far, the jury seems to be settling on a warming somewhat less than the global surface average, but still enough to tip glaciers toward melting. Across eastern Africa, the scanty data show little if any warming trend, which means that drying and other factors may be teaming up to produce Kilimanjaro's rapid ice loss (see p. 128).

Aside from temperature, there's a range of other processes that will help shape the future of tropical and midlatitude glaciers. These can differ from region to region and even on the scale of individual mountains, which leads to a globe-spanning patchwork of varied ice loss sprinkled with local advances in a few midlatitude areas.

* **Sunlight** can directly melt glacial ice even when the air remains well below freezing (as noted previously). Local increases in cloud cover can either enhance or detract from ice loss, depending on how they affect sunlight as well as the longwave (infrared) radiation that clouds send toward the ground.
* **Moisture** in the air can also help or hinder glacier growth, depending on location. In normally dry parts of the tropics, where glaciers

Picturing glaciers as they go

It may be the most ambitious round of time-lapse photography ever attempted—and certainly the most sobering. Since 2007, dozens of cameras deployed at glacier-studded sites across the globe—from Greenland and Iceland to Alaska, the Rockies, and the Alps—have each been taking daylight photos once per hour. The Extreme Ice Survey (**www.extremeicesurvey.org**) is the brainchild of U.S. photographer James Balog, a trained geologist and regular contributor to *National Geographic*. Balog teamed with glaciologists, computer scientists, and experts in still and video imagery to assemble the project's network of resilient cameras, nonpolluting power sources, and uplink facilities. The survey conveys ice loss to a broad audience through videos, Internet-based outreach, and a 2009 book and DVD. Balog and his work attained further prominence as the focus of a 2012 feature-length documentary, *Chasing Ice*.

James Balog installs one of his ice-profiling cameras at the Columbia Glacier. (Tad Pfeffer/Extreme Ice Survey)

typically sit at heights well below freezing all year round, any increase in dry conditions can allow for more snow and ice to slowly leave glaciers via sublimation (see p. 126). The melting process is different in the midlatitudes, where temperatures often rise above the freezing point in summer. If those balmy temperatures came with enhanced moisture and cloudiness, the clouds could send more infrared energy toward glaciers, thus hastening their melting.

As glaciers continue to shrink, their predicament gets increasingly dire. If a glacier is relatively flat, then the ratio of its surface area to its ice mass goes up. This gives the atmosphere more area to work on, causing the glaciers to change state more quickly.

Perhaps the only way for glaciers outside the poles to avoid death by warming would be through a boost in snowfall. That doesn't appear to be happening on a large scale, despite a general increase in atmospheric water vapor worldwide. The snowfall on glaciers atop midlatitude mountains is fed by strong, moist, upper-level winds, such as those near the west coasts of North America or Norway. Glaciers in these areas may stand the best chance at survival, as the winds reaching them could bear more moisture in years to come. Even then, temperatures would need to stay well below the freezing mark for the bulk of the year. At the elevations where temperatures are marginal, a small change can make a big difference. One study estimates that a 1°C rise in average temperature at the glacier's equilibrium line would need to be counteracted by a 25% increase in average snowfall—a massive amount when it comes to long-term climatology.

CHAPTER SEVEN

Oceans

A PROBLEM ON THE RISE

Glaciers calving like there's no tomorrow, sea ice dwindling to patches—it's easy to find drama in the warming of our planet's coldest waters. But only a small part of Earth's oceans ever produces ice. What about tropical and midlatitude waters, the ones that together span more than half the planet? If appearances can deceive, then these oceans are master tricksters. They might not look much different to the naked eye than they did a century ago, but they've gone through quite a transformation—warming up and growing more acidic, among other changes. It appears they've hidden the true impact of our energy choices quite skillfully, absorbing some of the heat and CO$_2$ that would otherwise have warmed the atmosphere.

Many scientists suspected as much for years, but the first solid numbers came from a team led by Sydney Levitus of NOAA. After assembling a variety of deep-ocean measurements for the period 1948–1998, Levitus announced that every major ocean exhibited warming down to at least 1000 m (3300 ft) and that the top 300 m (1000 ft) had warmed globally by about 0.3°C (0.54°F). You wouldn't be able to tell the difference in temperature by taking a quick dip, but the total heat added to the oceans represents as much as 90% of the energy trapped by human-produced greenhouse gases in those 50 years. Water expands as it warms, and the temperature increase translates to a rise in sea level of about 25 mm (1 in.). Again, that may not sound like much, but, as we will see, it's on top of the sea level rise from all other sources.

The undersea storage of vast amounts of heat has serious implications for humanity's future. Although the topmost layer of the ocean stays in balance with the atmosphere over timescales of a month or longer, the much deeper layer below the **thermocline** is more insulated. This means the heat it slowly absorbs from above will take a long time to work its way back out. Even if greenhouse gases magically returned to their pre-industrial levels tomorrow, it would take many decades for the heat tucked away in the deep oceans to work its way out of the climate system. This idea of "climate commitment" is a key theme in research on sea level rise. Some studies have estimated that if global greenhouse gas emissions had leveled off in 2000 (which they didn't), an additional warming of at least 0.5°C (0.8°F) beyond that year's global average would be guaranteed from twentieth-century emissions alone.

Of course, as climate change unfolds, the oceans will do far more than absorb and release heat. Ocean-driven storms, primarily tropical cyclones, are showing signs of strengthening as their fuel source heats up (see chapter 8). Sea levels are rising, too. Much of this is due to the expansion of ocean water as it warms, as noted above, but over time the rise will be increasingly enhanced by glacial melting. That melting, in turn, will shape ocean circulation in ways that push sea level even higher in some favored locations, as we'll see below. The uneven pattern of ocean warming may influence how, where, and when a variety of ocean–atmosphere cycles unfold. This could spell the difference between a life of plenty and a life of poverty for many millions of people living close to the land in drought- or flood-prone areas. The oceans are also gradually losing their natural alkalinity as they absorb carbon dioxide, a shift that threatens many forms of marine life. The warming and acidification of the world's seas, together with reduced oxygen in some waters due to changes in circulation, has led to a bleak four-word summary of our oceans' future: hot, sour, and breathless.

All of this has made oceans one of the favorite trump cards of climate activists. The possibility of islands disappearing under the sea does make for a potent image, and yet the image isn't as crystal clear as one might think. We still have limited data on the three-dimensional reality of oceans: how warm they are, where and how fast their currents are flowing, and so forth. There are big questions yet to be answered, including what "sea level" actually is.

From sticks to satellites: Measuring sea level

Mean sea level (MSL), the average height of the ocean surface at a given point, sounds as if it should be an easy enough thing to calculate. Subtract the astronomical tides, factor out the storms that come and go, and voilà—you've got your average. In truth, however, MSL is enormously difficult to nail down. For a start, the tides depend on the locations of the sun and moon within a set of cycles that takes 18.6 years to fully repeat itself. Even when you factor this out, the underlying MSL isn't the same everywhere, but varies across different sections of the open ocean by as much as 50 cm (20 in.), depending on the surrounding topography and the warmth of the underlying waters. The effects of global warming complicate things further.

Members of a 1926 Arctic expedition check a tidal gauge at Etah, West Greenland. (NOAA/C&GS Season's Report Rigg 1926–69)

The simplest and oldest method of measuring MSL relative to some point along the coast is through basic **tidal gauges**—essentially, sticks in the sand. A technological leap came in the 1800s with **wells** that used floating gauges to factor out the effects of waves. Similar techniques are still used at many coastal stations, though they've been joined lately by far more precise methods. Modern gauges measure the time it takes sound waves to bounce off the water's surface and back, converting that into distance and then to sea level height.

Radar altimeters on board satellites work in a similar way, typically sending microwave energy from space to the sea surface—the open ocean as well as coastal areas. One key advantage of satellites, when they're coupled with GPS data, is that they provide a measure of MSL that's independent of a land-based feature. The **TOPEX/Poseidon** satellite, a collaboration between the U.S. and French space agencies, began gauging ocean heights across much of the world in 1992, ending its long run in 2006. Its successor, **Jason-1**, scanned oceans from 2001 to 2013; **Jason-2** took to the skies in 2008, and **Jason-3** is expected to launch in 2015. **Jason-2** measures the height of virtually all of Earth's ice-free ocean surface with a vertical resolution as fine as 1 cm (0.4 in.). Even sharper-eyed is **GRACE (Gravity Recovery and Climate Experiment)**. This U.S.–European satellite system, launched in 2002, is expected to deliver useful data until the mid-2010s, with a follow-up mission planned for 2017. GRACE tracks MSL with an error of less than 1 mm (0.04 in.). Because it measures subtle changes in the gravitational tug exerted by the oceans as well as by Earth itself, GRACE helps scientists sort through the small-scale continental adjustments that may influence sea level.

Those geological tics can have a surprisingly big influence on sea level. Coastlines rise and fall with the clashing of tectonic plates. In some places, such as parts of Alaska and Japan, the motions are so pronounced that tidal gauges can't be used as a reliable index of long-term sea level change. Other coastal locations, such as Venice and New Orleans, are plagued by gradual subsidence. Still other regions are steadily rising, on the rebound after being squashed beneath kilometer-thick ice sheets for thousands of years. In some regions, such as the Gulf of Bothnia (bordering Sweden and Finland), the rise in MSL due to ocean warming is obscured by a much larger rise in the coastline itself due to postglacial rebound, much like a piece of dough springing back slowly from the push of a baker's thumb. The bounce-back from melting ice can be as much as 10 mm (0.4 in) per year, enough to more than cancel out sea level rise and produce a net drop in sea level from the perspective of some coastal towns.

Balancing the sea level budget

All of these various ups and downs have to be considered when evaluating changes in sea level. With this in mind, the Intergovernmental Panel on Climate Change has done its best to pool the data and calculate global sea level rise. The 2013 IPCC report estimates the average yearly rise in MSL from 1901 to 2010 at between 1.5 and 1.9 mm (0.059–0.075 in.) per year. The middle of that range, 1.7 mm (0.067 in.), corresponds to a sea level rise over 109 years of about 19 cm (7.5 in.).

It may seem as if the IPCC has set a fairly wide margin of uncertainty for something that's already happened. This is in part because, for many years, the factors known to raise sea level didn't add up to the actual gauge-measured rise since 1900. The Levitus study, noted above, acknowledged the problem. It proved that the oceans had warmed deeply and dramatically, but it also concluded that the ocean warming was responsible for only 25%–33% of the gauge-observed rise in MSL, or roughly 0.5 mm per year. Another 25%–33% was attributed to melting of land-based ice. That left the remainder—between 34% and 50% of the observed century-scale rise in global MSL—unaccounted for. This situation was dubbed the "sea level enigma" in a 2002 paper by Walter Munk (Scripps Institution of Oceanography).

Since then, researchers have floated a number of ideas on where the rest of the sea level rise may have come from. It now appears that the twentieth-century contribution from the melting of glaciers and the Greenland ice sheet may be larger than earlier thought, but still not enough to close the gap. That leaves Antarctica, where pre-satellite data are especially scarce, as the biggest question mark. A team led by Jonathan Gregory (University of Reading/Met Office) concluded in a 2012 analysis that water from the Antarctic ice sheet could easily account for the remaining unexplained fraction of global MSL rise. Still, the group noted that "[a] complete explanation remains to be achieved."

The pace of the sea's rise has increased since the 1800s, but with some noteworthy ups and downs along the way. Roughly in line with global air temperature, the rate of MSL rise appears to have increased from the 1920s to the 1940s, then slackened until the 1980s before increasing once more. Since the early 1990s, global sea level has risen at almost twice the rate of its twentieth-century average, or around 3.2 mm (0.13 in.) per year. Now that MSL is being tracked by the satellites mentioned above, it's possible to detect monthly and yearly trends with more precision, which has illuminated significant spikes and dips. For example, the pace of rise slowed without explanation between about 2005 and 2008, then picked back up. In 2010–2011, MSL actually dropped by 7 mm (0.3 in.), raising the eyebrows of experts. An analysis led by John Fasullo (NCAR) found the culprit: huge amounts of water vapor that had evaporated from the sea were squeezed out of the atmosphere by a rare confluence of weather patterns and dumped into Australia's vast, low-lying interior, where rainwater is partially blocked from flowing to the ocean. This process alone was enough to bring MSL down temporarily. Once those patterns shifted and Australia's record-setting torrents ended, global sea levels resumed their longer-term climb (see graphic).

How high will the sea get?

Assuming humans continue to burn fossil fuels, it's a safe bet that sea level will continue to rise over the long term. But by how much? And what does this mean for people on or near the coast?

As we'll see below, the amount of sea level rise is liable to vary from region to region, though the globally averaged value is still a useful and important index. Another thing to keep in mind is that such projections are typically calculated with the end of the twenty-first century as a benchmark, but some effects of sea level rise will be evident much sooner, and seas are expected to keep rising well into the twenty-second century (and probably far beyond it).

The 2013 IPCC report assigned likely ranges of sea level increase for each of four potential pathways of greenhouse gas increase. (See the box in chapter 1 for more on these representative concentration pathways, or RCPs.) The numbers below show how much MSL is projected to rise by 2081–2100 compared to 1986–2005.

* RCP2.6: 0.26–0.55 m (10–22 in.)
* RCP4.5: 0.32–0.63 m (13–25 in.)
* RCP6.0: 0.33–0.63 m (13–25 in.)
* RCP8.5: 0.45–0.82 m (18–32 in.)

Large as they are, these ranges actually suggest a gradual narrowing of what scientists believe could happen. This reflects improvement in the models that depict temperature rise, ocean expansion, ice melt, glacier behavior, and other key factors. The lowest-end projection is considerably higher than it was in the IPCC's 2001 assessment (0.09 m, or 3.6 in.), based in part on the increased rate of MSL rise since the 1990s. There's no guarantee on how long that faster pace will be sustained, but assuming it holds steady or grows, at least a few inches of global rise appear virtually certain. At the other end of the spectrum, the top of the highest-end projection is lower than it was in 2001, but just slightly (0.88 m, or 35 in.). Note that the differences between 2001 and 2013 also reflect changes in scenarios, so this isn't a totally apples-to-apples comparison.

The apparent constancy over the last few years of the high-end MSL projection for 2100 obscures some vigorous debate among climate scientists. One of the key points of contention is which type of model produces the best results: the complex, **process-based models** that aim to directly simulate atmosphere, ocean, and ice, or an alternative set of **semi-empirical**

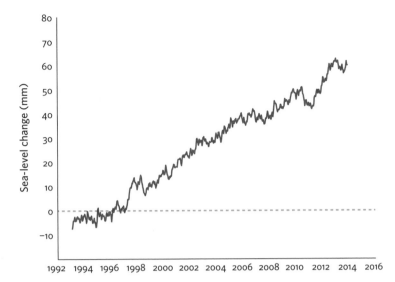

Sea level rise varies from month to month and year to year, but the long-term trend since 1993—shown here from satellite-derived data—is unmistakable, averaging about 3.2 mm per year. The normal seasonal cycle in sea level has been removed from this graphic. (Sea Level Research Group, University of Colorado Boulder)

models, which extend past behavior into the future by statistically relating increases in air temperature to sea level rise. Process-based models have been the favored tool for years, and their performance has steadily improved, but there's an appealing simplicity to the semi-empirical approach. After all, simply extrapolating recent trends would yield more than a foot of MSL rise this century, and it seems plausible that a warming world could accelerate those trends.

The schism between these two approaches broke open following the release of the 2007 IPCC assessment. That report's lower-than-expected MSL ranges excluded some aspects of rapid dynamic thinning of ice sheets that weren't considered well enough understood to incorporate in the projections. The IPCC did note that these factors could add another 100–200 mm to the top of the official range and stated that "larger values cannot be excluded." During the same week the IPCC outlook went public, it was joined by a *Science* paper from Stefan Rahmstorf (Potsdam Institute for Climate Impact Research). Using a semi-empirical model, Rahmstorf found that if sea level rise in the twenty-first century stays proportional to the increase in global air temperature (a relation that held fairly firm in the

Are the Maldives and Tuvalu doomed?

They sit thousands of kilometers apart, but the Maldives and Tuvalu have much in common. As two of the smallest, least populous nations on Earth, they're perched atop low-lying coral atolls, from which they watch the sea and wonder how long they can keep it from swallowing their countries whole. The world is also watching, as these two tiny, once-obscure countries have been at the fore-front of climate change coverage since the late 1980s. That's in part due to the eloquence and tenacity of their leaders, who have been driven onto the world stage by palpable fear of what the future may hold and anger at those deemed responsible. In 2002, Tuvalu threatened the United States and Australia with a lawsuit to be brought before the International Court of Justice, charging the two nations with reckless greenhouse gas emissions (the lawsuit never mate-rialized, though).

There's no doubt that water is lapping at the shores of both countries, and the IPCC projections of sea level rise offer no reassurance. The average elevation is around 1 m (3 ft) in the Maldives and around 2 m (6 ft) in Tuvalu, while the high-est terrain in each nation is around 3 m (10 ft) and 5 m (16 ft), respectively. The 2013 IPCC high-end projection for global sea level rise of 0.81 m (32 in.) toward the end of this century (2081–2100 would bring major impacts to both countries. Even now, storms are capable of flooding large parts of each island chain.

The 200 islands of the Maldives sit due south of India. They're so close to 0° latitude that they're at little risk from tropical cyclones, which need some distance from the equator in order to develop. But even far-off storms can send swells to worrisome heights, and flooding can also occur when intense mon-soon rains team up with tides. The Maldives' capital, Malé, is now fortified by a seawall 3 m (10 ft) high, built in the late 1980s and 1990s with the help of over US$60 million provided by Japan. However, on another island, Kandholhudhoo, over half of the residents had already committed themselves to leaving before the catastrophic Indian Ocean earthquake-generated tsunami of 2004 struck. The waters, topping around 1.5 m (6 ft), ruined almost every house and sped up plans to move residents to a nearby island, which will be bolstered by landfill.

Small as they are, the Maldives resemble an empire compared to Tuvalu, which is located well east of New Guinea and far to the north of New Zealand. Tuvalu's nine atolls only encompass around 26 km² (10 sq mi) in land area, and the total population barely tops 10,000. Stretching from 6° to 10°S, Tuvalu is far enough south to get battered by the occasional tropical cyclone. As with the Maldives, rogue waves and swells are an ever-present threat, and for both island chains the impact of such one-off events will only get worse as sea level gradually rises.

On the face of it, the residents of the two countries appear to have little choice but to either abandon their homelands or engineer their way to safety with expensive landfills and seawalls. But not everyone is convinced the islands are doomed. That's because atolls like the Maldives and Tuvalu form on the edge of coral reefs that sweep in narrow arcs around huge lagoons. The islands themselves are made up of the pulverized remnants of corals and other reef organisms. When the climate isn't changing and the coral are healthy, the islands appear to grow slowly over time, although there's a lot of uncertainty about the rate at which this occurs. Without humans intervening, big waves occasionally overwash the islands, scouring sediment from one side of the island and depositing it on the other. In theory, the island doesn't drown so much as it shifts backward. As sea levels rise, the islands should shift position on their reefs, generally moving toward their lagoons and growing taller.

This scenario is championed by geographer Paul Kench of the University of Auckland. In studying a variety of small landforms, such as tropical atolls as well as the barrier islands common along the U.S. Gulf and Atlantic coasts, Kench has come to believe that these lands are more resilient than they're given credit for. For a 2010 study published in *Global and Planetary Change*, Kench and colleague Arthur Webb (Pacific Islands Applied Geoscience Commission) examined changes in 27 atolls across Tuvalu, Micronesia, and Kiribati. They found that only four of the atolls lost at least 3% of their land area over the periods studied, which ranged from 19 to 61 years. Nearly half of them grew horizontally by at least 3%, largely due to coastal land reclamation that helped compensate for sea level rise.

Kench and colleagues don't foresee an easy path ahead for the Maldives, Tuvalu, and other reef-based populations. The problem is that the reefs' natural growth process, as described above, doesn't respect towns and buildings that are fixed in one spot. In addition, climate change may stunt the growth process itself. In a study partially supported by the World Bank, Kench and colleagues surveyed atoll and reef islands and adapted a model originally designed for sandy barrier islands in order to explain the life cycle of the coral-based atolls. Normally, the growing reefs provide extra sediment that helps build up the islands. As sea level rises and the ocean warms, Kench believes the heat-stressed corals (see p. 168) will devote most of their energy to self-preservation,

The Maldives lie uncomfortably close to sea level. (Nevit Dilmen/GFDL/www.gnu.org/copyleft/fdl.html/CC-BY-SA-3.0/Wikimedia Commons)

providing little or no sediment for the island. Thus, it seems, the residents will have to make do with the sediment already beneath their feet. For this reason, says Kench, "island nations should make conservation of island and nearshore sand resources a high priority."

Years of haphazard land use—including mining the reefs for building materials—by indigenous residents as well as nonresidents (such as the U.S. forces that occupied Tuvalu during World War II) have triggered chronic erosion and left a difficult legacy for islanders to overcome. The resulting problems along vulnerable stretches of coastline are sometimes erroneously tagged as climate

twentieth century), then an MSL boost of 500–1400 mm (20–55 in.) above 1990 levels could be on tap by century's end.

Subsequent papers by Rahmstorf and other experts—all using techniques that build on the correlation between past temperatures and past sea level rise—reiterated the idea that higher-end rises in MSL are quite

change impacts, especially when passing storms happen to push water onshore. While stressing the serious long-term threat posed by rising seas, Simon Donner (University of British Columbia) points to the risk of confusion along the way. "Instead of incorrectly attributing individual flood events or shoreline changes to global sea level rise, scientists and climate communicators can use such occurrences to educate the public about the various natural and human processes that affect sea level, the shoreline, and the shape of islands," says Donner.

As for the future, if Kench's model is correct, it may spell survival for some atoll islands—at least for the next century—but constant flooding and woe for their residents. Tuvalu has explored options for its citizens to resettle in Australia or New Zealand, where over 4000 Tuvaluans already live. In 2006 a retired Tuvaluan scientist, Don Kennedy, began lobbying to move the nation's 9000 remaining residents to the Fijian island of Kioa. There, ethnic ties with Tuvalu are strong—and the land is higher. However, Tuvalu prime minister Enele Sopoaga told Radio New Zealand in a 2013 interview that relocation "should never be an option because it is self-defeating."

With the Maldives' fate just as uncertain, Mohamed Nasheed became a high-profile figure in the global warming arena while serving as the nation's president from 2008 to 2012. He drew attention to the Maldives' plight in 2009 by joining 11 ministers to don scuba gear and hold a cabinet meeting underwater. According to Nasheed, "Climate change is a global emergency. . . . The world is in danger of going into cardiac arrest, yet we behave as if we've caught a common cold." Nasheed's efforts were profiled in the popular 2012 documentary *The Island President*. Under Nasheed, the Maldives also set an example for larger countries by setting out to become the world's first decarbonized economy by 2020. The goal isn't just to become carbon neutral (which could be achieved simply by offsetting their emissions) but to generate virtually all of the islands' power through such alternative sources as wind turbines, rooftop solar panels, and a biomass plant burning coconut husks, plus a complex of batteries to bridge gaps in power supply.

possible this century. Again using semi-empirical modeling, a 2012 study in *Nature Climate Change* led by Michiel Schaeffer (Wageningen University) asserted that a global temperature rise this century of 1.5°–2.0°C (2.7°–3.6°F) above preindustrial values, which falls toward the lower end of the latest IPCC temperature projections, could yield a twenty-first-century

MSL rise of 75–80 cm (30–31 in.), which falls near the IPCC's high-end sea level outlook.

The IPCC explicitly acknowledged the ongoing debate in its 2013 assessment. "Many semi-empirical model projections of global mean sea level rise are higher than process-based model projections (up to about twice as large), but there is no consensus in the scientific community about their reliability and there is thus low confidence in their projections," stated the report. Going forward, we can expect advances in both types of modeling, and the ocean's behavior itself should help reveal which approach might ultimately prove more accurate.

One of the main uncertainties that could yield higher-than-expected MSL rise this century is dynamic ice behavior, especially in the vast West Antarctic Ice Sheet (WAIS). Apart from Greenland and Antarctica, it's widely accepted that other glaciers and ice caps will continue to melt at a steady or increasing pace. A major 2011 analysis led by Valentina Radić (University of British Columbia) concluded that glaciers could add anywhere from 0.09 to 0.16 cm (3.5–6.3 in.) to MSL by 2100. Meanwhile, Greenland will be adding an increasing amount of water through surface melting of its huge ice sheet. However, this melting should increase at a pace more or less in line with regional temperature, so a catastrophic release of meltwater isn't expected over the next 100 years. And although some East Antarctic glaciers are accelerating toward the sea, rising temperatures over the frigid interior should lead to increased snowfall, so this part of the continent may actually gain ice and thus trim a small amount from the expected global rise in MSL rise over the next few decades.

It's the potential for sudden, large-scale ice loss from WAIS that's the greatest concern for many researchers. The land beneath much of WAIS actually slopes downward as you move inland; the sheet's leading edge extends into the ocean via tongues and shelves that serve to buttress the ice behind it. If the waters adjoining these extensions of oceanic ice warm up, or if shifts in ocean circulation happened to import warmer water, the buttresses might erode, which would potentially allow vast amounts of ice to flow seaward. The IPCC cited this concern as the major asterisk in its 2013 review of sea level projections: "Based on current understanding, only the collapse of marine-based sectors of the Antarctic Ice Sheet, if initiated, could cause global mean sea level to rise substantially above the *likely* range during the 21st century." The panel added—but only with medium confidence—that such an event would add no more than several tenths of a meter to the twenty-first–century MSL rise. Over the longer term, a full collapse of the WAIS could raise sea level by more than 3 m (10 ft).

Risks of a rising ocean

What if, by the year 2100, sea level were to rise on the order of 0.6 m (24 in.)? That's near the top of the 2013 IPCC midrange pathways and near the low end of semi-empirical projections. It wouldn't be enough in itself to inundate the world's major coastal cities, but it could certainly lead to an array of serious problems, especially for populous, low-lying regions. Some tropical islands could become virtually uninhabitable (see below) and, significantly, higher sea levels also mean a higher base on which to pile storm surges, tsunamis, and all other ocean disturbances. These are likely to be the real troublemakers as sea level climbs over the next several decades.

One region at particular risk is the Ganges Delta of eastern **India and Bangladesh**. While most of Bangladesh is expected to remain above water even if the ocean rises several meters, such a rise could still displace millions of people in this extremely low-lying country. Millions of other Bangladeshi people who remain near the coast will be vulnerable to catastrophic flooding. See chapter 8 for more on surges from tropical cyclones and coastal storms.

Another place set for serious problems from even modest sea level rise is the glittering string of hotels, offices, and condominiums along Florida's southeast coast from Miami to Palm Beach, as well as the more rustic Florida Keys. A sea level rise of 0.68 m (27 in.) would put some 70% of the Miami area—and nearly all of the Keys—underwater. Building effective barriers against the rising ocean will be difficult, if not impossible, because seawater will tend to sneak through the porous limestone that undergirds south Florida.

One crucial aspect of sea level rise is that it won't be uniform across the world's oceans, a point hammered home by recent studies. According to the 2013 IPCC report, close to a third of the world's coastlines may see an MSL rise that's at least 20% higher or lower than the global average. How could this be? As noted earlier, some coastlines are still rebounding from the last ice age, which would reduce their net sea level rise. Others are experiencing subsidence, which can occur through natural geologic processes as well as when oil, gas, and water are drawn from beneath the surface. Subsidence has allowed sea level to rise especially quickly in far southeastern Virginia, a critical nerve center for the U.S. military. At Norfolk, the average ocean height since 1990 has climbed 12 cm (4.8 in.), which is more than twice the global average.

The future of New Orleans is that it won't be there anymore." —Michael Schlesinger, University of Illinois at Urbana–Champaign

Parts of Florida susceptible to a 6-m (20-ft) rise in sea level are shown in red, with paler areas showing high-population zones. (Jeremy L. Weiss and Jonathan T. Overpeck, University of Arizona)

The northeastern U.S. coast is vulnerable to particularly large MSL rises due to potential shifts in ocean circulation. Should the Atlantic's northward flow be impeded by meltwater streaming from Greenland, then colder, deep water wouldn't form as readily. The net warming from this dynamic process would push sea levels even higher than projected across most of the North Atlantic, according to a 2009 study in *Nature Geoscience* led by Jianjun Yin (University of Arizona). The extra rise could total 10 cm (4 in.) in London and 20 cm (8 in.) in New York. Indeed, this dynamic process on top of global trends could end up producing some of the world's largest rises in sea level this century, focused on long-settled coastlines from the Mid-Atlantic north to the Canadian Maritimes.

Looking centuries ahead, the geographic picture of MSL rise could be influenced by the timing of the Greenland and West Antarctic ice sheet melts, because of the loss of large gravitational tugs exerted by each sheet. In both cases, the effect would be for water to move toward the tropics and away from the polar region where a given ice sheet is located.

With all of these changes unfolding, it's not out of the question—as odd as it seems to contemplate—that most of the world's coastal cities could be

Swimmers and sea level

If the water level in a bathtub goes up when you get in, couldn't people in the ocean be pushing up sea level? That's the semi-serious question posed by Gregory Pasternack of the University of California, Davis. While teaching short courses for high-school teachers in Maryland, Pasternack was inspired by a state politician who claimed that an excess of boats and boaters, rather than climate change, was behind the rise in global sea level. As part of a classroom exercise he developed, Pasternack encouraged students to measure how much water they displace in a bathtub, then to extrapolate that figure to a "world of swimmers." If everyone on Earth took a dip in the sea at the same time, they might occupy a volume on the order of a third of a cubic kilometer. But spread out over the vast area occupied by oceans, that translates to a sea level rise of a mere 0.0009 mm (0.000035 in.)—around a hundredth of the width of a human hair.

largely or completely underwater in a few hundred years. The risk is seldom put in such stark terms, but it's a logical outcome of the amount of warming expected to occur over the next several centuries unless major emission reductions take place. According to some estimates, sea levels ran 6–8 m (20–26 ft) above current MSL during the last interglacial period more than 100,000 years ago, enough to inundate every or nearly every low-lying coastal city. Global temperatures during this period are believed to have been only about 1°–2°C (1.8°–3.6°F) higher than today's. The warming back then was produced by changes in Earth's orbit around the sun (see p. 259), but greenhouse gases could bring as much or more warming in the next century and beyond.

It'll probably take more than a few decades for sea level to begin approaching such catastrophic values. A few outlier scientists, such as NASA's James Hansen, maintain that MSL rises of several meters are possible before this century is out. However, even the highest-emission pathway in the latest IPCC assessment fails to bring MSL beyond 1 m (1.3 ft). Several experts, including Tad Pfeffer (University of Colorado Boulder) and Jason Lowe (Met Office), have declared that an MSL rise of more than 2 m (7 ft) is almost certainly out of range by the year 2100. After that point, the risk of ultra-high seas will clearly rise more steeply. The catch is that the greenhouse gases we're now adding to the atmosphere at unprecedented rates will continue to warm the climate—and raise the seas—for much

The NAM, the NAO, and other climate cycles

After ENSO, the next-biggest cycles of climate variability are two circulations that play out over the North and South Poles, affecting ocean as well as land. The Northern Annular Mode and Southern Annular Mode (or NAM and SAM) refer to oscillations in the strength and structure of the upper-level winds that encircle the Arctic Ocean and Antarctica, respectively. Think of bracelets of strong wind, centered a few kilometers high, that loop around the North and South Poles respectively.

The SAM is the more straightforward of the two. Unblocked by land, its westerly winds howl above the Southern Ocean, circulating around a cold vortex locked over the South Pole. The circulation alternates between a tighter, faster-flowing ring (the positive mode) and a looser, more variable ring (the negative mode) that allows cold air to spill out from the pole across the Southern Ocean more easily. Since the 1960s, the SAM has trended more toward its tighter positive mode, which is likely related to ozone depletion (see chapter 2). This trend is projected by climate models to continue even after the ozone hole heals later this century. Among the implications for climate is a potential drop in winter rainfall for parts of Australia (see chapter 5).

On the other side of the globe, the NAM (also called the Arctic Oscillation) is a more complicated beast. The NAM encounters three continents, plus the thick ice sheets of Greenland and the warm currents of the North Atlantic and North Pacific—all of which interfere with the uniformity of the vortex and help to produce a more variable circulation. In and near the North Atlantic, the NAM manifests itself as the North Atlantic Oscillation (NAO), defined by pressure variations between Iceland and the Azores. It has a strong effect on European weather, especially in winter. The NAM and NAO tend to vary in tandem, shifting positive or negative for periods lasting from a week or so to a month or more. Their strongest effects on climate occur in the winter.

When the NAO turns negative, the polar jet stream tends to buckle northward across the Atlantic and then back south. This often throws northern Europe, eastern Canada, and the northeastern United States into extended cold spells, while milder, slow-moving storms drench the Mediterranean. A positive NAO, on the other hand, keeps Arctic air locked to the north and sends the pole-encircling winds more directly from the Atlantic across northern Europe, increasing the odds for mild and wet weather there as the Mediterranean stays dry.

Most climate models have projected a tendency toward a tighter, more positive NAM for the twenty-first century. However, since 2000 the NAM has actually trended toward its negative mode, with effects that sometimes contradict what people might expect from a warming climate. During the winter of 2009–2010, the NAM plummeted to unprecedented lows, with much of Europe and the eastern United States plunging into weeks of frigid temperatures and heavy snow. It was Britain's coldest winter since 1978–1979, and the U.S. corridor from Washington to Philadelphia set all-time seasonal records for snowfall. Not only did this winter onslaught afflict millions of people, it coincided with—and reinforced—a spate of skepticism about climate change (see chapter 13), despite the fact that the globe as a whole was experiencing near-record warmth. A strongly negative NAM/NAO was also in place during March 2013. That month ended up tied for the second-coldest March in more than a century of U.K. weather records, with huge snowdrifts reported in some areas even after the official start of spring.

It's hard to assess ocean cycles that span more than a few years from peak to peak, because sea surface temperature data before the advent of satellites in the 1970s is notoriously spotty. But many scientists are intrigued by the Pacific Decadal Oscillation (PDO), a seesaw of rising and falling temperatures across the North Pacific, and the Atlantic Multidecadal Oscillation (AMO), in which temperatures across much of the North Atlantic alternately warm and cool. The PDO typically switches from one phase to its opposite about every 20–30 years, with plenty of short-term variability along the way; the AMO takes closer to 30–40 years. In both cases, the physical driver isn't yet known, though there may be a link to the global oceanic conveyor belt (see p. 161). The PDO tends to reinforce the behavior of ENSO: when the PDO is in its positive phase, El Niños are more likely, while a negative PDO tends to correlate with La Niñas.

There's also the Indian Ocean Dipole. Like a sibling of the El Niño–Southern Oscillation (ENSO), the IOD involves abnormally warm and cool patches of ocean that build and decay along the equator, with each phase typically lasting a few months and recurring every three or four years. Positive IODs often occur with El Niño, while negative IODs are most common during La Niña. A positive mode of the IOD (i.e., warm surface waters focused in the western Indian Ocean) can help boost monsoon rainfall across India, and in Australia, a strongly negative IOD teamed with a La Niña event in 2010–2011 to help generate torrential rains.

Given that the physics driving some ocean cycles are still not fully understood, it'll take time to understand how climate change will affect each of them and how various cycles might interact in a warmer climate. The NAM/NAO is a case in point. Some research suggests that a chain of atmospheric events

This image from NASA's Moderate Resolution Imaging Spectroradiometer (MODIS) shows Britain encased in snow on January 7, 2010, midway through one of the harshest winters in recent times. (NASA Earth Observatory)

can lead from heavy autumn snowfall in Siberia to a buckling of the northern jet stream in winter and a subsequent negative NAM. Other scientists have found apparent links between reduced coverage of Arctic sea ice and a more circuitous, less progressive jet stream, again typically associated with a negative NAM. Research into these relationships is far from settled, though.

more than a century. One analysis led by the Potsdam Institute's Anders Levermann estimates that every degree Celsius of warming commits the planet to an eventual sea level rise of 2.3 m (7.5 ft) over the next 2000 years. In short, the greenhouse ship is becoming progressively harder to turn around, and the seas are rising around it.

Climate change and El Niño

The oceans play a vast role in shaping the vagaries of weather and climate, and much of that influence comes through a set of **ocean–atmosphere cycles**. Linked to arrangements of high and low pressure centers over various parts of the world, these cycles each alternate between two "modes," producing recognizable, repetitive weather patterns—drought, excessive rainfall, unusual warmth or cold, and so forth. These can unfold half a world away from their oceanic triggers, and they're natural parts of climate. It's possible that global warming will tamper with them, but right now, scientists have more questions than answers about whether and how this will happen.

Globally, the most important ocean–atmosphere cycle is the **El Niño–Southern Oscillation (ENSO)**, whose two modes are known as **El Niño** and **La Niña**. ENSO is based in the tropical Pacific Ocean, which spans a third of the globe from Ecuador to Indonesia. Trade winds blow more or less continuously across this huge area from east to west, pushing warm water toward Indonesia, where the resulting balmy seas help generate persistent showers and thunderstorms. The cold, upwelled water off Ecuador and Peru, meanwhile, stabilizes the air there and produces the region's legendary aridity (Peru's capital, Lima, gets about the same amount of rain each year as Cairo). Every two to seven years, the trade winds weaken or reverse, the surface layer of warm water deepens and expands into the eastern tropical Pacific, and an El Niño sets in, typically lasting one or two years. The flip side, La Niña, occurs when trade winds are stronger than average, pushing cooler-than-usual water westward into the central tropical Pacific for a year or two. About half of the time, neither El Niño nor La Niña is in progress and the Pacific is neutral.

El Niño increases the odds of drought across Indonesia, Australia, India, southeast Africa, and northern South America. It tends to produce mild, dry winters in Canada and the northern United States and cool, moist winters in the U.S. South and Southwest. It also lowers the chances of hurricanes in the North Atlantic and raises them for parts of the North Pacific. La Niña, in general, has the opposite effects.

Many of the year-to-year spikes and dips in global temperature are a result of ENSO. Earth's atmosphere warms by as much as several tenths of a degree Celsius during a strong El Niño, as warmer surface waters allow heat to escape to the atmosphere. Likewise, the atmosphere cools during a powerful La Niña, as cooler-than-average waters absorb heat from the air above. Indeed, as global temperatures were rising sharply in the 1980s and 1990s, five El Niños occurred, including the two strongest ones of the last century (1982–1983 and 1997–1998). However, La Niña began predominating after 2000, at the same time that the rate of global temperature rise dropped markedly. As noted in the 2013 IPCC report, there's little consensus on whether decadal variations in ENSO are modulated by human-produced greenhouse gases, mainly produced by natural variability, or some combination of the two.

Computer modelers are making progress in depicting El Niño and La Niña, and climate models can now simulate a variety of behaviors in the tropical Pacific's arrangement of ocean currents and rainfall patterns. However, due in part to the natural climate variability that affects the size and frequency of El Niño and La Niña events in both models and observations, it's not yet clear how global warming will affect the mix. Trade winds are expected to weaken, on average, and sea surface temperatures may warm more strongly near the equator than away from it. Yet as the various processes controlling ENSO evolve, their combined effects could go in either direction. Some models suggest an intensified ENSO cycle, while others point to a weaker one. The single most consistent trend among models that fed into the 2007 and 2013 IPCC reports is an eastward expansion of the western Pacific's warm pool, and its showers and thunderstorms, into the central Pacific. According to Gerald Meehl of the National Center for Atmospheric Research, it's possible that this could move the region's typical state closer to that of El Niño, but with individual El Niño and La Niña events still occurring on top of that new configuration.

Will the Atlantic turn cold on Britain?

Even as global warming appears to be proceeding full steam ahead, there are hints that the United Kingdom and parts of the European continent could be in for a countervailing force—a cooling influence that might inhibit human-induced warming in some areas. The storyline will be familiar to those who caught 2004's blockbuster *The Day After Tomorrow* (see p. 345), which took the idea to cartoonish extremes. The nugget of truth in the film is that the northward flow of warm water into the far North Atlantic could slow or even shut down entirely, which would plunge much of northwest Europe into a climate more akin to that of Canada, Russia, or other countries at the same latitude. In Hollywood's view, this occurs in a matter of days, flash-freezing almost everyone from Washington to Scotland. In real life, any cooling would be far less dramatic, unfolding over at least a few years, more likely decades to centuries. Even so, the possibility is worth taking seriously.

The Atlantic flow is part of the global loop of ocean circulation (see graphic above). It's a system whose mechanics are well mapped but not fully understood. One of the most important branches is often referred to as the **Atlantic thermohaline circulation**, because it's related to both temperature (thermo) and saltiness (haline). Many scientists now use a different name: the **Atlantic meridional overturning circulation** (AMOC). "Meridional" refers to its north–south orientation, and "overturning" refers to the somersault it performs in the far north. Before it gets to that point, the AMOC is a vast conveyor belt of warm water that flows northward from the Brazilian coast. In the North Atlantic, prevailing winds sculpt the sea surface flow into a clockwise gyre, sending warm water north along the U.S. coast in the Gulf Stream and bringing cool water south from Spain in the Canary Current. An extension of the Gulf Stream called the North Atlantic Current pushes warm surface water even farther northeast toward the British Isles. (Journalists often use the term "Gulf Stream" to refer to the North Atlantic Current, but strictly speaking, that isn't correct.)

The flip side of the conveyor occurs when the warm surface water approaches Iceland and splits around it, heading for the Labrador and Norwegian Seas. Along the way, it gradually chills and sinks. Then, at several depths ranging from about 2 to 4 km (6600–13,100 ft), the frigid deep water begins the return trip to the South Atlantic, thus balancing the warm northward conveyor and completing the AMOC.

By any measure, the AMOC is a serious workhorse. As it passes the latitudinal belt around 25°N, the circulation brings an average of roughly 1.3 petawatts of heat energy northward, day in and day out. That's enough energy to run almost 100 trillion 14-watt light bulbs, or more than 60 times the energy being used by all the homes, offices, and factories on Earth at any moment.

Scientists have speculated for more than a century that this giant natural turbine might have more than one speed setting. Since Earth began to warm up about 12,000 years ago, the AMOC has stayed fairly robust, except for a couple of dramatic weakenings around 11,500 and 8200 years ago, as glacial meltwater poured into the North Atlantic and suppressed

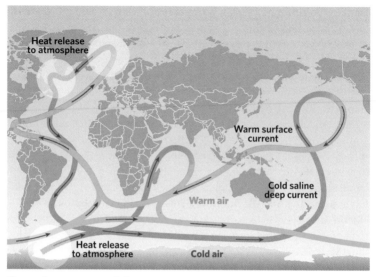

The great ocean conveyor belt.

the circulation. These events were enough to cool the winter climate in and around the present-day United Kingdom on the order of 10°C (18°F) below present-day readings. In other words, London in a typical January would have felt more like St. Petersburg in today's climate.

The main worry for this century and beyond is the increasing presence of warmer, less-dense water in the far North Atlantic, which has more trouble sinking and forming the Atlantic's cold, deep return flow. Increased melting from the Greenland ice sheet and glaciers adds to the freshening (i.e., reduced saltiness) of the waters, which reduces their density further. Analyses led by the United Kingdom's Centre for Environment, Fisheries and Aquaculture Science show that the upper 1000 m (3300 ft) of the Nordic seas—the main source of sinking water for the Atlantic—became steadily fresher from the 1960s into the twenty-first century. This doesn't seem to have had a pronounced effect on the AMOC thus far: in fact, the circulation actually strengthened during that period, perhaps due to natural variability that plays out over decades.

Computer models agree that evolving conditions in the Atlantic should put a substantial brake on the AMOC in the coming century. In response to this concern, the United Kingdom's Natural Environment Research Council funded a £20 million observational program called RAPID that ran from 2001 to 2008, with a follow-up called RAPID-WATCH continuing

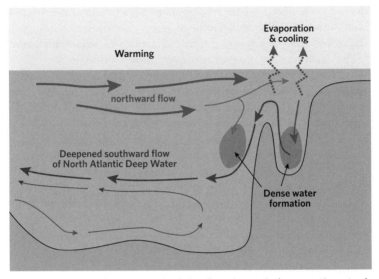

The Atlantic meridional overturning circulation shown in a vertical cross section extending from lower northern latitudes (left) to high latitudes (right).

through 2014. At the time of RAPID's launch, the Met Office Hadley Centre had produced simulations in which the AMOC was halted in its tracks. The results showed a net cooldown of around 1°–3°C (1.8°–5.4°F) by 2050 across most of the United Kingdom, Scandinavia, Greenland, and the adjacent North Atlantic, with warming elsewhere around the globe. However, as the models became more sophisticated because of RAPID and other research, they produced a more modest slowdown of the AMOC.

Nearly all of the modeling generated for the 2013 IPCC assessment points toward a gradual decrease in AMOC intensity by the end of this century. How much is still an open question: depending on how quickly you expect greenhouse gases to accumulate, the model consensus for the four IPCC emission pathways ranges from a slowdown of 11% to 34%. Even with the AMOC slackening, the newest models still produce a net warming in the British Isles and nearby countries, with greenhouse gases more than compensating for the ocean slowdown

On top of these big changes, there are signs of a natural speedup/slowdown cycle in the AMOC, with each phase peaking about every 60 or 70 years. Measurements in the deep Atlantic are too sparse to show this trend, and there's not yet an agreed-upon physical explanation for it. Nevertheless, sea surface temperatures across the North Atlantic show several distinct rises and falls since the late 1800s—a pattern referred to as the **Atlantic Multidecadal Oscillation**. Interestingly, global air temperatures have risen and fallen over a roughly similar timeframe, which has led some scientists to investigate how the two processes might be intertwined.

If this 60- to 70-year cycle is robust, then the AMOC ought to undergo a slight natural slowdown starting in the 2010s and lasting into the 2040s or thereabouts. The challenge will be to untangle this from any potentially larger human-induced slowdown. Improved measurements should help. In 2004, the RAPID project installed a set of moored instruments (maintained in collaboration with the United States) that monitors the AMOC continuously along 26.5°N. Data from the RAPID mooring has shown that the AMOC's strength can pulse up and down far more than previously believed. In fact, it can vary by a factor of nine within a single year. From early 2009 to early 2010, the AMOC flow dropped by an average of 30%. When this dip was examined in a modeling study led by the Met Office, the results suggested that the atmosphere itself may have put the temporary brakes on the AMOC at this subtropical latitude, especially when a strongly negative NAO pattern sent frigid air southward across the eastern United States, the North Atlantic, and Europe in the winter of 2009–2010.

With observations showing that the AMOC can vary naturally over a variety of time frames, it might take a while for any trends related to climate change to show themselves clearly. In the meantime, Britons might be best off planning for a warmer climate while learning to live with a bit of uncertainty about exactly how warm it might be.

Living in a changing sea

The creatures that make their home in the world's oceans might seem to have much more flexibility to deal with climate change than do their land-based cousins. But do they? Marine biologists are still sorting through the many interconnections among ocean species and are trying to map out how things might change as the sea warms up and as its chemistry changes.

As with life on land, there will be winners and losers, though not necessarily in balance. Migratory fish should in principle be able to shift their ranges relatively easily, provided they can still access the food and habitat they need, but not all components of the oceanic food chain will change in unison. A small rise in ocean temperature can affect the mix that exists among the tiniest marine plants and animals (including **phytoplankton**—which account for half of all the biomass produced on Earth—and the **zooplankton** that consume them). Changes in these could lead to mass die-offs of seabirds and fish further up the chain. Outflows of fertilizer-rich water can help produce huge blooms of phytoplankton near coastlines, but the picture may be different over the open ocean. A 2006 study using NASA satellite data showed a global-scale link between warmer oceans and a drop in phytoplankton abundance and growth rates over the preceding decade. This topic broke wide open in 2010 with the release of a bombshell study in *Nature* led by scientists at Dalhousie University. Combining satellite data with nearly half a million measurements of ocean color taken by ships, Daniel Boyce and colleagues concluded that phytoplankton had declined in most of the world's oceans—especially across the tropics and subtropics—at an astounding average rate of about 1% per year since 1950, in sync with rising ocean temperatures (as well as vastly expanded fishing). However, other work using model-based simulations has shown much smaller reductions in phytoplankton, and several critics have argued that Boyce's blending of datasets collected at different times may have biased the results downward.

Scientists can zero in on shorter-term links between ocean chemistry and climate by observing what happens during sharp regional warmings. Off the coast of Peru, the warm water of El Niño cuts off the upwelling of colder, nutrient-rich waters that support a variety of marine life. Back in 1972, an El Niño intensified the effects of overfishing and decimated Peru's lucrative anchovy industry. Many birds that fed on the anchovies died off, and in turn the local supply of guano—bird droppings used as fertilizer—plummeted. Such chains of events aren't limited to the tropics. In 1997, unusually warm waters developed in the eastern Bering Sea (which lies between Alaska and Russia), killing off the type of phytoplankton that normally dominate during the summer. The disturbance worked its way up to devastate local zooplankton, seabirds, and salmon.

Such studies imply that, as global warming unfolds, some of the most dramatic impacts on ocean life may occur through intense seasonal and regional spikes in water temperature driven by El Niño and other ocean cycles. It's somewhat analogous to the way climate change makes its presence most obvious and discomforting to humans during extreme heat waves. And there's more to consider than warming alone. Some ocean life is more dependent on seasonal changes in light than on temperature, especially at higher latitudes. Other species are more purely temperature-driven. Since cycles of sunlight aren't changing, the most light-dependent creatures haven't shifted their timing, whereas many species that are more temperature-dependent are starting their yearly growth cycle at earlier points in the warm season. The differences between these groups may lead to a large-scale mismatch, threatening the seasonal synchrony in growth and feeding times that supports much of the sea's food chain.

Looming over this entire scene is the gargantuan threat posed by **ocean acidification**, perhaps the most underappreciated risk posed by our long-time love affair with fossil fuels. This transformation is a direct result of the enormous amounts of carbon dioxide being soaked up by Earth's oceans. Each year a net influx of roughly seven metric gigatons of CO_2—which includes close to 25% of all the carbon dioxide produced by human activity—goes into the sea. In the well-mixed upper part of the ocean, where the absorption happens, the extra CO_2 is gradually making the water more acidic—or, to be more precise, the water is becoming less alkaline. A 2010 report by the U.S. National Research Council notes that average pH values at the ocean surface have dropped from around 8.2 in preindustrial times to around 8.1 today. That change may seem small, but it appears to be at least 10 times faster than anything the world's oceans have seen in the last few million years.

Because pH is measured on a logarithmic scale, every change of one point is equal to a tenfold increase or decrease. The 0.1-point drop in pH already observed thus translates to a 30% increase in hydrogen ions throughout the upper ocean. Depending on the pace of global greenhouse emissions, oceanic pH could drop to the range of 7.8–7.9 by 2100. A pH value of seven is neutral, so the oceans would still be slightly alkaline, but less so than they've been for at least hundreds of thousands of years.

Scientists are only beginning to probe the influence of acidification on marine life. As carbonate ions become scarcer, it affects the ability of many organisms, such as corals and oysters, to secrete their calcium carbonate shells. Ocean acidification has also been shown to affect the ability of young clownfish to sense chemicals in their environment that help them detect and avoid predators.

A growing body of work provides an unsettling preview of what acidification might have in store for marine life. Reefs located near natural oceanic vents for carbon dioxide, including sites in the Mediterranean and near Papua New Guinea, show dramatic declines in calcifying organisms and far less diversity in foraminifera when pH drops below 7.8. Among the winners in these regions are sea grasses, which take advantage of the additional CO_2 for photosynthesis. The regions where upwelling currents already bring cold, low-pH water to the surface are especially telling. Billions of oyster larvae have died off since 2005 during periods of upwelling at hatcheries along the Oregon and Washington coasts. The creatures' shell-building was hindered by pH that was unusually low, even for these locations. And in the Southern Ocean off Antarctica, a team of scientists found evidence in 2008 that pteropods—tiny creatures also known as "sea butterflies"—were suffering major damage to their shells in regions of upwelling.

Even if humans can turn the tide on greenhouse gas emissions, it will take tens of thousands of years more before the ocean returns to its pre-industrial pH levels. Some engineering fixes have been proposed—for instance, adding mammoth amounts of limestone to the ocean to reduce its acidity—but these are likely to succeed only on a local level, if even then. Any such technical fix is laden with ecological question marks: for instance, could an attempt to de-acidify the oceans cause something even more harmful to occur? Similarly, it's important to keep in mind that any scheme to tamp down global warming by blocking sunlight, as proposed in several geoengineering strategies (see p. 435), can't directly address ocean acidification, as these schemes don't reduce the amount of CO_2 entering the atmosphere and, by extension, the oceans.

Coral reefs at risk

The most famous oceanic victims of climate change may be the magnificent **coral reefs** that lace the edges of the world's subtropical and tropical waters. More than a quarter of the planet's coral reefs have already been destroyed or extensively damaged by various types of human activity and by spikes in water temperature that may be intensified by overall ocean warming. Nobody expects climate change to completely eliminate the world's reefs, but many individual coral species may face extinction.

It's amazing that coral reefs manage to produce the color, texture, and overall splendor that they do. After all, they grow in the oceanic equivalent of a desert: the nutrient-poor waters within 30 m (100 ft) of the surface. The reefs' success in this harsh environment is because of their unique status as animal, vegetable, and mineral. Coral reefs consist of **polyps**—tentacled animals that resemble sea anemones—connected atop the reef by a layer of living tissue. Inside the tissue are microscopic **algae** that live in symbiosis with the polyps. The algae produce the reefs' vivid colors and feed the polyps through photosynthesis, while the polyps provide the algae with nutrients in their waste. The polyps also secrete **limestone** to build the skeletons that make up the reef itself.

Coral reefs are durable life forms—as a group, they've existed for at least 40 million years—but they're also quite selective about where they choose to live. The microalgae need access to light in order to photosynthesize, and

Rich, diverse, and vulnerable—abundant life on the coral reef. (UCAR Digital Image Library, photo by Kathy Krucker)

the skeleton needs a foundation, so coral reefs form in shallow, clear water on top of a base that's at least moderately firm, if not bedrock. Finally, the waters around coral reefs need to be warm, but not too warm. Generally, the average water temperature should be above 27°C (81°F), with annual lows no cooler than 18°C (62°F). If the waters spend much time at temperatures more than 1°–2°C (1.8°–3.6°F) above their typical summertime readings, the algae can begin to photosynthesize at a rate too fast for the coral to handle. To protect its own tissue, the coral may expel the microalgae, allowing the white skeletons to show through the translucent polyps. When this occurs on a widespread basis, it's known as **mass bleaching** (although chlorine has nothing to do with it).

The record-setting El Niño of 1997–1998 led to unusually warm water in much of the tropical Pacific, triggering a bleaching disaster unlike anything ever recorded. Some 16% of the world's coral reefs were damaged in that event alone. A weaker event in 2002 bleached up to 95% of individual reefs across parts of the western Pacific, and another strong El Niño in 2010 produced record-warm oceans and a second round of global-scale bleaching. This time the victims included some groups of coral near Indonesia that had managed to survive the 2004 tsunami. Some coral species appear to be bouncing back from the 1997–1998 bleaching, although full recovery could take up to 20 years. Other species—especially those with lavishly branched forms and thin tissues—don't appear to be so lucky. Experiments hint that some coral may be able to adapt to bleaching over time by teaming up with different microalgae more suited to warmer temperatures. The upshot over the long term may be a thinning of the herd, as it were, leaving behind a hardier but less diverse and less spectacular set of coral reefs.

To complicate the picture, global warming isn't the only threat the reefs are facing:

* **Agricultural fertilizer** that washes offshore can deposit nitrogen and phosphorus that encourage the formation of seaweed or phytoplankton at the expense of corals.
* **Overfishing** can increase seaweed growth on top of reefs by removing many of the plant-eating fish that would otherwise keep the growth in line.
* **Sediment from rivers** can block needed sunlight, interfere with the polyps' feeding, or even bury a reef whole. Logging and other land-based activities have produced enough sediment downstream to put more than 20% of the coral reefs off Southeast Asia at risk, according to the World Resources Institute.

The ultimate impact of climate change on reefs depends in part on how extensive these other stressors become. By itself, the influence of global warming is a mixed bag. Sea level rise is a potential positive. Many older coral reefs have built upward to the maximum height that sea levels allow, so a higher sea surface would give these reefs more vertical room to grow. Meanwhile, the lowest-lying coral reefs could probably grow upward quickly enough to keep up with modest sea level rises—a typical vertical growth rate is about 1 m (3 ft) in a thousand years. As long as temperatures increase slowly enough to avoid coral bleaching, warmer temperatures appear to speed up the calcification process that builds reefs. And the global spread of warmer waters will add a sliver of territory around the current margins of coral reef viability, though that extra real estate is limited by other requirements such as a firm foundation and clean water.

Together, these pluses may not be enough to counter the negatives produced by the environmental stresses noted above, not to mention global warming itself. Along with causing bleaching, warmer waters may nourish the enemies of reefs as much or more than they help the reef itself. One study by Drew Harvell of Cornell University found that some pathogens tend to thrive at temperatures 0.6°C (1.0°F) or more higher than the optimal readings for coral themselves. Another risk is the ocean acidification discussed above, which can both limit the amount of carbonate available for reef building and perhaps erode the reef skeletons directly, in both cases slowing the rate of reef growth. Also, tropical cyclones may become more intense with time (see p. 183), although it appears that a healthy reef can recover from hurricane or typhoon damage fairly quickly. In short, we can expect some coral reefs to be around in the world of our grandchildren, but they may not be in the same locations or glow with the same brilliance that snorkelers and scuba divers now savor.

CHAPTER 8

Hurricanes and Other Storms

ROUGH WATERS

Spinning their way across the warmer parts of the globe, tropical cyclones throw fear into millions of coastal dwellers each year. Depending on which ocean they're in, these huge weather systems are known as hurricanes, cyclones, or typhoons. Whatever you call them, they've made their presence known across the tropics and midlatitudes, particularly in the new century. A record 10 typhoons raked Japan in 2004. Australia was hammered in early 2006 by a trio of cyclones, each packing winds above 250 kph (155 mph) either offshore or on the coast—the first time the region has recorded three storms of that caliber in a single season. Cyclone Nargis, which slammed into Burma in 2008, killed more than 100,000 people, making it the world's deadliest tropical cyclone in at least three decades. The Atlantic saw a major burst of activity in 2010, with five of the year's first six hurricanes reaching Category 3 status—twice the usual number for an entire season. And the super-sized, late-season storm known as Sandy, classified as a hurricane until just before it came onshore, ravaged large sections of the New Jersey and New York coastlines in 2013.

As terrible as Sandy was, the storm came and went in a matter of days, unlike the prolonged tropical onslaught borne by the United States in 2005.

Debris lines the streets of Tacloban, Philippines, after the catastrophic Supertyphoon Haiyan in November 2013. (Trocaire, Ireland/CC-BY-2.0/creativecommons.org/licenses/by/2.0/Wikimedia Commons)

On the basis of measurements of lowest pressure, that year brought three of the six most powerful Atlantic hurricanes on record up to that point: **Wilma** (the strongest), **Rita** (fourth strongest), and **Katrina** (sixth strongest). All three made devastating strikes on coastal areas. Wilma slammed Cancún and south Florida, Rita pounded the Texas–Louisiana border, and Katrina swept through Miami before bringing mind-boggling misery and more than 1800 deaths to the New Orleans area and the Mississippi coast.

Did global warming have anything to do with all this havoc? Several major studies have made a strong case that tropical cyclones worldwide are indeed getting more powerful. At the same time, a number of hurricane experts (many of them forecasters) have taken the position that global warming plays only a minor role at best in hurricane activity and that it will be far outweighed by other factors in the future. The legitimate science questions involved make this one of the liveliest arenas of climate change debate. The unexpectedly quiet Atlantic season of 2013—one of the weakest in decades—provided yet more intrigue.

To the north and south of the tropics, people tend to be more concerned with **coastal storms**. These wintry blasts of rain, snow, and wind sweep through places such as New England, British Columbia, and western Europe, packing less punch than hurricanes but often affecting much larger areas. Coastal storms appear to be shifting gradually poleward over time. Though the trends aren't highly dramatic, they could have major importance for the densely populated coastlines of North America and Eurasia.

Well away from the coasts, the most intense bouts of wind and rain are likely to occur with severe **thunderstorms**, which occasionally pack the tiny but powerful windstorms known as **tornadoes**. (Many hurricanes generate tornadic thunderstorms as well.) Although tornadoes are being reported more often in many parts of the world, it looks as if that can be pinned on greater awareness of them. Thankfully, there's no solid evidence that the most violent tornadoes and thunderstorms have become more intense or frequent because of global warming, though some hints of potential change down the road are showing up in computer models (see p. 186).

A taste of things to come?

Two main questions loom over any discussion of future tropical cyclones: will there be more of them in a warmer world, and will they be more destructive? In its 2001 report, the **IPCC** noted that no trend had yet been identified in tropical cyclone intensities, as measured by winds and pressures, adding that "past and future changes in tropical cyclone location and frequency are uncertain." The 2007 IPCC report went a step further, noting evidence that the number of intense hurricanes and typhoons had increased during recent decades in some regions. This shift remains difficult to nail down, and it isn't fully global, as reiterated by the panel's 2013 report. However, it's especially robust in the North Atlantic, where a distinct jump in frequency of the most intense hurricanes has taken shape since the 1970s.

The notion of stronger hurricanes and typhoons in a warmer world makes physical sense. **Warm ocean waters** help give birth to tropical cyclones and provide the fuel they need to grow. Generally, the cyclones can't survive for long over sea surfaces that are cooler than about 26°C (79°F). The larger the area of ocean with waters above that threshold, and the longer those waters stay warm in a given year, the more opportunity there is for a tropical cyclone to blossom if all other conditions allow for it.

Not all areas are conducive: within about 5° of the equator, the winds can't easily gather into a rotating pattern. The South Atlantic also doesn't seem to have the right combination of upper-level winds and warm surface water to produce hurricanes, although 2004 brought a major surprise (see box on p. 174). That leaves the prime breeding waters for tropical cyclones

A hurricane rolls down towards Rio

People in Brazil aren't used to hurricanes. In fact, no hurricane had ever been reported there until 28 March 2004, when a mysterious system packing winds up to 137 kph (85 mph) swept onto the coast of the state of Santa Catarina. Forecasters from Brazil and from the U.S. National Hurricane Center had been tracking the storm by satellite. Although no tropical cyclones had ever been officially recorded in the South Atlantic, the swirl of clouds had the clear eye and other telltale features of a bona fide hurricane. Despite the absence of a precedent, Brazil's meteorological service issued warnings and got people out of harm's way. The storm ended up destroying many hundreds of coastal homes, but only one death was reported.

Informally christened Hurricane Catarina for its landfall location, the storm continued to stir things up even after it died. Some meteorologists questioned whether it was truly a hurricane, and a few climate change activists pointed to Catarina as an illustration of a planet gone awry. In fact, it turns out that Catarina had some of the classic earmarks of a hurricane, but not all of them. For instance, the showers and thunderstorms circling its center were more shallow than usual. And Catarina can't be easily pinned on climate change, because the waters over which it formed were actually slightly cooler than average for that time of year. The storm was clearly unprecedented in its landfall impacts, but analysts poring over satellite records found at least one similar (if weaker) system in 1994 that stayed well out to sea.

In 2010, U.S. and Brazilian meteorologists tracked another system with tropical characteristics that spun for two days in the same offshore region where Catarina formed. It was dubbed Tropical Cyclone Anita by Brazil's meteorological service (though the World Meteorological Organization hadn't yet added the South Atlantic to the global naming system it manages). Researchers poring over satellite archives have found at least two other South Atlantic systems resembling tropical cyclones, as well as dozens of subtropical cyclones (storms that have some but not all characteristics of a bona fide tropical cyclone). All the attention has fed a growing awareness that tropical cyclones can develop in a region where few scientists were looking for them. However, in its sheer strength and impact, Catarina remains unique in modern records—a bizarre weather event that pushed the buttons of a world increasingly jittery about climate change.

between about 10° and 30° across the North Atlantic, North Pacific, South Pacific, and Indian Oceans (see map on p. 147).

The terrible 2005 trio of Katrina, Rita, and Wilma shows how important warm water is for hurricanes. Katrina and Rita both surged to their peak intensity—Category 5—just as they passed over a patch of the central Gulf of Mexico heated by an infusion of deep, warm water from the Caribbean. This channel, called the **Loop Current** for its arcing path into and back out of the Gulf, is often a hurricane booster. In 2005 it featured some of the warmest waters ever observed in the open Gulf, topping 32°C (90°F) in places. Similarly toasty waters had also overspread much of the Caribbean, where Wilma intensified in an astonishing 12 hours from Category 1 status to attain the lowest barometric pressure ever observed at sea level in the Western Hemisphere: 882 hectopascals (26.05 inches of mercury).

To keep a truly intense hurricane at top strength, the ocean's warmth needs to extend deep—ideally, at least 100 m (330ft). Otherwise, the powerful winds and waves can churn up enough cold water from below to dilute the fuel source and diminish a slow-moving cyclone in a matter of hours. Both Rita and Katrina began weakening after they left the Loop Current; they began to pull up cooler subsurface waters, while drier air and stronger upper-level winds from the United States started to infiltrate. By the time they struck the **Gulf Coast**, both storms were down to Category 3 strength. That didn't matter much, though, because the storms had already piled up fearsome amounts of water that slammed into the coast in the form of punishing storm surges. Katrina's surge was a U.S. record: at least 8.2 m (27 ft) above sea level at its worst, with waves on top of that, destroying the homes of thousands who thought they lived at safe elevations. The surge also helped lead to the failure of several levees in and around **New Orleans**, which in turn produced the cataclysmic flood that left most of the city submerged and paralyzed (see box, p. 144).

The wreckage of the 2005 Atlantic hurricane season extended far beyond U.S. shores. Ocean temperatures were markedly above average across vast stretches of the North Atlantic tropics and subtropics. All that long-lasting warmth, coupled with nearly ideal upper-level winds, led to 27 named Atlantic storms in 2005, which smashed the record of 21 set in 1933. Still another system was upgraded to tropical-storm strength posthumously, bringing the season's final total to 28. This swarm of storms was noteworthy not just for its sheer size but for its unusual geographic spread as well. Even Spain got its first-ever tropical storm, **Vince**, which made landfall on the southwest coast near Huelva.

Katrina, Sandy, and climate change

The damage and misery wrought by 2005's Hurricane Katrina topped anything produced by a single U.S. storm for many decades. More than 1800 people died, mostly in the New Orleans area, where public evacuation options outside of the wretched Superdome and convention center were nearly nonexistent. The storm's toll in property damage soared well into the tens of billions of U.S. dollars, and that doesn't include more than $14 billion spent on rebuilding and bolstering the New Orleans levee and water pumping system.

Some observers linked the Katrina debacle to climate change from the outset. In a *Boston Globe* editorial, journalist Ross Gelbspan declared, "The hurricane that struck Louisiana yesterday was nicknamed Katrina by the National Weather Service. Its real name is global warming." Others denied any connection between Katrina and climate change, a viewpoint that came to dominate U.S. media and legislative discussion before long. It's true that no weather event can be blamed solely on climate change, and certainly a storm like Katrina doesn't require global warming in order to flex its muscle. Though they're quite rare, hurricanes on a par with Katrina have developed in the Atlantic since records began. However, the gradual warming of tropical waters over the last several decades has made it easier for storms like Katrina to intensify when other conditions are right. Perhaps a more telling hint of a shift in climate wasn't Katrina itself so much as Emily, Katrina, Rita, and Wilma combined. It marked the first time four, or even three, Atlantic hurricanes had reached Category 5 strength in a single year in records that date back to 1851.

As for the landfall in New Orleans, that was simply the luck (albeit the bad luck) of the meteorological draw. Several near misses had grazed the city since 1965, when Hurricane Betsy produced flooding that killed dozens. Betsy prompted the U.S. Corps of Engineers to build the city-girdling levee system that was in place when Katrina struck. That system was built to withstand only a Category 3 storm, a fateful move sure to be analyzed by critics for years to come, especially since Katrina itself had weakened to Category 3 strength by the time it made landfall. Post-Katrina improvements to the system, designed to deal with a wide variety of scenarios, are expected to keep flooding minimal in New Orleans during a "100-year" storm (one with a 1% chance of occurring in a given year), and it should offer some protection against even stronger storms. However, the new system won't help outlying areas, where vulnerability may continue to rise.

Hurricane Katrina approaches the Gulf Coast on August 28, 2005. (NOAA)

In 2012, Hurricane Sandy brought climate change worries to the fore once again. The storm overtopped flood defenses in lower Manhattan and tore apart coastal towns from New Jersey to Long Island. (For the record, Hurricane Sandy was reclassified as a post-tropical cyclone a mere hour before landfall because its warm core was rapidly eroding—thus, the reason for the "Superstorm Sandy" moniker.) While the storm's death toll in the United States was far lower than Katrina's, the storm still caused more than 150 direct and indirect U.S. fatalities. Damage estimates hit $65 billion, making it the second most expensive American hurricane behind Katrina. The storm's psychological impact was no less devastating, as many residents didn't consider themselves hurricane vulnerable. "Our climate is changing," said New York City mayor Michael Bloomberg just after the storm. Though he claimed that Sandy and other extreme events may or may not be directly related to global warming, the magazine *Bloomberg Businessweek* dispensed with nuance in its front-page headline immediately after the storm: "It's Global Warming, Stupid." Sandy and its aftermath prompted Bloomberg to propose a $19 billion initiative to beef up the city's bulwarks against hurricanes and storm surges.

While hurricanes do reach the northeast U.S. coast from time to time, they're typically racing north or northeastward, which means they often sideswipe the coast from Virginia to Maine rather than hitting it head on. Sandy broke this mold in spectacular fashion. After striking Cuba and moving northeastward past the Bahamas, Sandy was blocked by an unusually intense zone of high pressure in the far North Atlantic. Meanwhile, a strong midlatitude storm dipped into the eastern United States. These features combined to shift Sandy toward the northwest and into the New Jersey coast—a trajectory never before observed for a U.S. hurricane in records going back to the 1860s. One analysis led by NASA's Timothy Hall found that a track this close to a right angle into New Jersey has a return period of roughly 600–700 years, assuming a steady-state climate. Sandy's square-on track also maximized the surge of high water pushed onto the coast north of its center. In lower Manhattan, a 9-ft (2.7-m) surge at the Battery teamed up with a high tide to produce the highest water levels seen there since records began in 1920. It's unclear whether several notable surges of the 1700s and 1800s were bigger than Sandy's; by some estimates, a Sandy-magnitude surge would be expected to recur in Manhattan an average of only every 400–800 years.

Did the hand of climate change help steer Sandy? Research led by Jennifer Francis (Rutgers, The State University of New Jersey) helped kindle a major debate over how the dramatic melting of Arctic sea ice could be affecting mid-latitude weather. In a paper published only a few months before Sandy, Francis and Steven Vavrus (University of Wisconsin–Madison) detected signs of weakening jet-stream flow and more elongated upper-level highs and lows since the 1980s. Francis also suggested in public remarks that high-latitude blocking highs such as the one that helped torque Sandy's path might be becoming more likely. However, some researchers questioned how and whether near-surface Arctic warming could influence conditions at the jet-stream level, and several competing analyses found little or no long-term trend in a variety of indices related to blocking. The main exception is in the October to December period, where studies tend to agree that at least some changes in blocking-related circulation over the North Atlantic may be under way. (Autumn is also the time when large areas once covered by Arctic sea ice remain open water months later than before.) The latest round of model simulations conducted in support of the 2013 IPCC assessment hints that blocking might actually decrease over the

A rollercoaster that thrilled generations of kids along New Jersey's Seaside Heights boardwalk became iconic in a different way after it was inundated and pulled offshore by Sandy. (FEMA/Patsy Lunch)

North Atlantic this century. However, such models also tend to underestimate the amount of blocking in today's climate, so a grain of salt may be in order.

There's no controversy over the fact that higher seas exacerbated the flood damage from Sandy. Mean sea level has risen roughly 28 cm (11 in.) at Manhattan over the last century—more than in many other parts of the globe—so it wouldn't be out of line to pin the topmost foot or so of Sandy's surge on a warming planet. Sandy's huge size also contributed to the destruction. Sustained gale-force winds sea covered a record-wide swath of up to 835 km (520 mi), which pushed vast amounts of water shoreward. At one point while Sandy was at sea, the kinetic energy in its waves and surge actually exceeded that of Katrina, which packed higher winds but covered less area. Several other recent hurricanes, including Ike (2008) and Isaac (2012), have also featured large circulations that delivered more storm surge than their peak wind speeds alone would imply. Time will tell whether the recent burst of jumbo-sized Atlantic hurricanes is part of a long-term trend or merely a passing quirk.

Keeping count: Will there be more cyclones in the future?

The sheer number of Atlantic hurricanes in 2005 led many to ask whether tropical cyclone frequency might already be on the rise due to climate change. One might think that a larger, longer-lasting zone of warm-enough water would be sufficient to spawn more and more tropical cyclones. And indeed, both 2005 and 2010 saw Atlantic waters that were unusually warm over vast areas for long periods, which helped seasonal forecasters accurately peg the action months ahead of time. Yet the 2013 season wasn't able to take advantage of relatively warm water. That year, the Atlantic produced a paltry total of two hurricanes—the smallest total in 31 years—and neither of these systems made it even to Category 2 status.

Both the 2007 and 2013 IPCC assessments concluded that, over the last century, the average number of tropical cyclones worldwide hasn't been significantly boosted by greenhouse gases. Making such an assessment of the frequency of past tropical cyclones, and projecting their future, is a complex task for several reasons:

* **Other factors.** Even over sufficiently warm water, only a small fraction of the weak circulations that form in a given year spin up to cyclone status. A number of factors can lead to their early demise: unfavorable upper winds, an influx of dry air, or competition from nearby systems, to name a few. Warm water is critical, but it isn't enough to guarantee that a hurricane will form.
* **Variable records.** It's hard to know exactly how many tropical cyclones roamed Earth before satellite monitoring became routine in the 1970s. Although ships did a good job of reporting stronger tropical cyclones in the pre-satellite era—at least across major shipping lanes—it's likely that some of the weaker, smaller systems over the open ocean were missed then but are being caught now. Better observations also help explain an apparent rise in the number of short-lived Atlantic cyclones, ones that last for two days or less. If these factors are taken into account (including an estimate of the cyclones that once went uncounted), the total number of Atlantic systems may not have risen significantly over the last century, even through the strongest have become more prevalent in recent decades.
* **Complex global interrelations.** For reasons not fully understood, the planet normally does a good job of conserving the total number of tropical cyclones. About 40–60 per year attain hurricane strength, and there has been no significant long-term trend up or down during

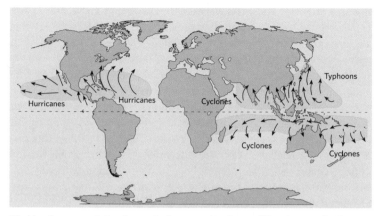

World regions prone to hurricanes, typhoons, and cyclones: different names for the same phenomenon.

the few decades of reliable data. The frequency stays fairly constant because the planet's ocean basins appear to trade off in their periods of peak production. Typically, when the Atlantic is producing hurricanes galore, parts of the Pacific are relatively quiet, and vice versa. Some of this seesaw effect is due to **El Niño** and **La Niña**, the cyclic warmings and coolings of the eastern tropical Pacific (see p. 159). The large-scale air circulations driven by La Niña encourage Atlantic hurricanes but suppress them in some parts of the Pacific; the opposite is true of El Niño.

Until relatively recently, computers weren't of much help in assessing whether we'll see more tropical cyclones in a warmer world. That's because most hurricanes and typhoons are small enough to fall in between the grid points of the comprehensive global models that map climate out to a century or more. It's also tough for a model to pick up on the meteorological subtleties that allow one system to become a hurricane and another not to—the same unknowns that plague tropical weather forecasters on an everyday basis. But over the last decade, researchers have made headway in using models to envision the long-term future of tropical cyclones. This is in part due to faster computers that allow new models to depict the entire globe in sufficient detail. An early sign of progress came in 2006, when a group of Japanese scientists led by Kazuyoshi Oouchi published results from a globe-spanning model with 20-km (12-mi) resolution, enough to follow individual tropical cyclones in a broad-brush fashion. The results suggested that in a midrange emissions scenario, the late twenty-first-century Earth could see fewer tropical cyclones but a higher fraction of intense ones. Other scientists have nested fine-scale regional models inside coarser global-scale ones or blended statistical and physical modeling.

The results from these varied approaches have largely been in sync with each other on the main take-home message: we may see fewer tropical cyclones worldwide over time—anywhere from 6% to 34% fewer by late this century, according to a 2010 roundup in *Nature Geoscience* by an all-star group of hurricane researchers led by Thomas Knutson (NOAA). The 2013 IPCC report echoed this conclusion, deeming it likely that global tropical cyclone counts will either drop or hold steady this century. It's not clear exactly why many models are tending to spin up fewer storms in the future. One possibility is a predicted drop in relative humidity aloft, which would threaten the cocoon of moisture in which a hurricane thrives. Another is a projected weakening of the global circulation through which air rises across the tropics and sinks over the subtropics. It's also possible that the SST threshold needed for tropical cyclone formation could rise hand in hand with overall ocean warming. It's worth noting that some of the most recent work with high-resolution modeling, including a 2013 analysis by Kerry Emanuel (Massachusetts Institute of Technology), suggests that tropical cyclones might actually become *more* prevalent in many regions, rather than less, but much more study is needed before such a potential paradigm shift could be confirmed.

Even if tropical cyclones become a bit less prevalent on a global scale, as most experts expect, they could still increase over some oceans while decreasing over others. Models haven't yet agreed on a collective portrait of region-by-region change, but a few intriguing threads have emerged. For example, a number of models suggest that a semipermanent El Niño–like state could develop. This could lead to a stronger warming of tropical waters in the Pacific than in the Atlantic, which in turn might trigger El Niño–like circulation patterns in the atmosphere. By examining 18 model simulations that fed into the 2007 IPCC report, Brian Soden (University of Miami) and Gabriel Vecchi (NOAA) found that upper-level winds became stronger (i.e., more hostile for hurricanes) in the northeast Pacific and Atlantic while relaxing in the Indian and western Pacific Oceans. However, given that global and regional models still struggle to depict the behavior of El Niño and La Niña, there remains a good deal of uncertainty in how hurricane counts will unfold from one basin to the next.

Regardless of whether or not we see more tropical cyclones, we can still count on most of them never reaching land. Even if a given ocean began to spawn more cyclones, it doesn't necessarily mean that more of them will hit the coastline, as evidenced by the busy Atlantic seasons of 2010 and 2011 that largely spared the United States. By the same token, even a slow season can bring destructive landfalls. In 1992, the Atlantic was far quieter than average, yet it still produced Florida's massively destructive **Hurricane Andrew**.

Packing more power

What about strength? Here, the tune is more ominous. Though the IPCC saw little evidence of change in its 2001 report, subsequent studies have found a jump in the destructive power of tropical cyclones on a global basis, as noted in the 2007 and 2013 reports. An increase in hurricane strength is both plausible on a physical basis and apparent in some of the most recent modeling.

Two key studies pushed this topic into the headlines in 2005, where it shared space with that year's bumper crop of Atlantic superstorms. The Massachusetts Institute of Technology's Kerry Emanuel examined tropical cyclones since the 1970s, when satellite monitoring became widespread. He looked not just at the raw wind speeds but at the destructive force they packed over their lifespan. This force, the **dissipation of power**, is proportional to the wind speed cubed. Thus, a wind of 100 kph would be twice as damaging as a wind of 80 kph. Small increases in wind speed, then, can produce major ramp-ups in destructive force. From this vantage point, Emanuel discovered that tropical cyclones in the Atlantic and northwest Pacific were packing nearly twice the power they did in the 1970s—the result of the strongest storms getting even stronger as well as lasting longer.

A team led by Peter Webster of the Georgia Institute of Technology found similar results to Emanuel's when they ranked tropical cyclones since the 1970s by **intensity category**. As expected, they found little change in the total number of tropical cyclones in all categories—a sign of the global balance noted above. The eye-opener was their discovery that the strongest cyclones, those at Category 4 or 5, were almost 50% more frequent globally in 1990–2004 than they were in 1975–1989. Put another way, Category 4 and 5 storms went from comprising about 25% of all tropical cyclones globally to comprising about 35% of them. A follow-up analysis in 2010 by coauthor Judith Curry showed the trend continuing (see graphic on previous page).

Predictably, these studies provoked a storm of controversy. They also exposed a schism between climate change researchers, few of whom had

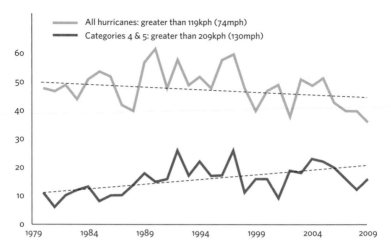

———	All hurricanes: greater than 119kph (74mph)
———	Categories 4 & 5: greater than 209kph (130mph)

Tropical cyclone totals for the period 1980–2009: while the total number of tropical cyclones around the globe ranked Category 1 to 5 didn't change much over these three decades, the proportion of Category 4 and 5 cyclones—those that produce the worst damage—rose markedly. (Judith Curry/IBTrACS)

done detailed studies of tropical cyclone behavior, and hurricane specialists, who knew the cyclones and the data by heart but weren't always experts on the larger-scale processes related to global warming. In his 2007 book *Storm World*, science writer Chris Mooney cast the battle as only the most recent example of a tiff between observationalists and theoreticians, with roots extending back to the 1800s.

One of the most vocal critics of the new findings was William Gray (Colorado State University), who founded the field of seasonal hurricane prediction. A long-time skeptic of climate change, Gray assailed the studies on various technical grounds. Also dismissive of a climate change link were several researchers at the **National Oceanic and Atmospheric Administration** (NOAA), whose **National Hurricane Center** (NHC) forecasts Atlantic storms. Individually, some NOAA scientists agreed that human-induced climate change has the potential for "slightly increasing the intensity of tropical cyclones," in the words of Christopher Landsea, who oversees research at NHC. However, Landsea and colleagues suggested that the apparent surge in top-strength hurricanes could be at least in part due to inconsistencies in the official record. For instance, lower-resolution satellites used in the 1970s may not have sensed the tiny areas of high, cold cloud tops associated with the strongest cyclones. Other observational tools, such as the instrument packages dropped from aircraft into hurricanes, have also evolved over the years, as have the analysis techniques that classify the storms. Landsea and colleagues are now in the midst of a long-term effort to reanalyze several centuries of Atlantic hurricanes, which has already resulted in some hurricanes being upgraded and others downgraded.

Even by restricting one's view only to years with the best data, it's apparent that change is afoot, as shown by James Elsner (Florida State University) and James Kossin (NOAA/University of Wisconsin–Madison) in studies published in 2008 and 2009. Acknowledging that observational techniques had changed, they set out to compare apples to apples by analyzing only those cyclones that occurred during the period of highest-quality satellite data, which began in 1981. Through this clarifying lens, the scientists found that the peak winds produced by the globe's strongest tropical cyclones had indeed grown stronger, at a rate of about 1–2 meters per second (2–4 mph) per decade. Even such a seemingly modest value adds to a storm's destructive power, as noted above. Over several decades, such a trend might be enough to push a substantial fraction of storms from one intensity category to the next.

The plot thickens even further when you consider the Atlantic Multidecadal Oscillation (AMO). It has long been known that, over the past century, sea surface temperatures in the North Atlantic alternated between warming and cooling for periods of about 30–40 years each. Many researchers have tied this waxing and waning of SSTs to presumed cycles in the Atlantic thermohaline circulation, also known as the meridional overturning circulation. This circulation is only now being measured with some precision (see chapter 7), and its dynamics are poorly understood. It's also possible that post–World War II pollution helped shade and cool the North Atlantic for decades. Nevertheless, Gray showed that the frequency and average intensity of Atlantic hurricanes has risen and fallen every few decades in sync with the AMO—lower from the 1890s to the 1920s, higher from that point through the 1950s, then lower again from the 1960s to the 1990s. On this basis, Gray predicted an increase in Atlantic hurricane action in the mid-1990s that would continue for several decades. Thus far, his vision has proven remarkably accurate.

Even if there is a strong physical basis for the AMO, it doesn't mean climate change can be ignored. The AMO only affects hurricane formation in the Atlantic, whereas the studies mentioned above suggest a strengthening trend across several other ocean regions as well. Moreover, the Atlantic uptick since the 1990s could be a combined product of the AMO and general ocean warming. For example, a 2006 study by Kevin Trenberth (NCAR) found that global warming had more than twice the influence of the AMO in producing 2005's extra hurricane-fueling warmth in the waters of the tropical North Atlantic. Thus, if and when the AMO cycle finally turns toward a quieter Atlantic—perhaps between 2015 and 2025—things may not calm down quite as much as one would otherwise expect.

Despite their limited ability to project tropical cyclones into the future, computer models tend to agree that overall warming ought to help induce a modest increase in strength. The IPCC's special 2012 report on climate extremes notes that the newer models consistently push typical winds in the strongest storms upward this century by anywhere from 2% to 11%. The latter would be nearly enough to bring a major hurricane from one Saffir–Simpson category to the next. This doesn't mean that more of these intensified storms will necessarily make landfall, but those that do could pack more of a wallop.

One question that remains largely unexplored is whether the size of tropical cyclones might be changing. The biggest cyclones can bring hurricane-force winds to 10 or more times greater area than the smallest ones, and they can drive enormous storm surges (see sidebar on p. 176). However, cyclone diameter isn't part of official global databases, and most climate change studies to date have omitted this important dimension, although greater attention is now being focused in this area.

Surges and downpours

Extreme **storm surges**—such as the one that swept across parts of the Philippines during Supertyphoon Haiyan in November 2013, killing thousands—will continue to loom as a catastrophic possibility from tropical cyclones in the century to come. However, the shape of that threat will shift as sea levels rise and the coast itself evolves. **Subsidence** and the loss of protective **wetlands** are already making many areas more vulnerable, including the Texas and Louisiana coastlines hammered in 2005.

More than any place on Earth, the **Bay of Bengal** exemplifies the risks of tropical cyclone surge. Fully half of the world's deadliest surge tragedies have taken place in eastern India, present-day Bangladesh, and Burma. Half a million Bangladeshi people died as a result of a catastrophic 1970 landfall (the event that prompted a famous relief concert by ex-Beatle George Harrison). In 2008, **Cyclone Nargis** took a track unusually far to the south, becoming the first major cyclone ever recorded in Burma's populous Irrawaddy Delta. A surge 3 m (12 ft) high struck at high tide and swept as far as 40 km (25 mi) inland. With no experience of such a storm in living memory—and with state-controlled media downplaying the storm's threat—delta residents had no inkling of what they would face. The official death toll topped 130,000, though many thousands more may have died.

Hurricane Danielle as seen from the International Space Station on August 27, 2010—a year in which 19 named storms raged across the Atlantic and Gulf of Mexico, including 12 hurricanes. The only prior year with more Atlantic hurricanes was 2005. (NASA)

Rising sea level only adds to the Bay of Bengal's flood risk, while soaring coastal populations make even more people vulnerable to a given flood. The intricate delta-style coastline would be extremely difficult to protect through engineering, even if the nations could afford such a fix. Using a regional model for the Bay of Bengal, Jason Lowe at the Hadley Centre found that extreme storm surges by 2050 could affect the India–Bangladesh coast in complex ways. Some areas become more vulnerable than others in the model, as storm patterns shift and as the higher sea level causes changes in tidal and surge behavior.

Landfalling tropical cyclones also produce a great deal of rain, on coastlines as well as far inland. That may get worse as moisture increases in the atmosphere and precipitation increases on a global scale. **Inland flooding** from heavy rain is already a serious threat in many hurricane-prone areas. With Katrina and Sandy as major exceptions, river flooding has been the leading cause of U.S. hurricane deaths in recent decades. It's also a major problem for agriculture. Rainfall in the average tropical cyclone may have already increased by 6%–8%, according to estimates by NCAR's Trenberth, and modeling associated with the 2013 IPCC report suggests that rainfall rates near the core of a typical storm could rise by 30% or more by later this century.

With all this in mind, it would seem sensible to prepare for the worst, despite the remaining gaps in knowledge. One major problem for planners to take on board is the explosive population growth and development taking place on many cyclone-prone coasts. Roger Pielke, Jr. (University of Colorado Boulder), argues that these factors could easily boost the human and financial costs of hurricane damage this century far more than any intensification related to global warming would—a reminder that demographics have everything to do with the ultimate impact of climate change. Researchers from across the spectrum of the hurricane–climate change debate crafted a landmark joint statement in 2006 stressing the human element of the problem: "We are optimistic that continued research will eventually resolve much of the current controversy over the effect of climate change on hurricanes. But the more urgent problem of our lemming-like march to the sea requires immediate and sustained attention."

Coastal concerns beyond the tropics

Hurricanes and typhoons may cause horrific damage, but coastal storms threaten the well-being of millions of people who live well away from the tropics. Unlike tropical cyclones, which need a cocoon of warm, humid air in order to grow, coastal storms—part of the larger group known as **extratropical storms**—thrive on the contrasts between cold continental air and warm marine air. Moving with the polar jet stream, in which they're typically embedded, coastal storms strike many of the major coastlines at mid- and high latitudes, mostly from autumn through spring. Along the U.S. Atlantic coast, they're often dubbed **nor'easters**, a reference to the strong onshore winds they bring. The lack of a mountain barrier in western Europe allows coastal storms to plow well inland with heavy rain and high wind. That was the case over the last week of 1999, when an especially vicious pair of post-Christmas storms dubbed **Lothar** and **Martin** pounded much of the region with winds as strong as 219 kph (136 mph). The storms killed more than 140 people and ruined centuries-old treescapes from France's Versailles to Germany's Black Forest.

As with hurricanes, the challenge in projecting coastal storms is twofold: will there be more of them, and will they become more intense? A few studies show that extratropical cyclones have tended to shifted poleward in both the Northern and Southern Hemispheres, a sign of the jet stream shifting in response to a warming atmosphere. Across Europe, this corresponds with the observed tendency in the late twentieth century for milder, stormier winters occurring to the north and drier-than-average winters near the Mediterranean—the calling cards of a positive **North Atlantic Oscillation**, or NAO (see chapter 7). As for intensity and frequency, the trends aren't as robust, and they vary by region. While extratropical storms appear to have become stronger and more numerous toward the top end of the North Atlantic, there's some evidence the opposite has occurred over the eastern Pacific and North America, according to the 2013 IPCC report.

What about the future? Here again, as the IPCC stresses, the outcome seems likely to vary by region, especially in the Northern Hemisphere. If the warm currents of the North Atlantic weaken as projected, then the resulting rearrangement of ocean currents may send stronger and more frequent midlatitude storms into western Europe. The latest round of high-resolution models also suggests a potential increase in the strongest extratropical cyclones along and near the U.S. East Coast. Globally, however, the models actually point toward a small decrease in midlatitude storm action. A multimodel review published in 2013 by Edmund Chang (Stony Brook University) finds a reduction of 6%–32% (depending on the time of year) in strong extratropical storms across North America, given a high-emissions scenario. This could spell trouble in the form of less rainfall for parts of the United States that straddle the zones between expected subtropical drying and higher-latitude moistening. Lower-latitude coastal regions such as southern California and the Mediterranean may find themselves experiencing more prolonged and/or intense dry spells in winter, as windy, rainy, fast-moving storms track farther north over time. Any such regional projections are framed by substantial uncertainty at this point. Smaller-scale effects that aren't well modeled or understood could overpower a global trend in particular locations, and although climate change models tend to make the NAO increasingly positive in the decades to come, that's hardly a done deal. There's plenty of natural variation in the NAO, both on a yearly basis (the index can slide back and forth several times in a single winter) and over longer time frames. In fact, the NAO has been trending away from its positive phase since the year 2000: the winter of 2009–2010 brought the most intensely negative NAO on record, along with frigid, snowy conditions to match (see chapter 7).

Coastal-storm flooding: A deepening problem

Even if coastal storms don't change drastically, the extreme high-water levels they generate will still be pushed upward by the long-term rise in sea levels. One region where high population meets up with a high risk for surge from coastal storms is the **North Sea**. The Netherlands have spent centuries defending themselves from the sea with a massive series of dykes. Much of coastal England, which sits slightly higher on average, gets along with a variety of dams and other flood-control devices. Both countries found themselves unprepared for the history-making North Sea storm of February 1, 1953, which sent huge surges into the British, Dutch, and Belgian coasts. Over 1800 people died in the Netherlands, where a powerful surge overpowered dikes and levees in the lower Rhine Valley. England saw more than 300 deaths on its east coast, including 58 on Canvey Island, from where more than 10,000 residents were safely evacuated.

The 1953 storm triggered a new age of massive efforts to tackle coastal flooding. The Netherlands spent more than US$5 billion on its **Delta Project**, completed over four decades. Its centerpiece is a pair of doors, each about 20 m (66 ft) high, that can close off the Rhine if a major coastal storm looms. Along the same lines, the 1953 storm inspired Britain to enhance seawalls along its east coast and to spend over half a billion pounds on the **Thames Barrier**, which was built between 1972 and 1982. Spanning the Thames in eastern London at a width of 520 m (1700 ft), it includes 10 massive gates, 6 of which rise from concrete sills in the riverbed and swing to form a single wall.

The Thames Barrier was employed only 11 times in its first decade of service. In 1999, the operating rules were changed, allowing for more frequent closures in order to keep tidal waters from blocking river runoff produced by heavy rain. This change makes it hard to assess how much climate change may have affected the closure rate. For whatever reason, the gates are swinging shut much more frequently than before—a total of 75 times in the decade 2000–2009, compared to 35 closures in the 1990s and only four in the 1980s. The record-wet British winter of 2013–2014 triggered an astounding 43 closures, and it's estimated that the barrier could be closing an average of 30 times per year by the 2030s. The region now has embarked on a 100-year plan to protect the Thames Valley from the increased surge risk related to climate change. One of the many options put forth is a far wider barrier across the Thames estuary that would span 16 km (10 mi). However, an environmental assessment released in 2009 recommended sticking with the current barrier, making repairs and improvements as needed, until at least the 2070s.

In another well-publicized initiative, Italy is building a set of 78 hinged, inflatable steel gates to protect **Venice** from rising waters. The project, dubbed **MOSE (Modulo Sperimentale Elettromeccanico)** in a tip of the hat to Moses' parting of the Red Sea, is due to be completed by 2016 at an estimated cost of more than US$6 billion. Normally lying flat on the seafloor, the gates will be pulled up as needed in a 30-minute process that would block three inlets from the Adriatic and help keep the city safe from increasingly threatening surges triggered by Mediterranean storms. Minor floods are now so frequent in this gradually subsiding city that visitors and locals alike often find themselves facing streets ankle-deep with water.

Venice's gates would be employed for any surge expected to exceed 110 cm (3.6 ft), and they're reportedly capable of handling surges up to 200 cm (6.6 ft). In 1996 alone, over 100 floods topped the 1-m mark, and a severe 1966 flood reached 180 cm (6 ft). Alas, it wouldn't take much of an increase in global sea level rise to push a repeat of that 1966 flood over the new barriers. The city also continues to subside by as much as 0.2 cm (0.08 in.) per year, which isn't helping. Some researchers have already pondered the idea of a "Great Wall of Venice" that might eventually surround the city in order to preserve it.

Tornadoes: An overblown connection?

One of the staple ingredients of many books and articles about climate change is a photo of a menacing twister, usually with a caption that goes something like this: "Tornadoes and other forms of extreme weather are expected to increase in a warming climate." Actually, there is no sign that the most violent **tornadoes** are on the increase globally, and most studies indicate that we shouldn't expect any dramatic increase in the strongest twisters as a result of human-induced greenhouse warming.

The popular confusion appears to arise because tornadoes often get lumped under the heading of "extreme weather." It's true that extremes of temperature and precipitation are already on the rise and are expected to increase further in a warming climate, but tornadoes are an entirely different animal. Twisters are spawned by **thunderstorms**, so they're dependent on the same warm, moist air masses that lead to thunder and lightning. But that's not enough in itself to produce a tornado. To get the most vicious type of twister—the kind that occur with regularity in only a few parts of the globe, most commonly in east India, Bangladesh, and the central and eastern United States—you need a certain type of thunderstorm known as a **supercell**. These long-lived storms thrive on a particular blend of conditions, including a layer of dry air about 2000–3000 m (6600–9800 ft) high as well as winds that increase and shift direction with height (wind shear). Supercells also feature an area of rotation, called a **mesocyclone**, from which tornadoes may emerge.

The rarity of all these conditions coming together in one place helps explain why so few parts of the world are prone to violent tornadoes (those

Aerial view of the MOSE project in Venice, Porto di Lido, late 2012. The project was criticized for being expensive, experimental, monolithic, and potentially obsolete even before it was completed. (Magistrato alle Acque di Venezia, Consorzio Venezia Nuova)

with winds topping about 320 kph/200 mph). Some are reported in Europe, but generally you need a location where warm, dry air (from areas such as the U.S. Southwest or India's Thar Desert) can easily clash with warm, moist air (from such sources as the Gulf of Mexico or the Bay of Bengal). Since the geographic variables that lead to supercells won't be changing any time soon, it's unlikely we'll see much change in the preferred stomping grounds of violent tornadoes, although it's conceivable that their U.S. range could shift northward and perhaps affect the southern tier of Canada more often. Another possibility is that the period considered tornado season (generally early spring in the U.S. South and late spring to summer in the Midwest) will shift a bit earlier in the calendar as the climate warms.

There's a bit more uncertainty with **non-mesocyclonic tornadoes**—the far weaker, far more common variety. These tornadoes don't require the exotic blend of ingredients above: as long as the updraft in a thunderstorm is powerful enough, it may be able to produce a weak, short-lived tornado (though very few do). With the advent of video cameras and the growth of weather hobbyism, people have been reporting more and more tornadoes. The 1997 Hollywood blockbuster *Twister* appears to have played no small part in this boom, said the late Nikolai Dotzek of the German Aerospace Center (DLR). Combing through these reports, he and other scientists confirmed that, in many places—including England, Ireland, South Africa, and France—tornadoes are more frequent than we once thought.

Whether the global incidence of all types of tornadoes is truly on the increase is difficult to know, but experts believe that most or all of the upward trend in reported twisters is simply because more people are reporting them. Statistics from the United States and Europe bear this out. The average yearly U.S. tornado count mushroomed from about 600 in the 1950s, just as the nation was implementing its watch and warning system, to 800 in the 1970s and around 1100 in the 1990s, when tornado videos became all the rage. The numbers have continued to climb, with the average for 2005–2013 running at more than 1400 twisters per year. However, U.S. reports of violent tornadoes—the kind rated EF4 or EF5 on the Enhanced Fujita Scale, which were hard to miss even before storm chasing became common—haven't changed significantly in the entire record, holding at around 10–20 per year (though it's possible that changes in damage survey techniques may be obscuring trends). As for Europe, the continent as a whole, including the United Kingdom, was once reporting around 160–180 tornadoes per year, according to a 2003 survey by Dotzek. That number ballooned with the advent of a quality-controlled database launched in 2006 by the European Severe Storms Laboratory. The continent now averages close to 300 reported tornadoes per year. As is the case in the United States, though, weaker tornadoes have accounted for most of the increase over prior years.

Scientists are now using leading-edge computer models to get a better sense of possible global changes in severe thunderstorms, the type that produce gale-force winds and hailstones (and, sometimes, tornadoes). A U.S.–Europe network of researchers led by Harold Brooks of NOAA is looking at how well global models reproduce the current patterns of severe local weather. These models are too coarse to pinpoint individual thunderstorms, so Brooks and colleagues have sifted through the global projections and hunted for the ingredients noted above, such as warm, moist air and **wind shear** (winds changing with height). Generally, they found, the models did a good job of reproducing where severe weather happens now. As for what's to come, the models' clearest signals appear over the storm-prone United States, where they point toward an increased prevalence of unstable air but a reduction in wind shear. All else being equal, this would be expected to tilt the balance of U.S. storms away from those that generate violent tornadoes and large hail and toward those more likely to produce localized bouts of strong wind and heavy rain. Both types of storms could qualify as "severe," and the total number of such storms

Researchers on a field project called VORTEX2 gathered extensive data on this tornado, which struck near La Grange, Wyoming, on June 5, 2009. Thankfully, there is little reason to believe that such twisters are increasing as the climate warms. (Robert Henson)

may rise, according to Jeffrey Trapp and colleagues at Purdue University. By implanting regional models inside global models to gain more detail on storm-scale weather, they have found that instability may increase enough in a future U.S. climate to compensate for weaker wind shear. As a result, they say, cities such as Atlanta and New York could see a gradual increase in the number of days each year conducive to severe thunderstorms—perhaps a doubling in some locations—by the late twenty-first century.

Of course, even when conditions are ripe, thunderstorms don't always form. Using a similar technique as above to gauge the risk for the central and eastern United States, NASA's Tony Del Genio and colleagues also project a future with higher instability and weaker wind shear overall. However, their work shows the most favorable zones of instability and wind shear coinciding more often, leading to a jump in the potential frequency of the most intense thunderstorms. This possibility was reinforced in a 2013 study led by Noah Diffenbaugh (Stanford University), which looked at the suite of computer model simulations carried out in support of the 2013 IPCC report. For the first time, these model results were detailed enough to allow researchers to scrutinize the interrelationships of severe weather ingredients on a day-to-day basis in future climate. Looking at a high-emissions scenario, they found a new wrinkle: the expected reductions in wind shear tended to occur on days when instability is low, thus muting their impact on storm behavior. On the other hand, an increasing number of spring days in the modeled future climate had the potent blend of high instability and high wind shear that can foster tornadic supercells.

There's still a lot of small-scale nuance in severe thunderstorm and tornado formation that current models can't fully capture. It's also possible that day-to-day and year-to-year variability might obscure long-term changes for many years to come—especially if the central United States sees dramatic swings from dry to wet periods and back again. Thanks largely to such climatic swerving, the nation veered from one of its most active multi-month tornado strings on record in 2011 to one of its least active in 2012. The big question is whether any long-term trend might emerge from the substantial year-to-year noise. If further work supports the latest modeling, then some of North America's most populous areas might have to add severe weather to their list of climate change concerns.

CHAPTER NINE

Ecosystems and Agriculture
THE FUTURE OF FLORA, FAUNA, AND FARMING

If all that global warming did was to make life a bit steamier for the people who consume the most fossil fuels, then there'd be a karmic neatness to it. Alas, climate change doesn't keep its multitude of effects so nicely focused. A warming planet is liable to produce a cascade of repercussions for millions of people who have never started up a car or taken a cross-country flight. Many animal and plant species will also pay a price for our prolific burning of fossil fuels.

Food, and the lack of it, could be where a changing climate exerts some of its most troublesome impacts for society. While the changes could affect ranching and grazing as well as arable farming, much of the research to date has focused on croplands. Because of longer dry spells, hotter temperatures, and more climatic uncertainty, the twenty-first century is likely to see major shifts in the crops sown and grown in various regions. Well-off countries might break even or even benefit from the changes at first, if they can keep a close eye on the process and adapt their agriculture early and efficiently. This is especially the case for large midlatitude countries where crop zones can shift yet remain within national borders. Sadly, the same may not be true of the tropics, where most of the world's food crops are grown. The most problematic impacts on agriculture may wind up occurring in the poorest countries, those with the least flexibility and the most potential for catastrophic famine. On top of all this, there's global population growth and the accompanying increase in food demand.

Humans use about a third of Earth's land surface for farming and other purposes. What global warming does to parts of the other two-thirds—the world's natural ecosystems—could be even more wrenching than the effects on managed lands. Pollution, development, and other side effects of civilization have already put uncounted species at risk, and now climate change threatens to make the situation far worse.

As with agriculture, the tropics appear to be particularly vulnerable when it comes to wildlife. The normally muted variations in temperature at low latitudes imply that it would take only a small warming to push organisms into conditions they've never faced before. This concept was put in stark terms by a 2013 study in *Nature* led by Camilo Mora (University of Hawai'i at Mānoa). Scrutinizing a blend of 39 models used for the latest IPCC assessment, Mora and colleagues calculated the point at which annual global temperatures would stay completely above the yearly range observed from 1860 to 2005. Globally, in a high-emissions scenario, the estimated date for this transition is 2047, but many tropical regions would pass the threshold as soon as the 2030s. With large areas entering uncharted atmospheric territory, many creatures would have to venture far in order to stay within a familiar climate regime. Along these lines, a 2009 study by S. Joseph Wright (Smithsonian Tropical Research Institute) studied tropical mammals whose present range is no more than a few dozen kilometers. Wright found that more than 20% of those species would have to travel at least 1000 km (600 mi) in order to get to a place where temperatures by the year 2100 will be comparable to readings they experienced in their habitats in the 1960s.

How many creatures could simply evolve in place? Many forms of life are quite adaptable in their behavior, and when faced with a pressing crisis, such as new competition for food, some species can even go through physical evolution remarkably quickly—as fast as 22 years, in the case of one type of Darwin's finch whose beak size changed to improve its access to seeds. However, most adaptation to past climate change unfolded over much longer time scales, and it's not yet clear which species could keep up with today's comparatively rapid warming. In a 2013 survey of more than 500 birds, mammals, amphibians, and reptiles, Ignacio Quintero (Yale University) and John Wiens (University of Arizona) found that the average rate of climate-driven adaptation for most groups corresponded to a temperature rise of less than 1°C per million years—which is thousands of times slower than this century's climate may demand.

One landmark study of the extinction risks from climate change, led by Chris Thomas of the University of Leeds and published in *Nature* in 2004, looked at regions that together span a fifth of the planet's land surface. The report found that, by the year 2050, 15%–37% of plant and animal species across these areas would be committed to eventual extinction should a middle-of-the-road scenario for increased emissions come to pass. If emissions are on the high side, the range jumps to 21%–52%. The IPCC's 2007 and 2014 reports reinforced the grim picture painted by Thomas and colleagues, with the 2014 report noting that the risks of species extinction rises under all of the scenarios examined. As models of species risk have incorporated more variables, the range of possibilities has grown. This means an even wider swath of species could be in jeopardy—or, on the other hand, if we're lucky, perhaps a smaller number than previously thought.

While climate skeptics sometimes note that species have come and gone in vast numbers long before humans entered the picture, that's not exactly a reassuring thought, especially since climate change appears to have played a role in some of the biggest mass extinctions in Earth's history (see chapter 11).

The canaries of climate change

Miners once brought canaries into their dank workplaces to test whether the air was safe to breathe; if the canary died, it was time to get out. Similarly, some creatures can't help but reveal the mark of a shifting climate through their very ability—or inability—to survive.

Amphibians and **reptiles** offer some of the most poignant examples of species at risk. Because these are ectothermic (cold-blooded) creatures, they must keep themselves within a fairly restricted range of air temperature, and they've got to crawl wherever they go. Those are tough constraints where the climate is warming and where highways and other human construction make it difficult to migrate. For many species of **turtle** (as

> " It will be difficult to anticipate future threats to biodiversity and ecosystem dynamics, even if we could know future climate change with perfect accuracy."
> —*Jonathan Overpeck,*
> *University of Arizona*

Golden toads were wiped out by a fungal disease during a La Niña spring in Costa Rica. (Charles H. Smith, U.S. Fish and Wildlife Service)

well as some other reptiles, amphibians, and fish), there's another complication: the ratio of females to males changes as the ambient temperature in their nesting area goes up. If a given species can't migrate quickly enough to avoid regions that are warming, then any gender imbalance could become a threat to survival. This only adds to the many other threats facing beach-nesting turtles, from commercial trade to pollution.

There's no doubt that amphibians are in serious trouble. More than 30% of amphibian species were found to be vulnerable, endangered, or critically endangered in the most recent comprehensive survey, the 2004 Global Amphibian Assessment. What's been more challenging is to pin down the role of a changing climate amid many other threats to amphibians. Temperatures in many key habitats haven't yet risen to dangerous levels, but there's some evidence that even modest warming can have significant impacts when it's accompanied by other risk factors, such as the presence of pesticides.

Frogs are among the species most at risk at many locations around the globe. One of the biggest emerging threats to amphibians is the skin-damaging, reflex-slowing fungus known as **chytridiomycosis**. It's been implicated in amphibian loss over the last couple of decades in many locations around the world, including the **Monteverde Cloud Forest Reserve** in Costa Rica. Twenty out of 50 species of frogs at the preserve vanished after the unusually warm, dry spring of 1987. One special species found nowhere else, the exotically colored **golden toad**, numbered at least 1500 during that La Niña spring. Within two years, it could no longer be found. Researchers were puzzled: **chytridiomycosis** tends to prosper in cool, moist weather, yet the Monteverde Cloud Forest Reserve was warming overall. However, it turned out that the bulk of the warming at Monteverde was occurring at night, whereas days had actually cooled slightly (probably due to increased cloud cover). The reduced day-to-night range could have kept conditions closer to optimal for the villainous fungus. However, the timing might simply have happened to coincide with the natural course of the fungal epidemic. A broader look at these relationships, led by Jason Rohr (University of South Florida), suggests that the main climatic culprit may actually be increased temperature variability—as during a shift from El Niño to La Niña years—which could act to weaken the frogs' defenses against fungal invasions.

Butterflies are another excellent indicator of climate change on the march, partly because these widespread and aesthetically pleasing creatures have been carefully studied for centuries. Both drought and flood can cause a butterfly population to crash, as happened to 5 of 21 species studied during a dry spell in the mid-1970s across California. The severe drought and record heat across the U.S. Midwest in 2012 took a major toll on the iconic swarms of monarch butterflies that grow there before migrating south to Mexico each fall. The number of acres occupied by monarchs in Mexican reserves dropped by 49% in the winter of 2012–2013 compared to the previous year. Dry, hot conditions not only impede the summertime growth of the butterflies, but they also reduce the prevalence of milkweed, where monarchs lay their eggs. Milkweed is also attacked directly by the herbicide glyphosate, which became common across the U.S. Midwest in the 1990s. Since then, monarch production across the Midwest has reportedly dropped by more than 80%. (Such is the power of the monarch on our imagination that writer and biologist Barbara Kingsolver employed a dramatic rerouting of its winter migration in her 2012 novel *Flight Behavior*, with climate change a major part of the story.)

Because climate zones are concentrated in mountainous regions, the effects on butterflies are especially evident there. A 2010 study found that many California butterflies are moving upward as the climate warms, with an average elevation gain per species of nearly 100 m (330 ft) over a 20-to-30-year period along the west slope of the Sierra Nevada. The Apollo butterfly, whose translucent wings are often seen across high-altitude areas in Europe, is no longer found at elevations below 838 m (2749 ft) in France's Jura Mountains. Although they're skilled fliers, the Apollo butterflies are

The Apollo, one of many butterfly species being pushed to higher altitudes by warming temperatures. (Wenkbrauwalbatros/Wikimedia Commons)

finding themselves with less and less livable terrain as they're forced to climb in elevation. A major survey of European butterflies in 2009 found that dozens of species will be unsustainable in their present locations if the warming expected by late this century comes to pass. "Most species will have to shift their distribution radically to keep pace with the changes," said chief author Josef Settele (Helmholtz Centre for Environmental Research).

Even a creature as humble as the **dormouse** can tell us something about global warming. Made famous by Lewis Carroll's *Alice in Wonderland*, this tiny, nocturnal mammal now ranges only about half as far across Britain as it did in Carroll's time, and more than half of roughly 3000 British species of dormouse are on the decline. Part of this is due to the fragmentation of the trees and hedgerows where it likes to live. However, mild winters interfere with its hibernation patterns, and warm summers cause additional stress. Were it not for so much civilization in the way, the dormouse might easily be able to migrate north to escape the warm-up. A program to reintroduce the creature in various parts of Britain has found some success.

Trouble brews when interdependent species respond in differing ways to a warming atmosphere. Those who use temperature as their cue to breed or migrate may shift their timing as warming unfolds, while those whose behavior is driven more by the seasonal waxing and waning of sunlight may not. At the Netherlands Institute for Ecology, Marcel Visser has been studying disrupted synchrony—the mismatched timing that can occur when creatures find that a food source has departed from its usual seasonal schedule. In studying a bird called the Dutch Great Tit, for example, Visser and colleagues found that the birds' egg-laying behavior hasn't kept up with the earlier springtime emergence of its main food source, caterpillars. Similarly, in California, a Stanford University team found that the progressively earlier seasonal die-off of plantains on a nearby ridge left the Bay checkerspot caterpillar without its favorite food. The butterfly eventually disappeared from the ridge. Such timing mismatches could also threaten the relationship between plants and the insects that pollinate them. NASA is working to relate the amount of nectar in beehives (easily measured by weighing the hives) to the satellite-detected timing of vegetation. The agency's HoneyBeeNet reports that the peak nectar flow in Maryland hives now occurs nearly a month earlier than in the 1970s.

It's not only insects that are threatened by asynchrony. Eric Post (Pennsylvania State University) has found that the plants eaten by **caribou** calves in Greenland are emerging earlier than they did in the 1990s, as depleted sea ice allows air temperatures to warm across the region. However, the seasonally driven birth dates of the calves aren't changing, so the plants may now be past their nutritional peak by the time the caribou start foraging. As Post puts it, "The animals show up expecting a food bonanza, but they find that the cafeteria already has closed."

For other animals, it's camouflage—or the lack of it—that puts them at risk. Scott Mills (North Carolina State University) has described the effect climate change may have on a number of mammal species across the globe that seasonally molt from brown to white to camouflage themselves against their background. In a 2013 paper in *Proceedings of the National Academy of Sciences*, Mills and colleagues found that **snowshoe hares** have little ability to shift the timing of their coat color from brown to white or white to brown when the winters are short. Since fall and spring snowfall is decreasing throughout the globe in temperate regions, Mills' work demonstrates that the odds are rising that hares will be caught with their white coats mismatched against bare ground during these transition seasons, a time when their mortality rates are already the highest.

Do examples like the ones above add up to a **global shift**? In an extensive survey published in *Nature* in 2003, Camille Parmesan of the University of Texas at Austin and Gary Yohe of Wesleyan University found that 279 out of 677 species examined in large-scale, long-term, multi-species research showed the telltale signs of human-induced climate change. On average, these species had moved roughly 6 km (3.7 mi) northward, or about 6 m (20 ft) up in elevation, per decade. One of the most telling clues was the "sign switching" of eight butterfly species in North America and Europe. All of them moved northward during the warm period of the early twentieth century, shifted back south during the midcentury cooldown (see graph, p. 10), and then headed north again most recently. None of the butterfly species contradicted this pattern. Further evidence came with an exhaustive global survey in 2009, led by NASA's Cynthia Rosenzweig and published in *Nature*, of more than 29,500 datasets that each showed a significant physical or biological change related to temperature. About 90% of the biological changes were in the direction one would expect from a warming climate.

While the movement of species and the risks of extinction have gotten much attention, climate change is affecting some animals in a less-publicized way: they're getting smaller. A 2011 survey in *Nature* led by Jennifer Sheridan (National University of Singapore) assessed dozens of studies that found shrinkage among various types of plants, fish, amphibians, birds, and mammals. Smaller size can be a direct result of stressors such as drought or ocean acidification, but it can also serve as an adaptation that helps some heat-stressed creatures survive. A lower weight generally means more surface area per pound, which helps animals dispel heat more readily.

The big squeeze

Simply keeping up with a changing climate will be a challenge for many animals and plants. It's hard to overstate how quickly the chemical makeup of the atmosphere is changing relative to Earth's history. The unprecedented increase in greenhouse gases translates into a warming projected for the next century that may seem gradual to us but will be warp speed by geological standards (see chapter 11).

In the Arctic, where the range of life forms that can survive the harsh environment is relatively narrow, it's easy to identify the animals most at risk, chief among them **polar bears** (see chapter 6). Elsewhere around the planet, ecologists are keeping a close eye on **biodiversity hotspots**, where climate and geography team up to nurture a vast variety of species—far more than one might expect in such small areas. The 34 hotspots identified to date by researchers are scattered throughout the tropics and midlatitudes, largely on coastlines and mountain ridges. The Monteverde Cloud Forest Reserve mentioned above is part of one of the largest hotspots, the Mesoamerican forests that cover much of Central America. Most hotspots feature large contrasts in temperature and moisture across small areas, which enables

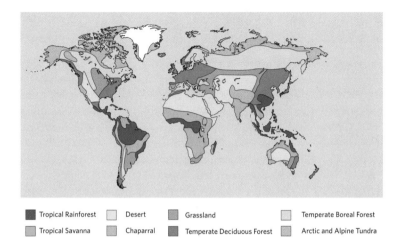

■ Tropical Rainforest	□ Desert	▨ Grassland	▨ Temperate Boreal Forest
▨ Tropical Savanna	▨ Chaparral	■ Temperate Deciduous Forest	▨ Arctic and Alpine Tundra

many different types of flora and fauna to find their niches. The flip side of this concentrated contrast is that a seemingly small shift in climate could have major effects on plants and animals accustomed to a very specific environment. Moreover, because many hotspots are localized—often surrounded by more uniform, human-dominated landscapes—plants and animals may not be able to easily shift their ranges outward.

What about the rest of the tropics and midlatitudes? These areas are often classified by geographers into **biomes**, regions that share similar traits of climate and landscape. Among the commonly used groupings for land areas are the following:

* **Forests.** Covering roughly a third of the planet, these include the great boreal forests of Canada, Alaska, and Russia; the evergreen and deciduous forests of midlatitudes; and the tropical rainforests that support amazing biodiversity despite nutrient-poor soil.
* **Grasslands.** The savannah that covers much of Africa, Australia, and southern South America, and the temperate grasses that sprawl across the mid-sections of North America and Eurasia as well as Argentina.
* **Deserts.** About a fifth of Earth's land surface. Many of the largest deserts are clustered in the zones around 30°N and 30°S, where air tends to sink and dry out. Most of Antarctica is also considered desert.
* **Tundra.** The cold lands of high altitudes and the Arctic that support only shallow-rooted, low-rising plants.

Up, up, and away?

For animals and plants in mountain homes that have grown too warm for comfort, there's nowhere to go but up—and that's not easy when you're already near the top. Some small mammals are at particular risk. For example, **pikas**—the endearing, hamster-sized creatures familiar to hikers across the high country of the U.S. West—have dwindled in number across both North America and Asia. With a metabolism that can't handle air much warmer than room temperature, pikas typically hide below rocks and ice to cool off, and they can't migrate easily. The U.S. Geological Survey reports that American pikas have vanished from more than half of 25 traditional haunts in the U.S. Great Basin since the early twentieth century. It's unclear whether heat has killed them directly or indirectly (by cutting down on their foraging hours) or whether other factors might be involved, such as a lack of insulating winter snow. In Australia, a comparable victim is the mountain pygmy possum, the only mammal native to that continent's high altitudes, whose numbers have dropped as warmer temperatures allow bush rats and feral cats to move upward into its terrain. The pika's decline prompted the U.S. Fish and Wildlife Service to investigate whether it should be declared a threatened or endangered species—which would have been the first such declaration tied to climate change in the 48 contiguous United States. In 2010, though, the agency concluded that pikas still had enough high-altitude habitat to remain viable as a species. Subsequent attempts to protect the pika at state (California) and federal levels have yet to gain traction.

When it comes to **plants**, the biggest losers could be those attuned to the highest, most rarefied conditions. As temperatures continue to climb, many of these species seem fated to die off. In Europe, high-mountain territory covers just 3% of the landscape but accounts for some 20% of all native plants. However, as in the Arctic, a warming climate could produce a larger zone of plant-friendly territory overall. Tree lines have risen some 150 m (500 ft) over the last century in Sweden's Scandinavian Mountains and up to 80 m (260 ft) in Russia's Ural Mountains. Swedish scientists were surprised to find large herbs at heights more than 100 m (330 ft) above their 1950s range, popping up in moraines only recently left behind by retreating glaciers. Some grasses may prove even more opportunistic, providing tough competition for other cold-adapted alpine plants. The potential "grassification" of high-altitude regions—hinted at by computer models, especially across the Alps—is a topic of keen research interest.

The mountain-dwelling American pika is already being challenged by rising temperatures. (Walter Siegmund/Wikimedia Commons)

Beyond anecdotal reports, there's a pressing need for a more thorough assessment of flora and fauna in a changing high-altitude world. That type of census is slowly taking shape from the Global Observation Research Initiative in Alpine Environments (GLORIA). Launched by the Austrian government with support from the European Union in 2000, this network of observing sites now includes more than 100 alpine observing sites in Europe, the Americas, Africa, Asia, Australia, and New Zealand. More than 75 papers have now drawn on GLORIA data, including a major survey in *Science* assessing changes in plant diversity on Europe's mountain summits. Between 2001 and 2008, plant species tended to move upslope across the region, with a richer array of species pushing onto most peaks. However, the new arrivals may put pressure on the more rarefied species found only at the highest altitudes. Meanwhile, several Mediterranean sites saw a decreased number of species, especially on the lower summits, where intensified dryness may already be taking a toll.

As climate change nudges flora and fauna upward, it could also open doors on plant and animal life of the past. Two tourists in 1991 discovered Ötzi, the Alps' now-famous 5300-year-old man, in a receding glacier. On a survey of Quelccaya in 2004, glaciologist Lonnie Thompson (see p. 127) found a small, shrubby plant that he expected might have been around 5000 years old. Carbon dating soon proved it to be a moss that had been under ice for most of the last 50,000 years—and probably twice that long, making it a likely remnant of life before the last ice age.

The edges of many of these biomes are shifting already. Some studies in Europe have shown a northward push of up to 120 km (75 mi) in the north edge of certain biomes over the last century. The tropics themselves—as demarcated by high-level jet streams—widened by roughly 180 km (110 mi) between 1979 and 2009, according to satellite data analyzed by scientists at the University of Washington and China's Lanzhou University. Other work suggests an even greater widening may have occurred. Implied in this expansion is a poleward push of subtropical deserts. Farther north, one of the largest-scale shifts expected over the next 100 years is the expansion of **boreal forest** poleward into fast-thawing tundra and, in turn, the transformation of the southern edge of some boreal forests into deciduous forests or grasslands. But forests take time to establish themselves, and climate may outrun the process, leaving vast areas in transition until soils and plants can adapt to a new climate regime. To make matters worse, any climatic state we envision for the future is itself a moving target, since the greenhouse gases we're adding to the air now will be warming the climate for a century or more. Will new forests be able to take root in these ever-shifting conditions?

As for land-based creatures in the tropics and midlatitudes, it might seem as if they can easily migrate poleward and upward as needed. The problem is that most of the planet's terrain outside hotspots and other protected areas is pockmarked with towns and cities and laced with highways, cultivated fields, and other potential barriers to migration. Those creatures adapted to higher altitudes, like the pika, may find themselves stranded on island-like patches of terrain, where climate will only constrict their range further over time. A growing focus of ecological research is the need for refugia—a term originally applied to locations where warmth-loving creatures survive ice ages and where those who prefer cold make it through interglacial periods. While such havens will remain critical for species over the very long term, there's a more immediate need for protection from human-produced climate change. This might be found in larger areas where species can migrate (macrorefugia) or in favorable pockets (microrefugia)—perhaps some as small as a few acres—that allow creatures to hunker down in place.

Whether it's ecological hotspots or the Arctic, if the exotic animals that favor these rarefied environments are potentially the biggest losers in climate change, the biggest winners may be the most mundane. Animals that are common and robust could spread even farther, such as the various species of deer now thriving across rural and semirural parts of North America and Eurasia. A landscape of shifting climate may also offer a nearly ideal platform from which nonnative plants, animals, and insects can spread like wildfire. In this era of mass global travel, it's easier than ever for **invasive species** to tag along and then move poleward with warming temperatures. **Kudzu**, a vine that grows as quickly as 22 cm (9 in.) per day, is a classic example. It was brought from Asia to Philadelphia in 1876 for the nation's centennial exposition, then used for decades across the U.S. South to stabilize roadsides and feed animals before its invasive qualities were recognized. Long since out of control, kudzu has been spreading steadily northward: in 2009 Canada's first-ever kudzu colony was found on the north shore of Lake Erie.

The most successful invasive species are often less picky than native species, able to tolerate a wide range of conditions, including heat and/or drought. One famous example is **tumbleweed**, or Russian thistle. Introduced to the U.S. Great Plains in the 1870s, tumbleweed quickly spread across the U.S. frontier, eventually becoming a symbol of its new home. Similarly, **rabbits** multiplied like, well, rabbits after 29 of them were brought to Australia in 1859 by rancher Thomas Austin. Today, the United States is dealing with the destructive and sometimes deadly **fire ant** (see sidebar, "March of the fire ants"), a major invasive pest that's ravaged both the countryside and populated areas and made its way to the Eastern Hemisphere. Where invasive flora and fauna are limited mainly by cold temperatures, a warming climate should only enhance their ability to spread.

Marine life is also likely to undergo major reshaping as the planet warms. For details on the effects of climate on coral reefs and other sea-based creatures, see chapter 7.

Bugged by a changing climate

It could be **insects**—adaptable creatures *par excellence*—that take the best advantage of changing climate in the coming century. Many bugs respond strongly to small shifts in temperature, rainfall, and humidity. The overall warming now under way may already be affecting insect distribution, and future warming should favor the spread of a number of species. In 2006, a spider-eating French wasp showed up in Sandy, Bedfordshire, the first of its species ever observed in Britain. On a more troublesome note, a landmark 2002 review by Drew Harvell and colleagues at Cornell University found that **parasitic diseases** are likely to become more widespread and/or severe as the planet warms and moistens. The challenge that remains is to determine how climate change that favors infectious agents will intersect with changes in where and how people live, among other factors.

In Central America, malaria is often transmitted by the mosquito species *Anopheles albimanus*. (James Gathany, U.S. Centers for Disease Control and Prevention)

History gives us a glimpse of what can happen when climate heats up on a large scale. The so-called Medieval Warm Period (see chapter 11)—the several centuries of relatively mild weather across much of the Northern Hemisphere between about AD 950 and 1300—may have helped rats and fleas to thrive across Europe, paving the way for the **Black Death** that killed a third of Europeans between 1347 and 1352.

More than two billion people live within the range of mosquitoes that transmit the parasites that cause **malaria**. These infect more than 200 million people and take more than 600,000 lives each year. Happily, protective measures such as insecticidal bed nets and mosquito spraying helped bring the global death toll downward by 26% from 2000 to 2010, according to the World Health Organization. It's obvious that any future rebound in infection rates would carry a huge toll in terms of human suffering, yet it's been difficult for researchers to nail down exactly how climate change will affect malaria, given the variety of other factors that shape the course of the disease's spread.

For more than 30 years, malaria has been moving into previously untouched parts of Africa's western highlands, possibly due to a scarcity of cool, mosquito-killing temperatures. It was long assumed that warming temperatures would expand the range of malaria in such a way, thus widening the swath of humanity threatened. However, limited data from the

March of the fire ants

It sounds like Hollywood horror at its most lurid: a voracious, supercharged pack of bright orange bugs chewing their way across the United States. This terrifying scenario isn't happening on the silver screen but in real life, and a warming climate appears to be contributing to the problem. The red imported fire ant, which arrived in the United States from South America in the 1930s, has conquered large swaths of the South, from North Carolina to Texas. Since the 1960s, the ants have expanded their territory each year by an area the size of New Hampshire, gradually pushing north and west toward their climatological limits.

The ants have many qualities going for them. Multiple queens can team up to help launch a colony (though workers eventually kill all but one of them). The ants reproduce like mad, often pushing out native ants completely and preying on bird hatchlings. They'll also attack mammals, stinging repeatedly rather than merely biting, and they've even killed a few people, most of whom have been unlucky enough to fall on or near a nest due to an accident or injury. The ants can even survive floods: one of the surreal spectacles from Hurricane Katrina was the sight of fire ants floating on top of the waters in dense clots, a technique they can sustain for up to two weeks.

Temperatures much below −12°C (10°F) spell doom for fire ants, but the gradual warming of winter nights could allow their northward U.S. spread to continue this century. They're already expected to head west into California and up that state's coast. One projection shows that if atmospheric carbon dioxide increases by 1% a year over the next century, the fire ant could increase its U.S. range by over 20%, populating a belt from Oklahoma to Delaware. Since the year 2000, the ants have also swept into many other parts of the world, including Australia, southern China, New Zealand, the Philippines, Taiwan, and Thailand. Genetic analyses indicate that most of these invaders originated in the United States rather than the ants' original South American homeland.

(Agricultural Research Service, USDA)

region left open the possibility that the highlands weren't actually warming. The intersection of climate and malaria in Kenya's western highlands proved to be a contentious topic during the first decade of the 2000s. "The spat paralyzed the discussion of climate and health at the highest policy level," states a 2011 *Nature* overview led by Madeleine Thomson (International Research Institute for Climate and Society). As it turns out, the most thorough study to date found that temperatures at the highland city of Kericho indeed rose roughly 0.6°C (1.0°F) between 1979 and 2009. However, "given the high complexity associated with emergence of malaria epidemics, drawing a direct correlation between temperature and malaria risk is unwise," notes a 2013 report issued by the International Institute for Sustainable Development.

Recent models suggest that the range of malaria-carrying mosquitos might actually shift rather than expand. Intensified drought, for example, may render some areas unable to sustain malaria-carrying mosquitoes even as the disease moves into other areas, such as African highlands, that were once malaria-free. A 2013 modeling study headed by Volker Ermert (University of Cologne) suggests that malarial risk should decrease by mid-century across much of tropical Africa while increasing in some other parts of the continent, including the East African highlands (especially at altitudes above 2000 m or 6600 ft) as well as the southern Sahel, where the population is growing rapidly.

What about other parts of the world? The Malaria Atlas Project found that the disease's range encompassed 58% of Earth's land area in the year 1900 but only 30% in 2007. Much of the progress is likely due to insecticides, vaccines, mosquito nets, and other advances in prevention, with urbanization possibly helping as well. In places like Europe and the United States, where malaria was endemic for centuries, the disease is considered unlikely to make a comeback—even with warmer temperatures—as long as control measures and public-health systems remain strong. Reassuring as this progress might seem, mosquito populations are booming across many of the developing regions where malaria is most prevalent. And as noted above, warming temperatures may be helping the virus spread into areas where people aren't accustomed to it and may not have the resources to deal with it. Together, these factors pose continuing challenges to public health efforts in chronically poor areas.

Dengue fever is already on the rise. It's caused by a group of five potentially fatal mosquito-borne viruses, one of them just discovered in 2013. Around 40% of the world's population now lives in affected areas, according to the World Health Organization. Anywhere from 50 to 100 million people are infected with dengue each year, of whom around 500,000 need to be hospitalized. Children account for a large proportion of these serious cases, including most of the fatalities. One study estimates that a 1°C (1.8°F) rise in average global temperature could raise the number of people at risk of dengue by up to 47%, but the complex factors that influence transmission make it hard to know if the disease would actually spread that quickly. A separate study found that increased economic development has the potential to more than counteract any climate-related boost in dengue's spread by 2050. Still, the disease has been pushing its way into developed countries: a few cases have been reported in far southern parts of Texas and Florida since 2009, and the first European outbreak in almost a century struck the Portugal islands of Madeira in 2012. Researchers are now investigating whether climate change and other factors might help dengue to spread from lower elevations east of Mexico City into the high-altitude urban core, where it would threaten close to 20 million people. The project has already found *Aedes aegypti*, the primary dengue-carrying mosquito, at altitudes of 2130 m (7000 ft), the highest on record for its presence in Mexico.

Unusual summertime warmth and drought helped another mosquito-borne illness make quick inroads into North America. The **West Nile virus** first showed up around New York City in 1999. Hot summers can be favorable for proliferation of mosquitoes that transmit the virus, which is hosted by birds in the wild and can spill over into humans. The virus spread regionally at first and then, during the record-breaking heat of 2002, swept across much of the United States and into Canada. Up to 2005 it had killed more than a thousand North Americans. By the end of the decade, the annual human toll had dropped rapidly, perhaps because birds were becoming immune to the virus or possibly because fewer humans with low-grade infections were being tested and counted. However, the disease again spiked during the torrid summer of 2012, with 286 U.S. deaths reported— the largest annual toll since the virus first reached American shores.

Insects can also cause trouble for people indirectly, such as by attacking the food we eat. The Colorado potato beetle, a striped, centimeter-long U.S. native, is a serious threat to potato crops in Europe. Each year a few of these beetles hitchhike into the United Kingdom, but to date the bug has failed to establish a foothold, thanks to vigilant inspections, public awareness efforts, and eradication campaigns by the U.K. government. However, a study by Richard Baker of the U.K. Central Science Laboratory found that if Europe were to warm by an average of 2.3°C (4.1°F) by 2050, this pest could more than double its potential range in Great Britain, putting virtually all potato-farming areas at risk.

Shrinking forests

As we've seen, rainforests and their loss can have a huge impact on the global climate (see p. 11), but how does global warming affect the rainforest itself? That's not so clear. In the 2009 study noted above, S. Joseph Wright found that the IPCC's midrange emissions scenario could push 75% of the world's tropical forests into annual temperatures higher than any of the yearly averages now observed in closed-canopy forests. On a seasonal scale, it's precipitation more than temperature that drives the rainforest, so anything that affects when and how hard it rains can have a major impact. El Niño tends to produce drought across much of the Amazon, so a key factor will be how El Niño evolves in the future, which is still an open question (see chapter 7). In addition, it appears that unusually warm Atlantic waters northeast of Brazil can help strengthen Amazonian drought. The Atlantic may also be supplying enhanced moisture to the Amazon during non-drought years, which could help explain why the region has become wetter on average since 1990, despite two intense droughts in 2005 and 2010.

The various global climate models considered in the 2013 IPCC report tend to support this trend toward a beefed-up hydrologic cycle across the Amazon. Under the highest-emissions scenario for the 2080s and 2090s, the maximum five-day rainfall expected in a given year rises by 10% or more, while each year's longest dry spell is extended by a week or more. Should climate change end up pushing the Amazon toward more bona fide drought, the trend could be exacerbated by land clearing, which causes local-scale drying due to loss of the moisture-trapping forest canopy. The fear is that these and other positive feedbacks could lead to massive rainforest loss across the Amazon, although that scenario is far from a given. Models that track climate together with the interactions of species within a rainforest canopy are still early in development, but they may eventually clarify how tropical forests as a complete ecosystem stand to deal with drought.

There's more certainty about what climate is doing to forests farther toward the poles. Mountain and boreal forests are one of the largest ecosystems already being strained by a rapidly changing climate. Permafrost melt (see chapter 6) is one big source of high-latitude forest stress, but there are other major factors. Across western North America, the intersection of pests and a warming atmosphere has led to one of the world's largest **forest die-offs**. A series of major droughts and warm winters since the 1990s from Mexico to Alaska has transformed the landscape across huge swaths, destroying millions of hectares of forest through **beetle invasions** and **forest fires**. The extent of both is unprecedented in modern records. In 2002 alone, fire and disease killed enough lodgepole pines across the Canadian

Trees in the Rocky Mountains near Granby, Colorado, show the effects of an attack by the mountain pine beetle. (Carlye Calvin)

province of British Columbia to cover the entire area of Belgium and the Netherlands. As of 2012, the total beetle-affected area had reached 180,000 km² (69,000 sq mi). Half of British Columbia's pines are now gone, and the total may approach 60% by the 2020s. Even as the epidemic slows in that province, it's been gaining ground toward the northeast into Alberta.

The main insect behind this devastation is the **bark beetle**, which comes in various species across North America. This hardy creature burrows into trees, lays its eggs, then girdles the tree from within, blocking the nutrient flows between roots and branches. Climate in recent decades has given North American bark beetles carte blanche for explosive growth. The spruce bark beetle, one of the most common varieties in colder forests, normally requires two years to reach maturity, but an extremely warm summer can allow a generation to mature in a single year. Conversely, it takes two bitterly cold winters in a row to keep the population in check. The combination of milder winters and warmer summers since the early 1990s has allowed the beetle to run rampant.

Another variety, the **mountain pine beetle**—once limited to lower-elevation tree species such as lodgepole and ponderosa—has moved uphill as temperatures warm. It has invaded whitebark pine in Utah at elevations up to 3000 m (10,000 ft). In Colorado and Wyoming, the beetles' territory doubled between 2007 and 2010, encompassing 14,000 km² (5400 sq mi). Though it's now peaked across the region, the epidemic has left swaths of dead or dying forest covering some of the nation's most famous scenery, with pine needles turning a sickly red before they fall off. Hikers must now keep a wary eye for dead trees that can tumble into campgrounds and onto trails. In Canada, the beetles are moving into boreal forests that were long thought to be less vulnerable than their southern counterparts. They've also begun to reproduce within **jack pine**, a critical move that could allow the insects to cross the continent and lead to massive infestations across the forests of eastern North America. At the same time, the southern pine beetle, endemic from the U.S. South into Central America, is moving north, already establishing a foothold in the pine forests of southern New Jersey.

Drought weakens the ability of trees to fight off these aggressive bugs, because the dryness means a weaker sap flow that's easier for beetles to push through. Across parts of New Mexico, Arizona, Colorado, and Utah, where a record-setting drought took hold in 2002, pine beetles and other drought-related stresses killed up to 90% of the native **piñon** trees. A drought in the 1950s had already taken many century-old piñons, but the more recent drought produced "nearly complete tree mortality across many size and age classes," according to a 2005 study led by David Breshears (University of Arizona). Even assuming the climate supports piñon regrowth, it will take decades to reestablish the landscape. Breshears sees the event as a sign that climate-driven landscape change could be far more rapid and widespread than even the experts expect.

Certainly, both insects and fire are part of the natural ecosystems of North America. Some species of trees actually rely on fire to propagate. Jack pine seeds are tucked into tightly sealed resinous cones that only open when a fire comes along and melts the resin, which then allows the seeds to spread and sprout in the newly cleared landscape. The question is how climate change might amplify the natural process of forest fires, especially across the high latitudes of North America and Asia. For example, some studies project an increase of up to 100% by the end of this century in the amount of land consumed by forest fire each year across Canada. These fires are a major source of pollution, and they pack a double whammy on the carbon budget: the fires themselves add significant amounts of carbon dioxide to the air (not to mention carbon monoxide, methane, and ozone-producing oxides of nitrogen), and the loss of trees means less CO_2 being soaked up from the atmosphere (at least until new growth occurs).

A wildland fire rages in the forest behind Great Sand Dunes National Park in southern Colorado on June 23, 2010. Many researchers expect such fires to become more frequent and severe as the planet warms. (UCAR Digital Image Library, photo by David Hosansky)

As for the beetles and budworms, most major infestations tend to run their course after a few years, followed by new epidemics a few decades later as the next crop of tree victims matures. But nobody knows how that natural ebb and flow will intersect with the climate-driven shifting of the insects' potential range across the U.S. and Canadian West. It's also not out of the question that a decades-long **megadrought** could set in. It's happened before, at least across the U.S. Southwest. Richard Seager (Lamont-Doherty Earth Observatory) warns that by 2050 that region may be in the grip of a warming-driven drought that's more or less permanent. Already, a large swath from New Mexico to California has been struggling with on-and-off drought. Most years from 2000 to 2013 fell short of average rainfall in Los Angeles, and the period from July 2006 through June 2007 was L.A.'s driest in more than a century of record keeping, with a mere 82 mm (3.21 in.) of rain. That set the stage for catastrophic wildfires in and around the city that autumn. Both Los Angeles and San Francisco, along with several other cities, went on to smash all-time calendar-year records for low precipitation in 2013. Lake Mead, which slakes the thirst of the growing Las Vegas area, dipped to water levels in the early 2010s unseen since the lake was created in the late 1930s. The long-term risk to Las Vegas

The northward flow of maple syrup

The shift of a forest can be just as significant psychologically and economically as it is biologically. Take maples, for example. They're so closely identified with Canada that they're the centerpiece of the national flag, but for generations of Americans, it's New England that's synonymous with maple syrup. This piece of Americana appears to be in jeopardy, according to the U.S. National Assessment of Climate Change. To produce the best syrup, maples need a series of freezing nights and milder days, together with a few prolonged cold snaps in late winter. But since the 1980s, winters across New England haven't lived up to their past performance. There

(Pete Markham/Wikimedia Commons)

have been fewer stretches of bitter cold, and nights are staying above freezing more often, even in midwinter.

Farther north, maple production in Quebec was long limited by the lack of daytime thaws, as well as by the deep, sustained snow cover that kept maple harvesters from their trees. Now, improved technology allows the Quebecois to gather syrup more easily in deep snow, and the climate appears to be shifting in the Canadians' favor as well. Not only is maple production moving out of New England, but the long-term survival of maple trees themselves is in question across much of the Northeast, according to the U.S. National Assessment. That would be a potential hit to the spectacular autumn foliage that attracts thousands of tourists each year.

Perhaps New England's best hope of keeping its stunning foliage, and its syrup, is the chance of a shift in North Atlantic currents (see chapter 7) large enough to chill the region—or at least keep it from warming as much as other parts of the globe. The chances of this happening, however, are slim.

became dire enough to trigger construction of a third intake valve—a new "straw" that will allow the city to continue drawing water even as lake levels reach new lows.

All of these issues complicate long-running debates on how best to manage ecosystems across western North America. The overarching question is how to allow natural processes to play out while fending off the kind of large-scale, fire- or bug-produced devastation that could permanently alter the landscape.

Crops and climate: A growing concern

In recent years, a variety of factors (no doubt including the influence of factory farming) has led to an enormous global reliance on a tiny number of staple crops. Out of some 50,000 edible plant species, a mere three—wheat, rice, and maize (corn)—make up more than half of the world's current food supply. To deal with the demands of climate change, especially over the long haul, food producers around the world will need to demonstrate flexibility in the timing, location, and nature of their planting. That won't be easy when there is no single, fixed "new" climate to plan for, but a climate in constant flux. The challenges will be especially great for subsistence farmers whose options are restricted even in the best of times.

The complex effects of climate change on plants are often reduced to a simplistic, and misleading, equation: plants take in carbon dioxide and use it to grow, so increasing the amount of CO_2 in the air can only be a good thing. One of the more prominent U.S. organizations stressing the potential benefits of a world with more carbon dioxide was the now-defunct Greening Earth Society. From the 1980s into the 2000s, this group promoted the view that the extra CO_2 in the air would help crops and other vegetation to grow more vigorously. As they pointed out, the tendency of global warming to be most pronounced in the winter and at night should only lengthen growing seasons on average across much of the world.

By itself, the fertilization effect of carbon dioxide isn't controversial; it's been supported by research in the laboratory and in the field, albeit with a few caveats (see sidebar "Will greenhouse gases boost plant growth?"), and underscored in many climate reviews, including the IPCC's. To cite just one prominent study, the Maryland-based Smithsonian Environmental Research Center reported in 2010 that a set of 55 plots of trees monitored there for 22 years grew at up to twice the expected rate. Among other things, the enhanced CO_2 triggers a shrinkage of the pores on plant leaves, allowing them to retain moisture more easily. A 2013 analysis headed by

Will greenhouse gases boost plant growth?

Scientists have long recognized one potentially positive side effect of fossil fuel use: the stimulation of forests, crops, and other vegetation by the carbon dioxide we're adding to the atmosphere. Initial work in the 1980s suggested that a doubling of CO_2 could increase crop yields by as much as 35%. However, much of the initial work involved plants in open-top chambers, which allow fresh air in through the top but don't fully replicate how plants would grow outdoors. A more true-to-life model would be to add CO_2 to the air around vegetation in a natural setting and see what happens. This is the idea behind a series of projects called Free Air CO_2 Enrichment (FACE). More than a dozen large-scale FACE sites have been operating in Germany, Italy, Japan, New Zealand, Switzerland, and the United States since the early 1990s. At each site, a circle of vent pipes injects CO_2 into the atmosphere surrounding a plot of land 8–30 m (26–100 ft) in diameter. Through computer-controlled settings, the CO_2 levels can be adjusted for winds and other factors.

In a 2005 review of more than a hundred FACE studies, Elizabeth Ainsworth and Stephen Long of the University of Illinois at Urbana–Champaign found that the crop benefits from enhanced CO_2 fell short of expectations. For example, in studies where CO_2 was increased over present-day levels by about 50%, the yields rose by about 7% for rice crops and 8% for wheat. This is only about half the effect one would have expected by extrapolating from the doubled-CO_2 levels in the earlier chamber studies, according to Ainsworth and Long. Other FACE studies indicate that the benefits may tail off for at least some crops as CO_2 levels continue to increase. On the other hand, trees have fared better than predicted in the FACE studies. Ainsworth and Long found that the extra CO_2 caused the total amount of dry biomass produced above ground by trees to increase by an average of 28%.

Perhaps the biggest bonus from enhanced CO_2 could be an increase in the ability of some plants to deal with drought. Ainsworth and Long found

David Keenan (Harvard University) found that trees at more than a dozen spots across the Northern Hemisphere had become increasingly efficient in their use of water over the last two decades—even more than models had predicted. Again, enhanced CO_2 and its effect on plant pores turned out to be the best explanation for the trend.

Increased carbon dioxide also speeds up photosynthesis rates for so-called C_3 plants, including wheat, rice, and soybeans. In one experiment, the yields of C_3 crops increased by an average of 14% when carbon diox-

that when conditions were dry, the boosted CO_2 levels improved yields by an average of 27%; during wet conditions, the extra CO_2 had no significant effect. A leading explanation for this is that the extra CO_2 constricts a plant's pores, which also helps it retain moisture.

In order to take advantage of higher CO_2 levels, plants will need to draw on other nutrients to support their increased growth. That's possible but by no means certain. Nitrogen is a particular question mark. It's a critical part of industrialized agriculture: crops are fertilized with it so commonly, and so inefficiently, that most of the nitrogen bypasses the crop and enters the atmosphere, where it can travel thousands of kilometers before it's taken up by the soil. Nitrogen is also generated by microbes within the soil itself. It's unclear whether these two sources will provide enough nitrogen to allow natural vegetation to thrive on the extra carbon dioxide. Research to date suggests that, for grasslands at least, extra nitrogen would be critical in order for plants to take advantage of any benefit from enhanced CO_2.

In this 1995 photo, FACE study team leader Bruce Kimball adjusts wind sensors used to control the release of CO_2 over wheat plots near Phoenix, Arizona. (U.S. Department of Agriculture)

ide was boosted to 590 ppm compared to 370 ppm. (Livestock that graze on C_3 grasses, which are mainly at higher latitudes, could also benefit.) In contrast, C_3 crops, including maize and sorghum, gain relatively little benefit from the extra CO_2, as they don't get the boost in photosynthesis that C_3 plants do.

Among nonagricultural plants, woody vines such as kudzu (see p. 209) appear to thrive on the extra CO_2. This is in part because their clingy mode of growth allows the plant to devote more energy to photosynthesis as

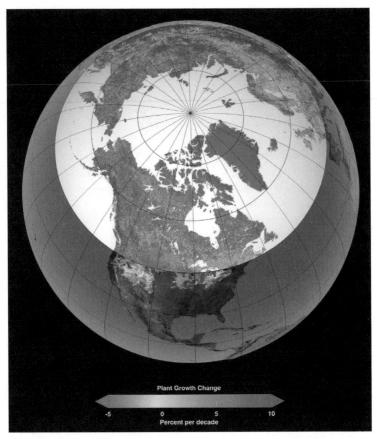

Data collected by NASA satellites between 1982 and 2011 reveals trends in vegetation across higher northern latitudes. Between 34% and 41% of vegetation-supportive terrain showed an increase in plant growth (green and blue), while 3%–5% showed a decrease (orange and red). (NASA Goddard Space Flight Center Scientific Visualization Studio)

opposed to building a structure to keep it standing. A 2006 study at Duke University found that one noxious vine—poison ivy, which sends over 350,000 Americans to the doctor each year—grew at three times its usual pace with doubled CO_2. To make matters worse, the ivy's itch-producing chemical became even more toxic than before.

The most recent analyses have brought down the overall agricultural benefit we can expect from carbon dioxide increases, in part because the negative influence from **near-surface ozone** will be larger than scientists once thought. One of the open-air studies carried out as part of the Free

Air CO_2 Enrichment program, or FACE (see sidebar "Will greenhouse gases boost plant growth?"), tested the effects of enhanced carbon dioxide and ozone on soybean fields. A doubling of atmospheric CO_2 boosted soy yields by about 15%, but that benefit went away when ozone levels were increased by as little as 20%. Another open-air study, this one in Wisconsin, obtained similar results for aspen, maple, and birch trees. Ozone may also cut the nutritional value of some crops, although research is still in its early days.

Global warming itself will have an impact on crops, of course, some of it favorable. Modeling that fed into the 2013 IPCC assessment shows how **frost days** (the number of 24-hour periods when temperatures drop below freezing) might evolve over the next century. The number of such days drops by more than a dozen by the end of this century under the midrange RCP4.5 emissions pathway, with the largest changes across far western North America, eastern Europe, and central Asia. That, in turn, should lead to somewhat longer growing seasons across much of the planet's midlatitudes. Already, plant hardiness zones across the contiguous United States have shifted north in many areas, according to maps produced in 1990 and 2012 by the U.S. Department of Agriculture.

Across the globe, scientists expect a hotter climate to enhance the overall potential for cereals and some other crops across large parts of northern Europe, Russia, China, Canada, and other higher-latitude regions. This should lead to an overall expansion of the world's land areas conducive to agriculture. However, that alone doesn't guarantee tomorrow's plants will be just as healthy as today's. The protein level in grain crops can decrease if nitrogen doesn't increase in tandem with carbon dioxide, and a number of studies to date show that CO_2-enhanced crops can be significantly depleted in zinc, magnesium, or other micronutrients, perhaps because there aren't enough trace elements from the soil entering the plant to keep up with the photosynthesis boost from CO_2.

Winners and losers in farming

A battle royal is setting up for the coming decades across the world's farmlands, where the benefits of extra CO_2 and longer growing seasons will fight it out with intensified drought, spikes of extreme heat, invasive species, and other negatives. There are both encouraging and worrisome signals. On the down side, the fertilization effect of CO_2 for crops like wheat and rice tends to decrease once CO_2 is boosted beyond a certain point. While some developed areas may benefit from climate change, at least for a while, these gains could be offset by thirsty soils and a resulting spike in irrigation demand. Even if the world is lucky enough to get a few decades where agricultural winners match or even outnumber losers, the resulting imbalance could put huge stresses on agricultural trade and food security, including relations between developed and developing nations—and that's without considering the impact of converting food crops to fuel (see chapter 16), or the relentless growth in meat consumption, which increases the amount of land and water devoted to food production.

On the plus side, the world's food production system has done an impressive job in recent decades in keeping up with the ever-increasing number of people on Earth. World population has ballooned from three to seven billion since 1960, but global harvests have more than tripled, even though the amount of land being cultivated has increased by only about 10%–15%. Technology and innovation should continue to help boost total global yields for the foreseeable future, and the effects of global warming aren't expected to cause a literal drop in productivity anytime soon. The main concern is that even a small climate-related dent in expected yields (especially if other problematic factors come into play at the same time) could have an outsized impact on a growing world that's come to expect a bountiful supply of food that increases year over year. The question then becomes: how much will tomorrow's agriculture be held short of its potential as the planet warms up?

It now looks as if the net global effects of carbon dioxide and climate change may shift from positive to negative sooner than expected. The 2014 IPCC report on impacts, adaptation, and vulnerability (Working Group II) brought this uneasy message home. It found that the lion's share of studies now indicate that climate change will have an overall negative effect on crop yields as soon as the 2030s or 2040s (see graphic), with decreases of 25% or more appearing by that time period in some studies. The verdict right now is hardly unanimous: as can be seen in the graphic, a substantial percentage of studies show that climate change may actually help *increase* crop productivity. Taking all this into account, the panel's consensus outlook is that median global yields can be expected to drop by 0%–2% per decade for the rest of this century, compared to the yields that would have occurred without a changing climate. The panel notes that adaptation could make a serious difference. If farmers can adjust what they grow and when they grow it, working to match the evolving climate in their location, then some could actually come out ahead—at least in some areas, and assuming that local temperatures don't rise much more than 2°C over preindustrial values.

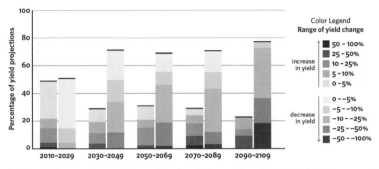

Projected changes in crop yield as a function of time, with and without adaptation, incorporating more than 1000 studies from across the world. For each time interval, each colored slice indicates the percentage of studies for that interval showing a given change in crop yields. (IPCC)

The 2014 IPCC report also took a region-by-region look at where climate change is likely to have the biggest impacts on agriculture. Rising temperatures and changes in rainfall may trim the potential productivity of rice crops in Asia and cereal crops across Africa, possibly dealing a major blow to African food security. Europe is a mixed bag, with climate change likely serving as a net boost to grains across northern Europe and perhaps eastward into parts of central Asia, while cereal crops toward the Mediterranean take an overall hit. Much like Europe, North America could see net benefits to crops toward the north and detriments toward the south, although farmers across much of the continent will need to adapt to an enhanced risk of heat waves and precipitation swings. That's even truer for Australia, where recurring floods and droughts already exert a huge impact on agriculture.

Regional projections of future agriculture must reckon with the potentially destructive role of extremes, such as the damage produced by hot, dry summers. For example, the **groundnut**—a staple across western and southern India—produces far less fruit when temperatures consistently top 35°C (95°F) over a few days. Many other crops have similar temperature thresholds. **Wheat** that's exposed to 30°C (86°F) for more than eight

> The relevant question at the global scale is how much of a headwind climate change could present in the perpetual race to keep productivity growing as fast as demand."
> —*David Lobell and Sharon Gourdji, Stanford University*

Wheat harvests at higher latitudes (such as shown here in the Akershus region of Norway) may get at least a temporary boost from more CO_2, but climate change will stress crops across many parts of the globe, especially in the tropics. (Øyvind Holmstad/ Wikimedia Commons)

hours produces less grain, and **rice pollen** become sterile after only an hour of 35°C heat.

Scientists are increasingly incorporating the mix of crops and climate into computer models. Most studies of individual crops have been relatively small scale, while analyses of future climate are at their strongest in depicting patterns across large sections of the globe. A natural intersection point is regional climate models, which zero in from global models to portray climate across a specific area in finer detail. The expanding set of global and regional efforts is now being evaluated through the Agricultural Model Intercomparison and Improvement Project, which involves participants from around the globe active in crop, climate, and economic modeling. As of 2013, the effort had already entrained dozens of models, including 13 for rice, 19 for maize, and 27 for wheat. The project's ultimate goal is to improve the collective understanding of world food security and to enhance the ability of both developing and developed nations to adapt to changes down the road.

Meanwhile, agricultural specialists are working on hardier strains of key crops that can deal with potential climate shifts. Half the world's population eats rice as a staple, so its fate on a warming planet is critical to world food stocks. In the Philippines, the International Rice Research Institute

(IRRI)—famed for its work in bolstering yields in the 1960s Green Revolution—is using non–GMO-style crossbreeding to develop strains of rice that can resist heat, drought, and flood. The institute recently unveiled a variant that can produce a decent crop even after being fully submerged for two weeks. One challenge in developing heat- and drought-suited crop strains is that they may prove less productive than traditional strains during good years. Another critical step in formulating crops for a warming world will be to ensure that they stay as nutritious as their predecessors, given the effects of CO_2 on micronutrients noted above. Otherwise, the silent threat of "hidden hunger"—nutrient deficiencies that are now estimated to affect as many as two billion people—could only get worse.

Technological gains may amount to little unless social and financial arrangements are in place to help even the most humble farmer take advantage of the progress. Many subsistence farmers stick to tried-and-true crops and techniques, preferring to minimize risk given their lack of safety cushions. International alliances such as the Consultative Group on International Agricultural Research are looking into policy options such as new forms of crop insurance that could give risk-averse farmers more room to adapt to a changing climate. Farmers throughout the tropics could also benefit from locally tailored information on climatic shifts.

The Science

HOW WE KNOW WHAT WE KNOW ABOUT CLIMATE CHANGE

A composite view of Asia and Australia at night from NASA's Suomi NPP satellite, April and October 2012. (Robert Simmon/ NASA Earth Observatory)

CHAPTER TEN
Keeping Track
TAKING THE PLANET'S TEMPERATURE

It's customary to see polar bears, forest fires, and hurricane wreckage when you turn on a TV report about climate change. What you're unlikely to see are the dutiful government meteorologist launching a weather balloon, the grandmother checking the afternoon's high temperature in her backyard, or a satellite quietly scanning Earth's horizon from space.

These are among the thousands of bit players—human and mechanical—who work behind the scenes to help monitor temperature and other variables that reveal the state of our planet. There's no substitute for this backstage action. Day by day, month by month, year by year, the people and instruments who take Earth's vital signs provide bedrock data for the tough decisions that individuals, states, and companies face as a result of climate change.

The rise of climate science

The practice of assigning a number to the warmth or coolness of the air has royal roots. Galileo experimented with temperature measurements, but it was Ferdinand II, the Grand Duke of Tuscany, who invented the first sealed, liquid-in-glass thermometer in 1660. The Italian Renaissance also gave us the first barometers, used for measuring air pressure. By the 1700s, weather observing was all the rage across the newly enlightened upper classes of Europe and the U.S. colonies. We know it was a relatively mild 22.5°C (72.5°F) in Philadelphia at 1:00 p.m. on the day that the Declaration of Independence was signed—July 4, 1776—because of the meticulous records kept by U.S. cofounder and president-to-be Thomas Jefferson.

Some of the first sites to begin measuring temperature more than 300 years ago continue to host weather stations today. Most reporting sites of this vintage are located in Europe. A cluster of time-tested stations across central England provides an unbroken trace of monthly temperatures starting in 1659—the longest such record in the world—as well as daily temperatures from 1772 (see graph for annual averages). Like the remnants of a Greek temple or a Mayan ruin, these early readings are irreplaceable traces of a climate long gone. Such stations represent only a tiny piece of the globe, so on their own they don't tell us much about global-scale changes. However, they do help scientists to calibrate other methods of looking at past climate, such as tree-ring analysis. They also shed light on the dank depths of the **Little Ice Age**, when volcanoes and a relatively weak sun teamed up to chill the climate across parts of the world, especially the Northern Hemisphere continents. A few regional networks were set up in Europe during those cold years, starting in the 1780s with Elector Karl Theodor of Bavaria, whose meteorological society in Mannheim, Germany, provided thermometers and barometers to volunteers across Europe and collected and analyzed the annual data.

The more thorough assembling of temperature data began in earnest after the arrival of the telegraph in the mid-nineteenth century, which made possible the rapid-fire sharing of information. Overnight, it seemed, weather mapping changed from a historical exercise to a practical method of tracking the atmosphere from day to day. The embryonic art of weather forecasting, and the sheer novelty of weather maps themselves, helped feed the demand for reliable daily observations. **Weather services** were established in the United States and Britain by the 1870s and across much of the world by the start of the twentieth century. Beginning in the 1930s, the monitoring took on a new dimension, as countries began launching weather balloons that radioed back information on temperatures and winds far above ground level. Most of these data, however, weren't shared among nations until well after World War II.

One of the products of the postwar drive for global unity was the **World Meteorological Organization**, founded in 1951 to link national weather agencies and later folded into the United Nations infrastructure. In 1961, U.S. president John F. Kennedy urged the UN General Assembly to consider "further cooperative efforts between all nations in weather predic-

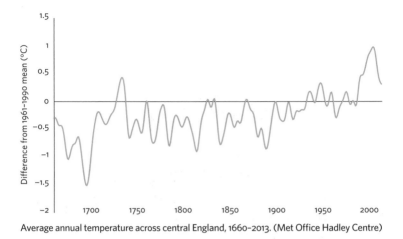

Average annual temperature across central England, 1660–2013. (Met Office Hadley Centre)

tion and weather control." The latter goal proved to be far fetched, but the WMO did begin work on getting the world's weather services to agree on observing practices and data protocols. Before long, a global weather observing system was in place, a model of international collaboration and data exchange that stood firm against the tensions of the Cold War and continues to this day.

During this midcentury whirlwind of progress, the focus was on improving daily weather forecasts. What about the temperature of the planet as a whole? Amazingly, hardly anyone was tracking it. The pioneering climatologist **Wladimir Köppen** took an early stab at it in 1881. There was a brief flurry of activity in the 1930s when, in separate studies, the U.S. Weather Bureau and British engineer Guy Stewart Callendar drew on the limited data available—mostly from North America and Europe—and found that Earth had warmed since the late 1800s. Little did they know that the warming itself happened to be focused in North America and Europe, a fact that emerged only decades later. "But for this accident, it is not likely that people would have paid attention to the idea of global warming for another generation," notes climate biographer Spencer Weart.

> Climate was described as 'average weather' and climatology was looked upon merely as the dry-as-dust book-keeping branch of meteorology." —*Hubert Lamb, 1959, referring to the period between World War I and II*

Though World War II interrupted observing efforts at many weather stations, it also brought new instruments into vogue, including the radiosonde (balloon-borne instrument package), used here in 1944 by U.S. Army Air Force meteorologists in Iceland. (NOAA)

Interest in global temperature had cooled off—as had the planet it-self—by 1961, when Murray Mitchell of the U.S. Weather Bureau found a cooling trend that dated back to the 1940s. His was the most complete attempt up to that point at compiling a global temperature average. How-ever, like his predecessors, Murray was tricked by the wealth of U.S. and European weather stations. In the sparsely observed Southern Hemisphere, the midcentury cooldown was far weaker than it was in the north. Still, the trend was evident enough by the 1970s to prompt widespread atten-tion (see chapter 13). It took until the late 1970s for temperatures to begin inching back up and concern about global warming to resurface. Finally, in the early 1980s, several groups began to carefully monitor the annual ups and downs of global air temperature across the entire planet, eventually including the oceans. These datasets are now among the most important pieces of evidence brought into the courtroom of public and scientific opinion on climate change.

Measuring the global warm-up

Among those who find human-induced warming not much to worry about, there's a fairly wide spectrum of thought. Many now accept that

So what is Earth's temperature, anyway?

If you live in Rome, Melbourne, or Shanghai, you might not realize how close your climatic experience is to that of Earth as a whole, at least in the average. Each of these cities has an annual mean air temperature within 2°C of 14.4°C (57.9°F), which is roughly the planet-wide average. A hundred years ago, the global average was closer to 13.6°C (56.5°F).

Earth is warming up but argue that the rise isn't catastrophic and that any further rise should be well within our ability to adapt. At the other end of the "don't worry" spectrum, a very few diehards maintain that the planet really hasn't warmed up measurably in the last century—a position dismissed by virtually all climate scientists.

There's plenty of physical evidence beyond temperature measurements to indicate that a warm-up is under way, from sea level rise to melting glaciers. Still, the diehards' position begs an interesting question: how do the experts come up with a single global surface temperature out of the thousands of weather stations around the world? It's a far more involved process than simply adding up the numbers and averaging them out. For starters, you need to include the best-quality observations available. Modern weather stations are expected to fulfill a set of criteria established by the WMO. For example, thermometers should be located between 1.25 and 2 m (49–79 in.) above ground level, because temperature can vary strongly with height. On a cold, calm morning, you might find frost on the ground while the air temperature at station height is 2°C (36°F) or even warmer. Also, an official thermometer should be housed in a proper instrument shelter, typically one painted white to reflect sunlight and with louvers or some other device that allows air to circulate. Anyone who's been in a car on a hot day knows how quickly a closed space can heat up relative to the outdoors.

Consistency is also critical. Ideally, a weather station should remain in the same spot for its entire lifetime, but as towns grow into cities, many of the oldest stations get moved. That means the station is sampling a different microclimate, which can inject bias into the long-term trend. One of the biggest agents of station change in the mid-twentieth century was the growth of air travel. Countless reporting stations were moved from downtown areas to outlying airports, where regular observations were needed for flight purposes and where trees and buildings wouldn't interfere with wind and rainfall measurement.

Every day, thousands of volunteers across the world take weather reports that become part of the long-term global database. (© UCAR)

Even when stations stay rooted, the landscape may change around them and alter their readings. Many urban locations are vulnerable to the heat-island effect (see "Bright lights" sidebar). Large buildings help keep heat from radiating out to space, especially on summer nights, and this can make temperatures in a downtown area as much as 6°C (10°F) warmer than those in the neighboring countryside.

On top of all this, gaps can emerge in the records of even the best-sited, longest-term stations. Natural disasters, equipment or power failures, and wars can interrupt records for days, months, or even years, as was the case at some locations during World War II. Problems may also emerge after the data are sent, as electronic or software glitches may corrupt the numbers in ways that aren't immediately obvious. Even seemingly innocuous changes, such as moving to a different time of day for once-daily measurements, can have a real impact. If highs and lows are recorded in the late afternoon, the long-term averages will end up higher than if the measurements are taken at midnight, since an especially warm afternoon and evening could then spread across two 24-hour observing periods instead of just one.

In a perfect world, every temperature record would be free of the blemishes noted above. That's not the case, of course, but climate scientists have found clever ways to work with what they've got. For instance, they make use of the fact that temperature departures have large spatial scales: if Washington, D.C., is unusually warm on a given day, then nearby Baltimore should be unusually warm as well. This allows biases at individual sites to be teased out through station-to-station comparisons. By using stations with long records, and weighting each station by the area it represents, researchers can produce a reasonable estimate of long-term worldwide trends in the form of a **time series**—an accounting of global average temperature from one year to the next.

One major task in building such a time series is to determine how best to come up with a spatial average of global temperature, given that some parts of the globe are much better monitored than others. Another is to find out how much the global reading for a given year has departed from the recent average (which is typically calculated from data covering a 30-year period, long enough to transcend shorter-term dips or peaks). While individual scientists have taken stabs at time series of global temperature for more than a century, three major climate centers now carry out yearly analyses of global temperature, each with its own methods of tackling the challenge.

* **University of East Anglia.** Phil Jones and colleagues at the Climatic Research Unit of the University of East Anglia (UEA) apply a technique called the "climate anomaly method." As the name implies, this method starts with the anomaly observed at each station in a given year—how much warmer or cooler it is relative to a 30-year base period, such as 1961–1990. UEA does this calculation for about 4200 stations worldwide. It averages those station anomalies across grid boxes that span 5° each of latitude and longitude, then by hemisphere, and finally for the globe as a whole, with the Met Office Hadley Centre providing the ocean analyses.
* **The National Aeronautics and Space Administration.** A group at NASA originally led by James Hansen devised the "reference station method." Using temperatures observed at about 6000 stations, it calculates averages across a much smaller number of much larger grid boxes than UEA uses (about 80 in all). The longest-term record in each grid box is dubbed the reference station and considered the gold standard for that box.
* **The National Oceanic and Atmospheric Administration.** NOAA uses a "pairwise" method that automatically identifies and compares temperatures at nearby stations. The technique hunts for red flags such as a gradual drift or an abrupt jump at one site that's not

Bright lights, big cities, bogus data?

One of the most calcified critiques of climate change science is the accusation that the warming indicated by global observations is largely an artifact of urban growth. Nobody doubts the existence of the heat-island effect, by which the dense buildings and paved areas of cities absorb and trap heat in various ways. Indeed, urban heat islands are a form of climate change in themselves. Those who live in an expanding urban area would probably experience a gradually warming climate even if we weren't pumping the atmosphere full of greenhouse gases. One example: at a group of stations scattered across fast-growing metropolitan Shanghai, the average temperature climbed by more than 0.5°C (0.9°F) per decade between 1981 and 2007. Almost half of that rise can be attributed to urban effects, according to a 2011 study led by Xuchao Yang (China Meteorological Administration).

Only a tiny fraction of Earth's surface is urbanized, so heat islands would tend to melt into the background of global temperature even if urban stations weren't culled. Still, mainstream climate scientists don't consider it kosher to mix the localized effects of heat islands with the planet-wide warming produced by fossil fuel emissions. Thus, the big three data compilers described in this chapter use a variety of methods to filter out urban effects from their datasets. The first challenge is working out which stations should count as urban and which should count as rural.

James Hansen's group at NASA started out by using population data to identify urban stations, but they soon discovered this approach couldn't always distinguish whether a reporting station was sited in the heart of a big city or just outside town. In 2001 they began identifying urban areas through the lights detected by U.S. defense satellites on clear nights (especially during new moons). Hansen found that many rural U.S. areas generated enough nighttime light to put their free-of-heat-island status into question. Eventually, the group removed about 80% of their U.S. reporting stations from the global average. That left roughly 200 bona fide rural sites—still enough to provide a reasonable picture of U.S. climate, according to Hansen.

In other parts of the world, where rural analogs for urban stations are scarce or where development covers entire regions, it's harder to weed out the city influence. However, urban heat islands may not be affecting the measured global average as much as scientists once thought. Thomas Peterson of the U.S. National Climatic Data Center found that, from 1880 to 1998, the rural subset of the center's 7200-station global record actually warmed a touch faster than the

The lights of humanity in a nighttime composite image of the world as seen from orbit. (Data courtesy Marc Imhoff, NASA/GSFC, and Christopher Elvidge, NOAA/NGDC. Image by Craig Mayhew and Robert Simmon, NASA/GSFC.)

full record—hardly a sign of urban contamination. And if a weather station has always been located within a large, well-established city, temperatures there might be rising at a pace not too different from their rural counterparts, albeit from a warmer baseline. Taking these and many other factors into account, the 2013 IPCC report deemed it unlikely that any residual heat-island effect has boosted the long-term global temperature trend by more than 10%.

Urban heat islands are strongest on calm nights, when there's no breeze to help disperse the city-trapped air. This inspired David Parker (Met Office Hadley Centre) to see whether calm nights might be warming up more than windy nights globally, which would presumably be a sign of heat-island bias in the climate record. As he reported in *Nature*, that doesn't seem to be the case. For a set of 264 stations across the globe during the period 1950–2000, he found identical warming trends for nights categorized as windy, calm, or lightly breezy. Parker's study added to the overall body of work showing that heat islands are a poor scapegoat for planet-wide warming. One other clue: the recent warming has manifested not only on land, but over the oceans—which aren't known for their urban sprawl.

reflected by its neighbor, then adjusts the data for these effects. Among the three groups, NOAA calls on the largest number of stations: about 7200, with that number expected to rise severalfold in the mid-2010s. Similar to UEA, it calculates anomalies for each station relative to a 30-year base period, with a separate calculation done for each month of the year. Data-sparse areas are filled in on the basis of the long-term patterns of temperature variation observed within grid boxes that span 5° of latitude and longitude.

Each one of these techniques contains a lot of redundancy—on purpose. According to some experts, it takes no more than about a hundred strategically placed points around the world to create a decent first cut at global temperature. Everything beyond that refines and sculpts the estimate; important work, to be sure, but nothing that would change the result drastically.

Which of the three major groups' methods is best? That's an impossible question to answer, because there's no absolutely perfect source to compare the results against. A better question might be, how sensitive is the final result to different approaches? If only one approach were used, we really wouldn't know, but with several techniques in play, we can begin to assess how solid the results are. Each center's approach is designed to address the issues it believes are most critical. Trends among the three groups can vary depending on what years you start and end with, but reassuringly, there doesn't seem to be all that much difference among the results. What may look like important contrasts—for instance, NASA ranked 2013 as the seventh-warmest year on record, while NOAA put it in fourth place—relate more to tiny differences in particular years (though some skeptics have interpreted the occasional reshufflings of the rankings as a sign of sloppy research and/or insignificant warming).

There's no disagreement among the three groups on the big picture of long-term warming of the atmosphere at Earth's surface. As reflected in the 2013 IPCC report, the consensus shows a global temperature rise, considering both land and ocean areas, of **roughly 0.89°C (1.60°F) from 1901 to 2012**, with 90% confidence that the actual value is in the range of 0.69°–1.08°C (1.24°–1.94°F). Most of that net warming has occurred in the last few decades—**about 0.72°C (1.30°F) from 1951 to 2012**, with a 90% confidence interval of 0.49°–0.89°C (0.88°–1.60°F).

The solid nature of this consensus was reinforced in a highly touted independent study led by Richard Muller (University of California, Berkeley).

Amassing the figures

It was 1979 when Phil Jones and colleagues at the Climatic Research Unit of the University of East Anglia (UEA) came up with the notion of a global temperature index. The group knew that others had tried to take Earth's temperature in a systematic way but hadn't achieved truly global coverage. With support from the U.K. and U.S. governments, Jones and Tom Wigley (now at NCAR) led the charge to sift through stacks of data, gather readings from weather services around the world, and correct as many sources of bias as possible. "It's the best thing we've ever done," Jones said in 2005.

In New York, James Hansen and colleagues at NASA's Goddard Institute for Space Studies were on the same wavelength. Hansen, a pioneer climate modeler, set out to maximize the value of climate data in poorly sampled parts of the globe. His group's technique allowed the most isolated stations to represent conditions within a radius of 1200 km (750 mi), which helped expand the analysis across the station-scarce Southern Hemisphere and near the poles.

As the 1980s unfolded, the NASA and UEA groups took turns making headlines with a series of landmark papers. NASA led off in 1981 with the startling announcement that the world was once again warming—and, in fact, that it had been since the 1970s. UEA concurred in a 1982 paper, and in 1986 Jones and colleagues detailed their techniques in the most thorough global analysis to date. By the time Hansen testified before the U.S. Congress in 1988 (see chapter 13), the National Oceanic and Atmospheric Administration had begun its own program to monitor global temperature, led by Thomas Karl.

Little did the researchers know how politicized their products would become. Climate skeptics besieged Jones and others with requests for data (some of which Jones couldn't legally share) and hunted for issues and inconsistencies that might weaken the case for global warming. Then, in 2009, Jones found himself at the center of the unauthorized release of emails dubbed Climategate (see chapter 13). A few months later, although some concerns remained about data access, all inquiries had cleared Jones and colleagues of data-tampering accusations.

"There are many flaws in the input data for the temperature change analyses, especially in the early years, and the effect of these flaws cannot be fully removed," noted Hansen in 2005. "Despite these problems, however, the reality of global warming in the past century is no longer at issue."

The work garnered support from the Charles Koch Foundation, whose stated mission is to "advance social progress and well-being through the study and advancement of economic freedom." Given the well-publicized conservative leanings of oil magnate Koch, some skeptics assumed the Berkeley Earth Surface Temperature (BEST) project would find less global temperature change than the big three groups and thus provide the vindication they'd so long sought. Muller and colleagues drew on the same 7200-station network as NOAA, but added several thousand stations (many with shorter records) from other databases. They also approached local discontinuities in a way that differed from the three groups above, separating before-and-after climate datasets as if they'd come from independent stations.

In a series of peer-reviewed papers in 2012 and 2013, the BEST team burst the collective bubble of skeptics. With a confidence level of 95%, they estimated land-only temperatures in the decade 2000–2009 to be 0.85°–0.95°C (1.53°–1.71°F) warmer than the 1950s. As it happens, this closely paralleled the trends found by the other three groups for the last century.

Although the Berkeley group focused on land-based readings, most estimates of full-globe temperature include temperatures over the **oceans**, which are a bit tricky to incorporate. Since oceans represent 70% of the planet's surface, it's critical that surface temperature is measured over the sea as well as over land. There's no fleet of weather stations conveniently suspended above the sea, unfortunately, but there are many years of sea surface data gathered by ships and, more recently, by satellites. Because the sea surface and the adjacent atmosphere are constantly mingling, their temperatures tend to stick close to each other. This means that, over periods of a few weeks or longer, a record of sea surface temperature (SST) serves as a reasonable proxy for air temperature.

Until the mid-twentieth century, SSTs were mainly measured by dipping uninsulated buckets (originally made of wood, then canvas) into the ocean and hauling them back to deck, with the water often cooling slightly on the trip up. Around the time of World War II, it became more common for large ships—especially U.S. ones—to measure the temperature of water at the engine intake. This tended to produce readings averaging about 0.4°C (0.7°F) warmer than the canvas-bucket method. Some ships measured air temperature directly on deck, but these readings had their own biases: the average deck height increased through the last century, and daytime readings tend to run overly high due to the sun heating up metal surfaces on the ship. It wasn't until the 1990s that routine measurements from ships and buoys were gathered promptly, processed for quality control and coupled with newer estimates derived from satellites, which can infer SSTs by sensing the upward radiation from the sea surface, especially where it's not cloudy. Together, all these sources now help scientists to calculate temperatures for the entire globe, oceans included. Researchers adjust the SST data so that the pre–World War II bucket observations are consistent with the more recent engine-intake readings. As with the land readings, there are differences in how the three leading groups compile and process ocean data, but these don't appear to influence the long-term trends for global temperature too much.

Even if it's clear the world is warming, it's equally evident that the current observing network is far from perfect. That point was brought home in 2009 by an energetic blogger, Anthony Watts, who rallied more than 650 volunteers through Watts Up With That?, his hugely popular, skeptic-oriented website. The investigators in Watts's **surfacestations.org** project found hundreds of weather stations across the United States that were improperly located near water treatment plants, pavements, air conditioners, and other sources of artificial heat. A photo-studded report, "Is the U.S. Surface Temperature Record Reliable?" was published by the Heartland Institute, a nonprofit outfit that promotes free-market approaches to problem-solving (see sidebar, "Pushing against climate change action"). Though the report wasn't a peer-reviewed study, it made quite a splash in the U.S. media and the blogosphere. It found that nearly 90% of the 860 stations examined from NOAA's U.S. Historical Climatology Network (USHCN) fell short of the agency's own protocol that stations should be at least 30 m (100 ft) from artificial objects that could produce, reflect, or radiate heat. The report also claimed that the adjustments used by NOAA to homogenize the datasets simply expanded the impact of the siting problems. If the much-vaunted U.S. system was this flawed, Watts argued, perhaps the global database was just as compromised. The findings prompted a group of U.S. congresspeople to ask whether NOAA was adequately ensuring the quality of its climate data. A 2011 inquiry by the U.S. General Accounting Office found a lower percentage of problematic stations (42%) than the Watts report, while recommending improvements in how the USHCN network is managed.

It didn't take long for researchers to demonstrate that the station-location issues, while in need of addressing, had virtually no impact on the big picture. In a 2010 paper, a NOAA team headed by Matthew Menne compared the adjusted data from stations rated by Watts to those from a gold-standard group of well-sited stations called the U.S. Climate Reference Network. The trends were nearly identical, even for those stations whose siting was rated "poor" by Watts—an illustration of how even less-than-ideally-positioned instruments can still capture a trend if it's distributed widely. As for the

Improving the global thermometer

As critical as they are to climate change science, traditional weather stations and balloon-borne radiosondes were never designed to measure the subtle trends of long-term climate change. Just as a hi-fi that's fine for heavy metal might not convey all the nuance of a symphony orchestra, a weather station might capture large day-to-day temperature changes well enough but contain tiny biases that become evident only after years have passed. To address the need for better long-term data, several UN agencies teamed with the International Council for Science in 1992 to launch the Global Climate Observing System (GCOS). Through a broad web of activities, GCOS fosters the improvement of all types of data collection on climate change and its impacts, particularly on ecosystems and sea level.

With more and better data in the queue, the job of connecting these observational dots looms larger. In 2005 nearly 60 governments and the European Commission endorsed a 10-year plan to build a Global Earth Observation System of Systems (GEOSS). The name itself signals the multilayered complexity of the task. GEOSS has been looking for ways to incorporate upcoming satellite systems and new ground-based tools, while working to maintain the integrity of the observational network already in place. That network suffers from multiple ailments: gaps in coverage across space and over time, inadequate archiving, and the growing risk of discontinuities from one satellite to the next as tight budgets threaten new launches. Some countries don't take or share many observations, and many older records in paper form haven't yet been digitized.

Even the best datasets can't help much if they're inaccessible. In 2010 the World Meteorological Organization began considering a proposal from the Met Office to build an open-access archive of global temperature records. By 2014 the WMO was moving toward a new policy that would encourage greater sharing of historical climate data. Right now much of the best-quality data, including hourly temperature reports, is locked up by the weather services of nations who turn a profit by selling their data to users. If a new storehouse managed to pool these observations, researchers would have more than twice the holdings than are now at their disposal. The archive would also maintain details on all adjustments and corrections in a transparent fashion, thus bolstering its usefulness while helping to boost confidence in its data. (Even without thermometers, of course, there's ample evidence that the world is warming, from a moistening atmosphere to shrinking glaciers to poleward-migrating wildlife.)

unadjusted data, they did show a bias, but not in the expected direction: the poorly sited stations actually ran cooler overall. It's mainly because those sites were more likely to have newer thermometers whose installation had led to cooler daily highs, due to the effects of a concurrent change in instrumentation as well as a switch in the timing of once-per-day temperature checks. Adjustments already in place accounted for these biases well, noted Menne and colleagues. They stressed the need to wade through actual numbers—"photos and site surveys do not preclude the need for data analyses," they wrote—while also praising the work done by Watts and his volunteers to document problems in station placement. Berkeley's Richard Muller, who has also voiced support for the station analyses carried out by Watts's group, confirmed the lack of significant impact of such biases on long-term trends in a 2013 paper for *Geoinformatics and Geostatistics.*

Heat at a height

When people refer to global warming, they're usually talking about the air in which we live and breathe—in other words, the air near ground or sea level. But the atmosphere is a three-dimensional entity. Conditions at the surface are affected profoundly by what's happening throughout the **troposphere**—the lowest layer of the atmosphere, often called the "weather layer." The troposphere extends from the surface to about 8–16 km (5–10 mi) above sea level. It's tallest during the warmer months and toward the tropics, wherever expanding air is pushing the top of the troposphere upward.

There are two ways to measure temperatures high above Earth's surface. The first is to use **radiosondes**, the balloon-borne instrument packages launched from hundreds of points around the globe twice each day. Sending data via radio as they rise, the devices measure high-altitude winds, moisture, and temperatures. As useful as they are, radiosondes have their limits for tracking climate change. They're launched mainly over continents, with large distances separating their paths; they're produced by more than one company and thus don't behave uniformly; and they're prone to small biases that can make a big difference when assessing long-term trends. The second technique is to use **satellites**. These provide a more comprehensive global picture, but they were also at the center of a now-resolved issue that was, for years, the subject of one of the most heated scientific debates in climate change science.

Shortly after global warming burst into political and media prominence in 1988, Roy Spencer and John Christy of the University of Alabama in Huntsville began using data from a series of satellites to infer

How the greenhouse turns

The troposphere's name stems from the Greek *tropos*, meaning to turn—which is exactly what the troposphere does. Heated largely from below, as sunlight strikes Earth, this layer of air churns and bubbles like a boiling pot of water, with calm conditions the exception more than the rule. Usually the air is blowing faster horizontally than vertically, but no matter how you look at it, the troposphere is an ever-changing domain. All this motion—up, down, and sideways—helps give rise to the variety of weather we experience at ground level.

Because the troposphere is so well mixed, there should be a close connection between how much it warms near the surface and higher up. There's not a one-to-one relationship, however, because greenhouse gases aren't spread through the troposphere as evenly as you might expect, and their effects vary with height.

* **Shorter-lived greenhouse gases**, such as low-level ozone, are chemically transformed or washed out by rain or snow within a few days. Thus, they tend to stay within the same general latitudes as their sources—often riding the jet stream from Russia to Canada or from eastern North America to Europe during their short time aloft.
* **Water vapor** is most prevalent at lower altitudes, especially above the oceans, but it's been increasing higher up, thanks in part to high-flying aircraft. (Planes also add fossil fuel emissions to the thin, cold air.)
* **Longer-lived greenhouse gases**, such as carbon dioxide, are thoroughly mixed across the troposphere, both horizontally and vertically. That's why a single station high atop Hawaii's Mauna Loa (see p. 39) can tell us how much CO_2 the global atmosphere holds and how that's changing.

In general, the higher in the troposphere a greenhouse gas is, the more powerful its effect on the troposphere as a whole. This is because the amount of energy a greenhouse molecule emits is directly related to its temperature. At high altitudes, where temperatures are low, the molecules emit less of the heat

global temperature in three dimensions. Throughout the 1990s and into the new millennium, their findings roiled the world of climate research and lent ammunition to global warming skeptics. The satellite data suggested that the troposphere wasn't warming much if at all, despite the observed warming at ground level. Unless the fundamental understanding of lower atmosphere was off base, then one set of measurements had to be wrong.

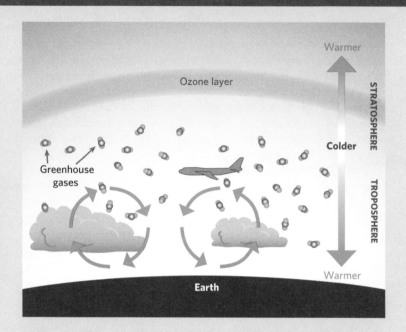

Warmer

Ozone layer

STRATOSPHERE

Colder

Greenhouse
gases

TROPOSPHERE

Warmer

Earth

energy they receive from Earth into space, resulting in more heat remaining in the troposphere.

This picture reverses in the stratosphere, where—oddly enough—human actions have led to *cooling* rather than warming. Carbon dioxide doesn't absorb much heat energy from Earth at these heights, but it continues to radiate heat to space, thus acting to cool the stratosphere. Meanwhile, the partial loss of sunlight-absorbing ozone over the last 30 or so years has exerted a cooling influence. As a result of these and other factors, temperatures in the lower stratosphere have plummeted to record-low levels.

Spencer and Christy's data are gathered primarily by instruments called **microwave sounding units** (MSUs), stationed on board NOAA satellites. The MSUs measure the temperature of gases at various altitudes below them by measuring the energy that they emit. Unlike a surface station or weather balloon, the MSUs provide data that's inherently averaged over large areas. The formulas are complicated, but in general, each individual temperature measurement covers anywhere from about 8000 to 38,000 km² (3100–14,700 sq mi) through a depth of a few kilometers and a time period of a few hours. All told, the satellites cover more than 90% of the planet's surface.

The fireworks began with the first Spencer–Christy paper, published in the journal *Science* in 1990. It showed no discernible temperature trend since 1979 in the lowest few kilometers of the atmosphere. By comparison, surface temperatures had warmed roughly 0.2°C (0.36°F) during the 1980s. Updated through the 1990s, the Spencer–Christy data continued to show virtually no warming. What's more, radiosonde analyses tended to agree with the satellites, with relatively minor warming in northern midlatitudes and a slight cooling in the tropics.

Many people inclined to disbelieve global warming cited these reports as proof that the world wasn't heating up, but most climate scientists weren't convinced. In 1998, two scientists in California, Frank Wentz and Matthias Schabel, discovered that atmospheric drag might be affecting the MSU readings. They claimed that air molecules were slowing down the satellites, increasing the effect of gravity on their paths and causing them to drop slightly, hence changing their viewing angle and skewing their measurements. Spencer and Christy adjusted for the error caught by Wentz and Schabel, but also made a separate correction in the opposite direction for another newly discovered bias, so the result was little net change. Meanwhile, the global surface temperature continued to climb, so the gulf between surface- and satellite-measured upper-air trends widened even further. The persistent disagreement left policymakers and the public wondering what to believe.

In 2000, the U.S. National Academies weighed in with a high-profile special report. It agreed that the overall surface warming trend since 1979 was "undoubtedly real" but noted the possibility that high-altitude air had warmed more slowly than the surface, thanks to factors such as sulfate pollution, cool air filtering down from the record-cold stratosphere, and volcanic emissions from 1991's eruption of Mount Pinatubo. Since then, events have conspired to further diminish the satellite data as an argument against global warming. For one thing, the years since 2000 have been no-tably warm compared to the 1980s and 1990s, both at the surface and aloft. Meanwhile, several other groups gained access to full MSU data, and their analyses (including new techniques for separating stratospheric cooling from tropospheric warming) tended to show even stronger satellite-based warming trends, closer to surface observations. Radiosonde trends have also crept upward in some studies.

Overall, these and other corrections have brought the surface and satellite estimates into much closer agreement with each other. The 2013 IPCC report examined seven different analyses of temperature trends in the lower stratosphere from 1979 to 2012, each drawing on radiosonde or satellite data. The best estimates all fall into the range of 0.13°–0.16°C (0.23°–0.29°F) per decade. That's in the same ballpark as surface-based trends for this period from the big three groups, which range from about 0.15° to 0.16°C (0.27°–0.29°F) per decade. With the dust now settled, it's clear that the atmosphere is warming up miles above us, as well as on the streets and farms where we live.

Global dimming?

To get a clear idea of how our climate is changing, you've got to measure more than temperature. **Rain and snow** are an important part of the story, and they're uniquely challenging to assess (see p. 77). Other variables, such as **water vapor** and the nature of **cloud cover**, have their own complexities. Satellites can measure both, but only with strict limitations. For example, some can't sense lower clouds if high clouds are in the way. A set of globe-spanning satellites called COSMIC (Constellation Observing System for Meteorology, Ionosphere, and Climate), launched in 2006 by a U.S.–Taiwan partnership, has provided fuller three-dimensional portraits of temperature and water vapor. Other satellites are keying in on cloud cover. Even so, it's likely to be years before we can assess with confidence how these features are changing.

The droplets and particles in our atmosphere have everything to do with how much sunlight reaches Earth. Sunshine and its absence weren't part of the dialogue on global warming until relatively recently. A 2003 paper in *Science* gave the phenomenon of **global dimming** its first widespread exposure, and by 2005 it was the star of a BBC documentary.

Global dimming is real—or at least it *was* real. From the 1950s through the 1980s, the amount of visible light reaching Earth appears to have tailed off by a few percent. In terms of climate, that's quite significant, though it would be difficult for the average Jane or Joe to perceive the dimming against the much more dramatic on-and-off sunlight cycles that occur naturally each day. The dimming also wasn't enough to counteract the overall global warming produced by greenhouse gases. In fact, the timing of global dimming's rise to fame was a bit paradoxical, since the phenomenon began to reverse in the 1990s (and as we will see, the recent brightening brings its own set of worries).

The effects of clouds on climate change are complex and not fully understood. (Robert Henson)

Global dimming offers a good example of how a set of little-noticed measurements, collected diligently over a long period, can yield surprising results and can help answer seemingly unrelated questions. Back in 1957, as part of the worldwide International Geophysical Year, a number of weather stations across the globe installed **pyranometers**. These devices measure the amount of shortwave energy reaching them, including direct sunshine as well as sunlight reflected downward from clouds. Slowly but surely, the amount of energy reaching the pyranometers dropped off. Gerald Stanhill (who later coined the phrase "global dimming") discovered a drop of 22% across Israel, and Viivi Russak noted similar reductions in Estonia. Although they and several other researchers published their findings in the 1990s, there was hardly a rush to study the problem. With the spotlight on global warming by that point, scientists might have found it counter-intuitive to investigate whether less sunlight was reaching the globe as a whole. Climate models, by and large, didn't specify such dramatic losses in radiation.

Yet the data cried out for attention, especially once they were brought together on a global scale. Using a set of long-term readings from the highest-quality instruments, Stanhill found that solar radiation at the surface had dropped 2.7% per decade worldwide from the late 1950s to early 1990s. Two other studies examining much larger datasets found a smaller global

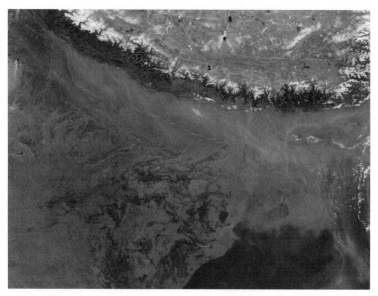

On February 5, 2006, the skies over northern India are filled with a thick soup of aerosol particles all along the southern edge of the Himalaya Mountains and streaming southward over Bangladesh and the Bay of Bengal. These toxic human-produced aerosols are certainly hazardous to life, but they may also act to slow global warming. (NASA)

drop—around 1.3% per decade from 1961 to 1990. Even these studies failed to jolt the climate research community until 2003, when two Australians, Michael Roderick and Graham Farquhar, linked the dimming to a steady multi-decade reduction in the amount of water evaporating from standardized measuring pans around the world. Warmer temperatures ought to have evaporated more water, not less, they reasoned (as suggested by global measurements of water vapor). However, Roderick and Farquhar realized that a reduction in sunlight could more than offset this effect. It's possible that global evaporation actually rose in spite of the pan-based data, as noted in the 2007 IPCC report. Pan-based sensors can only show how much water might evaporate from a completely wet surface, but a warmer, wetter, and dimmer world could still send more total moisture into the air—through the leaves of plants, for example. The lifetime of moisture in the atmosphere, which could change with the climate, is another factor that makes the evaporation issue especially hard to unravel.

Still, the notion of dimming through much of the latter twentieth century is now fairly well accepted, and it's a small but significant part of the picture when scientists try to balance the global energy budget of incoming

The plane truth about contrails

Can those sleek, wispy lines of cloud that trail behind jets really affect the future of our climate? Conspiracy theorists point to contrails—which they've dubbed "chemtrails"—as an existential threat, a shadowy government plot to inject mysterious chemicals into the air. Reality-grounded scientists don't have to invoke such scheming to find ways in which contrails actually do influence the workings of the atmosphere.

Contrails are a special type of **cirrus**, which are high, thin clouds composed mainly of ice crystals and often resembling streaks or sheets. Contrails form as the water vapor and particles spewed out by an airplane's exhaust stream create water droplets that soon crystallize into ice. Pollutants in the exhaust plume serve as focal points for additional ice crystals within the contrail as well as in neighboring clouds. One of those pollutants, nitrous oxide, is a powerful greenhouse gas in itself—but at flight level, it reacts with ozone, another greenhouse gas, diminishing the net effect.

Most types of clouds cool Earth's surface as they screen out sunlight. However, cirrus clouds have an overall warming effect. That's because cirrus usually allow a good deal of sunlight through but effectively trap radiation flowing up from Earth. When a contrail first forms, it behaves more like a lower-level cloud, as the water droplets reflect ample amounts of sunlight and exert a cooling effect. Larger aircraft tend to generate thicker, longer-lived contrails. Gradually, each contrail spreads, thins out, and freezes, and its net effect soon shifts from cooling to warming. Many contrails evaporate within an hour or so, but some evolve into much larger cirrus clouds that can last 3–12 hours or more, boosting their collective warming influence severalfold.

Contrails can reflect a good deal of sunlight, partially offsetting their 24/7 warming influence noted above. At night, they provide only a warming effect, since there's no sunshine to be reflected (which argues for flying by day rather than by night; see chapter 19). One of the first analyses to combine a high-end climate model with hour-by-hour flight data was led by Chih-Chieh Chen (NCAR). Published in 2013, the study found that night flights contributed 88% of contrails' total impact on radiative balance across the eastern United States and 69% across central Europe, well above a previous global estimate of 60%.

Of course, contrails aren't uniformly distributed around the world; they're focused where people happen to be flying. Right now contrails are most prevalent across North America, western Europe, and eastern Asia. In these regions the effect of contrails is much larger than the global total. There's some uncer-

(© UCAR)

tainty in contrails' vertical distribution, because they can hide in or beneath other clouds and escape the view of cloud-detecting satellites. A 2010 study by the University of Leeds and the Met Office found that unrelated cloud cover negates almost half of the potential effect of contrails on climate. Conversely, the German study above found that contrail-related cirrus can itself inhibit the formation of other clouds.

Aviation only represents about 3% of the carbon dioxide humans put into the atmosphere each year, but that fraction is rising, and contrails add to the climatic impact of flying. Studies have pegged the total warming effect from contrail cloudiness as anywhere from half as large to fully as much as greenhouse forcing from aircraft-emitted CO_2. As for the future, it seems inevitable that increasing affluence will spawn more contrails above China and India, although it's unclear how cut-rate airline traffic in Europe and North America will evolve amid increasingly steep fuel costs. Whatever the total impact of contrails may be, it looks set to rise as global air travel continues to grow, adding further to the implications of what is already a very climate-unfriendly form of transport (see sidebar, chapter 3).

and outgoing radiation at Earth's surface over the last few decades. Dimming might even help explain why some Arctic tree-ring records show a divergence between temperature and tree growth since the 1950s. If these high-latitude trees are especially light-sensitive, then even a small dimming might be enough to limit their growth.

Some areas—especially urban, industrialized parts of the Northern Hemisphere—showed several times more dimming than other parts of the globe. This regional patchwork makes sense given what we know about the causes of dimming. The prime suspects are soot and other airborne particles, or aerosols, produced when fossil fuels are burned. For hundreds of years, people in London had first-hand experience of the sky-darkening effects of soot. Aerosol pollution does this type of darkening in several ways:

* **Attracting water.** Many aerosols are hydrophilic, or water-attracting. When conditions are right for clouds to form, a highly polluted air mass can produce many times more cloud droplets. The moisture is spread more thinly, though, so each droplet is smaller on average. The upshot is that this larger array of droplet surfaces can reflect more sunlight upward before it has a chance to reach the ground. From space, the clouds look brighter; from Earth, they're darker.
* **Extending cloud lifespans.** In general, it takes longer to get rain or snow out of a cloud full of many small particles, as opposed to fewer, larger ones. For this reason, the addition of aerosols seems to lengthen the lifespan of clouds and thus increase their reflective power over time.
* **Reflecting light.** Even on a cloudless day, aerosols themselves reflect sunlight to space. A major study of pollution across India and the adjacent ocean found that the subcontinent's thick veil of winter pollution can trim the amount of sunlight reaching the surface by as much as 10%.

Global warming itself may have been responsible for some of the dimming—perhaps around a third—by increasing the amount of water vapor in the atmosphere (which makes more moisture available for cloud formation) and perhaps rearranging the distribution of clouds (lower and/or thicker clouds tend to reflect more than higher, thinner ones). As noted above, it's not entirely clear how cloud cover is evolving on a global scale, though water vapor increased on the order of 1% per decade during the strong global warm-up of the 1980s and 1990s.

It's also unclear whether global dimming will be a major player in the decades to come. Aerosols are more visible and noxious than greenhouse gases but more practical to control, so their reduction can be an easier political sell. Most of the world's highly industrialized nations began cleaning up their smokestacks and exhaust pipes by the 1970s, and the economic downturn of the 1990s across the former Eastern Bloc further reduced aerosol pollution. Meanwhile, the breakneck growth across southern and eastern Asia has put aerosol emissions on the rise in those areas since the 1990s. The result is a more complex patchwork, with the net global effect appearing to be an overall drop in aerosol emissions in the 1990s, a slight rebound from about 2000 to 2005, and a slight drop later in the decade.

The amount of sunlight reaching Earth's surface has evolved in line with these trends, combined with other factors. At some locations, solar radiation jumped anywhere from 1% to 4% from the early 1990s to the early 2000s. Beate Liepert (Lamont-Doherty Earth Observatory) suspects this may have reflected a recovery from the sun-shielding effects of the 1991 Mount Pinatubo eruption and the extensive cloudiness generated by the frequent El Niño events of the early and mid-1990s. A slight dimming resumed in the early 2000s.

Our clean-up of aerosols comes at an ironic price. The transition away from global dimming helps explain why temperatures rose as sharply as they did in the 1990s. Few would complain about a world that's brighter and less fouled by aerosol pollution, yet if we manage to reduce the global prevalence of aerosols in a sustained way, we'll also lose some of the overall cooling they provide as a bulwark against greenhouse warming. In gaining a cleaner atmosphere, the world thus stands to lose one of its stronger buffers against climate change. That may be one reason why the idea of consciously dimming the atmosphere through geoengineering has attracted attention (see chapter 16).

CHAPTER ELEVEN
The Long View
A WALK THROUGH CLIMATE HISTORY

During the four and a half billion years of our solar system, Earth has played host to an astounding array of life forms. Much of the credit goes to our world's climate, which is uniquely equable compared to conditions on the other major planets. Even if they had water and other necessities, Venus would be too hot to sustain life as we know it, and Mars and the planets beyond it too cold. However, there's evidence that both planets once harbored oceans and may have been mild enough to support life before their climates turned more extreme. Likewise, during its lively history, Earth's climate has lurched from regimes warm enough to support palm-like trees near the poles to cold spells that encased the present-day locations of Chicago and Berlin under ice sheets more than half a mile thick.

The changes unfolding in our present-day climate make it all the more important to understand climates of the past. Thankfully, over the last 50 years, technology and ingenuity in this field of study have brought dramatic leaps forward in our knowledge. The clues drawn from trees, pollen, ice, silt, rocks, and other sources answer many questions, while leaving many others still open. Taken as a whole, though, these data paint a remarkably coherent picture. We now have a firm grasp of the changes that unfolded over the past three million years—a period ruled mainly by ice ages, with the last 10,000 years of interglacial mildness being one of the few exceptions. We also know a surprising amount about what transpired even earlier, including many intervals when Earth was a substantially warmer place than it is now.

Could those ancient warm spells serve as a sneak preview of what we might expect in a greenhouse-driven future? We know that there's already more carbon dioxide and methane in our atmosphere than at any time in at least 800,000 years, and perhaps much longer than that. Carbon dioxide has increased by more than 30% in only a century, an astonishing rate by the standards of previous eras. At the rate we're pumping greenhouse gases into the atmosphere, they could be more prevalent by the year 2050 than they've been in at least 10 million years. That's one reason why **paleoclimatologists** (those who study past climates) are taking a close look at how climate and greenhouse gases interacted during previous warm intervals. The analogies are admittedly less than perfect. As we'll see, Earth's continents were positioned quite differently 200 million years ago than they are now. Even 10 million years ago, India was still drifting across the Indian Ocean, and North and South America had yet to join up. As a result, the ocean circulations were very different as well. Also, most of the species that were then swimming, crawling, or flying across our planet are now extinct. Many other aspects of the environment at that time are hard to discern today.

In spite of our less-than-ideal hindsight, we've gained enormous ground in our understanding of how climate works by piecing together the past, and major advances continue to arrive in the still-young field of paleoclimatology. In 2004, a new ice core from Antarctica provided a record that was almost twice as long as the previous one (see "Core value" sidebar). There's hope that a future core could yield data going back nearly 1.5 million years, including the point when a switch occurred in the pace of the on-and-off tempo of ice ages. If there's one message that's become increasingly clear from the ice cores, it's that natural forces have produced intense and rapid climate change in the past. That's something worth thinking about as we add carbon dioxide to the air at a rate far beyond natural processes.

What controls our climate?

The great detective story that is paleoclimatology hinges on a few key plot devices. These are the forces that shape climate across intervals as brief as a few thousand years and as long as millions of years. Greenhouse gases are a critical part of the drama, of course, but other processes and events can lead to huge swings in climate. These vary greatly in terms of timespan. The sun-blocking particles spewed into the stratosphere by an unusually large volcano may cool the planet sharply, but most of those impacts are gone after a couple of years. In contrast, climate change due to slight shifts in the sun's orbit plays out over tens or hundreds of thousands of years. That's still relatively fast in geological terms. The slow drift of continents, another major player in climate variation, can operate over hundreds of millions of years.

The overlap among these various processes creates a patchwork of warmings and coolings at different time intervals, some lengthy and some relatively brief. It's a bit like the mix of factors that shape sunlight in your neighborhood. On top of the dramatic but dependable shifts in sunlight you experience due to the 24-hour day and the transition from summer to winter, there are irregular shifts produced as cloudiness comes and goes. Below is a quick summary of the biggest climate shapers. We'll come back to each of these later in the chapter.

* **The sun is literally the star of the story.** At its outset, the sun shone at only about three-quarters of its current strength; it took several billions of years for its output to approach current levels. One fascinating question (discussed later) is how Earth could be warm enough to sustain its earliest life forms while the sun was still relatively weak. In another five or six billion years the sun—often referred to as an "ordinary middle-aged star"— will reach the end of its life. At that point it should warm and expand dramatically before cooling to white-dwarf status.

* **Variations in Earth's orbit around the sun are hugely important.** These small deviations are responsible for some of the most persistent and far-reaching cycles in climate history. Each of these orbital variations affects climate by changing when and how much sunlight reaches Earth. As shown in the graphics on the opposite page, there are three main ways in which Earth's orbital parameters vary. Several visionaries

James Croll. (J.C. Irons/photo courtesy NOAA)

in the mid-1800s, including James Croll—a self-taught scientist in Scotland—suspected that these orbital variations might control the comings and goings of ice ages. In the early 1900s, Serbian mathematician Milutin Milankovitch quantified the idea, producing the first numerical estimates of the impact of orbital variations on climate. While his exact calculations have since been refined, and there's still vigorous debate about what other factors feed into the timing of climate cycles, the basic tenets of Milankovitch's theory have held firm.

✳ **The changing locations of Earth's continents also play a huge role in determining climate.** Francis Bacon observed in 1620 how easily South America and Africa would fit together if they were puzzle pieces. Many schoolchildren notice the same thing long before they're taught about **continental drift**, Alfred Wegener's bold 1915 hypothesis that Earth's continents moved slowly into their present positions. Today, satellites can monitor continental drift by tracking minuscule land movements, but scientists are also able to deduce the long-ago locations of continents thanks to various techniques—including analyzing traces of magnetism generated by Earth's geomagnetic field over the millennia and preserved in volcanic magma. These and other clues tell us that much of Earth's land mass once existed as **supercontinents**. The first (and most speculative) is **Rodinia**, a clump that bears little resemblance to today's world map. Formed more than a billion years ago, Rodinia apparently broke up into several pieces. Most of these were reassembled as **Gondwana**, which drifted across the South Pole between about 400 and 300 million years ago. Gondwana became the southern part of **Pangaea**, an equator-straddling giant. In turn, Pangaea eventually fragmented into the continents we know now. The locations of these ancient continents are important not only because they shaped ocean currents but because they laid the literal groundwork for ice ages. Only when large land masses are positioned close to the poles (such as Gondwana in

The Earth, circa 250 million years ago, showing the supercontinent Pangaea.

the past, or Russia, Canada, Greenland, and Antarctica today) can major ice sheets develop. However, high-latitude land isn't sufficient in itself to produce an ice age. The southern poles harbored land for most of the stretch from 250 to 50 million years ago, yet no glaciation occurred, perhaps because CO_2 concentrations were far higher than today's.

✳ **Volcanoes** can spew enormous amounts of ash, soot, and other particles and gases—some of them greenhouse gases—into the atmosphere. A single large eruption can be powerful enough to loft sun-shading particles into the stratosphere, as was the case when Mount Pinatubo blew its top in 1991. When this happens, the high-altitude debris can cool the planet by more than 1°C (1.8°F) for a year or more, until gravity pulls the particles back to Earth. Tropical volcanoes are especially effective at this cooling, since their strato-spheric debris can easily spread across both Northern and Southern Hemispheres. Smaller volcanoes have less of an individual impact, but they can make a difference collectively. The 2013 IPCC report concluded that a series of minor eruptions after 2000 played at least some role in the relative hiatus in global warming observed since the late 1990s. In the much longer term, volcanoes act as a warming

Mount Pinatubo erupts on June 12, 1991. (U.S. Geological Survey/Cascades Volcano Observatory)

Round and round we go (and so does our climate)

Everyone who lives outside the tropics knows the power of the four seasons, but a surprising number of people—including most of the Harvard graduates interviewed at random in the classic documentary *A Private Universe*—can't explain why the seasons occur. Many assume it's because Earth is closer to the sun in summer and more distant in winter, but that doesn't wash: it's true that Earth's orbit is slightly asymmetric, but we're actually closest to the sun each year between January 2 and 5, when it's summer in Sydney but winter in London. The real reason for the seasons is that Earth is tilted about 23.4° relative to its orbit around the sun (see graphic). That tilt aims each hemisphere toward the sun in an alternating pattern that produces summers, winters, springs, and autumns.

What does this have to do with climate change? Each of the factors just alluded to—Earth's tilt, the eccentricity of its orbit, and the timing of its closest approach to the sun—change slightly over time in a cyclical manner. This is due mainly to the intersecting gravitational effects of the sun, the moon, and the other planets in our solar system. Together, the orbital cycles produce dramatic swings in climate over tens of thousands, or even hundreds of thousands, of years, with carbon dioxide greatly boosting the atmosphere's response. By studying how these natural climate shifts come and go, scientists also find clues about how Earth might respond to change that's human induced.

* **Earth's tilt** goes up and down, ranging from about 21.8° to 24.4° and back over about 41,000 years (the range widens or narrows a bit with each cycle). The tilt is now just over 23.4° and on the decrease. When the tilt is most pronounced, it allows for stronger summer sun and weaker winter sun, especially at high latitudes. Ice ages often set in as the tilt decreases, because the progressively cooler summers can't melt the past winter's snow. At the other extreme, increasing tilt produces warmer summers that can help end an ice age.

Minimum tilt: 21.8°
0° Current tilt: 23.4°
Maximum tilt: 24.4°

* **Earth's orbit** around the sun is not precisely circular, but elliptical, with the sun positioned slightly to one side of the center point. Currently, this brings Earth about 3% closer to the sun in early January (perihelion) than in early July (aphelion), with about 7% more solar energy reaching Earth at perihelion. The **eccentricity** or "off-centeredness" of the orbit varies over time in a complicated

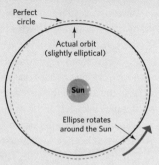

Perfect circle

Actual orbit (slightly elliptical)

Sun

Ellipse rotates around the Sun

way. The net result is two main cycles, one that averages about 100,000 years and another that runs about 400,000 years. When the eccentricity is low, there's little change through the year in the Earth–sun distance. When eccentricity is high, the sunlight reaching Earth can be more than 20% stronger at perihelion than at aphelion. This either intensifies or counter-acts seasonality, depending on the hemisphere and the time of year that perihelion and aphelion occur. The last eight ice ages have come and gone roughly in sync with the 100,000-year eccentricity cycle, but it's not yet clear how strongly the two are related (see p. 282).

* **Earth's axis** rotates slowly around an imaginary center line, like a wobble in a spinning top. The main cycle of this process—known as **preces-sion**—takes about 26,000 years, and it shifts the dates of perihelion and aphelion forward by about one day every 70 years. Thus, in about 13,000 years, Earth will be closest to the sun in July instead of January. This will intensify the seasonal changes in solar energy across the Northern Hemisphere and weaken them in the south. The stronger summer sun is likely to bolster the African and Asian monsoons, as was the case around 10,000 years ago (see p. 78).

Precession (rotation of the axis itself): 26,000 years

0°

Rotation of Earth on its axis: 24 hours

influence, since they add perhaps 0.1–0.3 metric gigatons of carbon to the atmosphere per year on average. That's a substantial amount, though it's less than 1% of what humans now add through fossil fuel burning. At present, it appears that the volcanic addition is balanced by other natural forces and feedbacks that remove carbon at roughly the same pace. However, it's possible that a frenzy of volcanism early in Earth's history pumped the atmosphere's greenhouse gas levels far above anything that's occurred since. Later on, undersea eruptions associated with the spreading of continental plates may have been the main source of the carbon dioxide that led to Earth's remarkable warmth from about 250 to 50 million years ago (see p. 274).

✳ **Celestial bodies** such as **asteroids** make cameo appearances in Earth's climate saga. Most are minor. Only the very rarest of giant asteroids, which typically arrive millions of years apart, can produce climatic effects comparable to those from a major volcanic eruption. As with volcanoes, a huge asteroid might cool climate for a year or two by blasting dust and soot into the stratosphere. However, a big enough asteroid could also produce an explosive wave of heat as it enters the atmosphere, enough to set much of Earth's vegetation ablaze. Such fires would send enough carbon dioxide into the atmosphere to cause a substantial global warming that could last centuries, far longer than the brief post-collision cooling. This lurid scenario apparently played out around 65 million years ago, when an enormous asteroid struck Earth. The most likely point of impact is an area on the north end of Mexico's Yucatán Peninsula now covered by a giant crater of pulverized rock. Scientists combing the globe have found widespread deposits of iridium and other substances that are far more prevalent in asteroids than on Earth, all apparently deposited at the same time. Also dating from the same period are clumps of charcoal-like carbon, a hint that widespread fires occurred. More than half of all species are thought to have perished in this cataclysm and its aftermath, including most dinosaur species. (A few that survived may be the predecessors of modern-day birds and crocodiles.)

How scientists read the clues

The technology used to analyze materials and glean climate clues from them has advanced at a phenomenal rate in recent decades. The traces of past climates can show up in animal, vegetable, and mineral forms, providing a wealth of evidence for paleoclimatic detective work. Below are the four main types of paleoclimate **proxy data**—records that can be used to infer atmospheric properties such as temperature or precipitation:

* **Biological:** tree rings; pollen; corals; and fossils of plants, animals, insects, and microscopic creatures such as diatoms. Their locations, concentrations, and conditions point toward the atmospheric and oceanic conditions under which various life forms prospered. Many biological materials can be dated using radioactive decay rates or the presence of stable isotopes (more on these later). The chemical makeup of these samples can also yield important information.

* **Cryological:** cylindrical samples called ice cores, collected by drilling through dense, deep ice sheets or glaciers where air bubbles and dust are trapped within yearly layers of snowfall. The bubbles preserve the greenhouse gases and other atmospheric constituents present at a given time, and the ice cores can be analyzed through radiometric dating (see p. 266). The dust layers and the character of the ice itself also reveal climate processes at work.

* **Geological:** rock, sand dunes, ocean sediments, glacial debris, stalagmites, and other materials sampled from sea and lake beds and dry land. Again, radioactive decay within these materials helps establish their age. Volcanic rock can also be dated by the traces of Earth's geomagnetic field. The location and condition of telltale geological features helps establish when sea levels rose and fell or when glaciers scoured the land surface. Many ocean sediments are laden with the remnants of shells built by creatures the size of a grain of sand, including bottom-dwelling **foraminifera** as well as **plankton** that prefer near-surface waters. The prevalence and type of these shells can reveal the amount of carbon and other elements that entered the sea at various points.

* **Historical:** written records of atmospheric conditions and biological phenomena tied to climate. The latter might include the timing of spring blooms for a given type of tree or the poleward extent of a particular species of insect. In some areas, including England and Japan, such records go back hundreds of years with almost daily precision.

Establishing dates for climate proxies can be a challenge. Trees are among the easiest: each ring indicates a year's worth of growth, and its width hints at the state of that year's climate. For trees that are still alive, or ice cores drawn from the surface downward, it's a simple matter to count backward from the present and establish dates in the past, using multiple samples to

What trees tell us about climate

It was an astronomer—who happened to be looking down instead of up—who launched the science of **dendrochronology**. In the 1920s, Andrew Ellicott Douglass pondered the rings in tree stumps he encountered while hiking through the forests of northern Arizona. Douglass discovered that the width of the tree rings in this arid area was correlated with the precipitation amounts that occurred during the trees' lifetimes. He went on to found the Laboratory of Tree-Ring Research at the University of Arizona in 1937. Among his noteworthy accomplishments was developing a method using tree rings to determine the age of wood beams in prehistoric structures across the U.S. Southwest.

From its roots in Arizona and elsewhere, dendrochronology has branched out across the world, with past climates now profiled at thousands of sites. To create a typical dataset, the ring patterns obtained from several trees in the same area are matched, cross-dated, and analyzed to produce an annual, localized set of tree-ring growth indices known as a site chronology. For some types of living trees, such as bristlecone pines, these chronologies can go back thousands of years. (Tree lovers, take note: the slender cores can be extracted from live trees without harming them.) If a series of progressively older dead trees can be found whose lifespans overlap with living trees, the sequence can be extended back even farther. In Germany, the University of Hohenheim has created a 12,480-year record using oak and pine trees.

It takes far more than counting in order to use tree rings as guides to past climates. Each site has a unique blend of characteristics—temperature, precipitation, soil conditions, and the like—that affect how trees grow. For example, trees at arid forest borders are limited by moisture availability, so the variations in their rings will reflect ups and downs in precipitation. It's also possible to deduce catastrophic events, such as when a tree is shunted by an advancing glacier or partially knocked over by a storm. One clue is the asymmetric ring pattern produced as the tree attempts to grow upright after the injury.

ensure accuracy. For many other proxies, such as fossilized trees, you also need a benchmark, something you can use to establish the time period in which the proxy was created. A typical starting point is **radiometric dating**, which revolutionized the paleoclimatic dating game in the 1950s.

Radiometric techniques rely on the fact that elements such as **carbon** and **uranium** exist in several different forms, or **isotopes**, each with a

An ancient bristlecone pine from the White Mountains of California. (Gnarly/Wikimedia Commons)

Some of the most dramatic findings of dendrochronology relate to the region where the discipline was born. Connie Woodhouse of NOAA went to the Arizona lab in the early 1990s and later made headlines when she and colleague Jonathan Overpeck compiled evidence, based on tree rings and other proxy records, of what Overpeck dubbed "megadroughts." These persistent dry spells, lasting several decades each, occurred across large portions of the central and western United States prior to 1600. Megadroughts can affect the course of entire civilizations (see p. 286), even though there may be occasional years of near-normal rain amid the dry years.

different number of neutrons. Some of these isotopes are unstable, and they decay over time at measurable and predictable rates. For instance, neutrons hitting the upper atmosphere are continually converting nitrogen into an unstable carbon isotope known as 14C. This sifts down into Earth's atmosphere and gets absorbed by living things, decaying over time as new 14C isotopes are created higher up. Given a certain amount of 14C, half of it will have decayed back into nitrogen after about 5700 years (its **half-life**). Knowing this, scientists can measure how much 14C has decayed in a given substance and use that information to determine how old the substance is. Although these unstable isotopes are rare compared to stable carbon atoms, they're well distributed throughout Earth and its ecosystems. This allows geochemists to date a wide variety of substances, from carbon dioxide in an ice-core air bubble to the shells of marine creatures. One limit to using 14C is its relatively short half-life. Because there's very little carbon left to measure after a few half-life cycles, most carbon dating only goes back about 50,000 years. For older substances, **mass spectrometry** dating is used. Uranium is the material of choice for this, because it decays into lead at a very slow rate. Uranium's half-life is about 700 million years for the isotope $_{235}U$ and an astounding 4.5 billion years—the age of Earth itself—for $_{238}U$.

As with any measurement technique, errors can creep into radiometric dating. This happens mainly when a substance is contaminated by other materials of a different age through chemical reactions, erosion, or some other process. Careful cross-calibration with similar materials from other locations can help alleviate this problem.

Isotopes don't have to be unstable to be useful in dating. Stable isotopes provide another window on the climate's past. The most extensively studied is $_{18}O$, a form of oxygen. Because $_{18}O$ is heavier than standard oxygen, it condenses more readily and evaporates more slowly. The difference between the two rates depends on temperature, ocean salinity, and other factors. For example, coral reefs hold differing amounts of $_{18}O$ depending on the sea surface temperature present when they formed. In ice cores (and other places where ancient water is trapped) the relative amount of $_{18}O$ can help reveal the temperature at which the moisture condensed. This can help delineate annual layers in the ice as well as reflecting longer-term temperature trends.

Computer models are another important part of the paleoclimate toolbox. In theory, any model that simulates future climate ought to be able to reproduce the past as well, given enough information to start with. Indeed, many of the most sophisticated global models designed to project future climate are tested by seeing how well they reproduce the last century. Another useful test is to see how well the models handle modern-day seasonal transitions: in midlatitudes, at least, going from summer to winter can serve as a microcosm of a long-term overall climate shift. For prehistoric times, there's far less data to check a model against, but even in these cases, simple models can be helpful in assessing what climates might be plausible.

The really big picture: From Earth's origins to the expansion of life

It takes quite a mental stretch to fully comprehend the total lifespan of Earth's atmosphere—unless you're a biblical literalist, that is (see p. 270). Most of our planet's history unfolded long before the appearance of life forms that we're now familiar with. To get a sense of this, picture Earth's 4.5-billion-year lifespan as if it were compressed into a mere 24 hours. In this imaginary day, dinosaurs wouldn't come on the scene until after 10:00 p.m. Even more impressive, the period since the last ice age ended—the time when most human civilizations emerged—would occupy only the last fifth of a second before midnight. Deducing the complete history of the atmosphere is thus a major challenge.

Earth itself offers few clues to help us figure out what climate was up to during the planet's first two billion years or so. In fact, a good deal of our knowledge about Earth's earliest climate comes from space. By studying the behavior of stars similar to the sun, astronomers have deduced that the solar furnace (a continuous nuclear reaction that forms helium out of hydrogen) took a long time to reach its modern-day strength. It appears the sun's luminosity started out only about 70%–75% as strong as it is now, gradually increasing ever since. With the sun relatively weak for so long, you might envision early Earth as a frozen, lifeless wasteland. Amazingly, this wasn't the case. Fossils of tiny bacteria show that life existed as far back as 3.5 billion years. Rocks from around that time show the telltale signs of water-driven erosion, another sign that Earth wasn't encased in ice. How did Earth stay warm enough for life during those early years? That's the crux of what paleoclimatologists call the **faint young sun paradox**, discovered by astronomers Carl Sagan and George Mullen in 1972.

Explanations for the paradox have been floated ever since. In 2010, Danish scientist Minik Rosing and colleagues proposed that widespread oceans and relatively thin clouds allowed the early Earth to absorb much more heat than it would today, but greenhouse gases are the most commonly proposed solution to the paradox. Put enough of these gases in the air, and Earth can stay warm enough to sustain life even with a weaker sun. The amount of carbon now stored in fossil fuels and in the oceans is more than enough to have provided such a greenhouse boost if it were in the air during the weak-sun period. Such a boost might not have come from carbon dioxide alone, though, as recent geochemical analysis of ancient soil suggests that there was far too little CO_2 in the air to do the trick by itself. Sagan and Mullen suggested that ammonia might be a lead player, but it doesn't hold up well under the ultraviolet radiation of the sun. Methane

Earth's age: 4.5 billion years or 6000 years?

Virtually all physical scientists accept the basic chronology of Earth's history summarized in this chapter. However, millions of Americans beg to differ. These Christian fundamentalists generally take the Bible at its word: that Earth was created in six 24-hour days about 6000 years ago. A cottage industry of experts, including some Ph.D. scientists, promulgates this view through such enterprises as the Institute for Creation Research. They often employ the Bible's story of a giant flood to explain the formation of the Grand Canyon and other geological features. In recent years, many U.S. school boards have pushed for the teaching of intelligent design (ID), a somewhat watered-down version of creation science that seeks evidence for the existence of a creator without specifying whom that might be. Most of the ID movement's energy is focused on biological rather than climate science, although its proponents seem to be philosophically inclined toward skepticism about human-induced global warming. (That's not the case for all U.S. evangelicals—see p. 340.) Public opinion polls suggest a cognitive split among many Americans that allows them to accept Earth science that contradicts the Bible even as they draw the line at the more discomforting idea of human evolution. A report by the U.S. National Science Foundation shows that more than 80% of Americans surveyed in 2010 agreed with the statement, "The continents have been moving their location for millions of years and will continue to move." However, U.S. surveys have also shown consistently that almost half of Americans believe God created humans in the last few thousand years.

is now becoming a key piece of the picture. In a 2013 study published in *Astrobiology*, Eric Wolf and Brian Toon (University of Colorado Boulder) simulated Earth's early atmosphere with a relatively complex three-dimensional model. By setting carbon dioxide at about 50 times its current concentration, and methane at about 1000 times the values observed today, Wolf and Toon obtained global temperatures consistent with geological indicators suggesting that the early Earth atmosphere was at least as warm as today's.

Volcanoes can inject huge amounts of greenhouse gas into the atmosphere in a short amount of time, so the intense volcanism of Earth's earliest days might have added enough to keep the planet from freezing. However, that begs another question: how did the greenhouse gas levels gradually decrease over billions of years at just the right tempo to keep Earth warm, but not too warm? It's possible, but unlikely, that volcanic activity diminished at exactly the pace needed. Another possibility is that some other climate factor kicked into gear to keep greenhouse gases within a range tolerable for life on Earth.

One candidate is the steady expansion and elevation of continents during much of Earth's history. **Land masses** reduce the amount of carbon dioxide in the air through a multimillion-year process known as **chemical weathering**, which occurs when rain or snow fall on rocks that contain silicate minerals. The moisture and silicates react with carbon dioxide, pulling the CO_2 out of the air. Carbon and water then flow seaward in various forms, and most of the carbon ends up stored in ocean sediments that gradually harden into rocks. Over billions of years, this process has been responsible for some 80% of the carbon now stored underground. (Rain and snow also draw CO_2 out of the air when they leach limestone, but in that case, the carbon heads seaward in a form used by marine creatures to build shells, a process that returns the carbon to the atmosphere.) Chemical weathering is such a powerful process that the growth of the Himalayas over the last 55 million years might have helped to pitch Earth into its recent series of ice ages, as we'll see later.

Apart from these vast changes, chemical weathering can also influence the atmosphere through a negative feedback process, one that tends to counterbalance climate change in either direction. In general, the warmer it gets, the more rapid chemical weathering becomes and the more carbon dioxide is pulled from the air. This is in part because the chemical reactions behind weathering operate more rapidly when it's warmer. Another reason is that warmer oceans send more water vapor into the air, and more rain and snow falls onto continents to stimulate weathering. Thus, the warmer the air gets, the more carbon dioxide is drawn out of the atmosphere through weathering, and that should help produce a cooling trend. Conversely, if ice starts to overspread the land, weathering should decrease and carbon dioxide levels ought to increase, thus working against the glaciation.

Earth's profusion of **plant life**, which began around 400 million years ago, serves as another brake on greenhouse gases. Plants and trees take in carbon dioxide through photosynthesis and release it when they die. Generally, plant growth increases on a warmer, wetter Earth—so the hotter it gets, the more plants there are to help cool things down. Indeed, the concept of Gaia positions life itself as the main negative or balancing feedback on the entire Earth system (see box above).

Of course, there are also many **positive feedbacks** between CO_2 and climate, which tend to amplify rather than dampen a change. For instance, as carbon dioxide increases and warms the planet, more water vapor—

Gaia and global warming

Though often associated with New Age thinking, the Gaia hypothesis (the idea that Earth and its life forms act as a self-regulating entity) isn't necessarily warm and fuzzy. The name itself, a reference to the Greek goddess of Earth, was suggested in 1969 by William Golding, author of the dystopian novel *Lord of the Flies*. The father of Gaia theory, scientist **James Lovelock**, has been profoundly concerned about the direction in which global warming—or as he calls it, "global heating"—is pushing the Earth system. His 2006 book, *The Revenge of Gaia*, warned that "we live on a live planet that can respond to the changes we make, either by cancelling the changes or by cancelling us." Even bleaker was his 2009 follow-up, *The Vanishing Face of Gaia*. Billed as his "final warning" to humanity, it raised the notion that humans might render themselves extinct via climate change. "We do not seem to have the slightest understanding of the seriousness of our plight," he says. More recently, Lovelock pulled back a bit, calling some of his previous writings "alarmist" in a 2012 MSNBC interview. Yet he remained concerned: Lovelock told the United Kingdom's *The Guardian* newspaper in 2012 that he believes "the climate situation is more complex than we at present are capable of handling, or possibly even in the future."

Lovelock was inspired in the 1960s by the contrast between the lack of evidence for life on Mars and the fact that life evolved and prospered on Earth despite the slow ramp-up of solar energy (see "The really big picture"). "Together, these thoughts led me to the hypothesis that living organisms regulate the climate and the chemistry of the atmosphere in their own interest," wrote Lovelock. He sees Gaia as a single, self-regulating entity that encompasses Earth's living things and the chemical and physical backdrop that sustains them. Though Lovelock doesn't view Gaia as an evolving organism in classical Darwinian terms, he does see the web of interacting feedbacks among greenhouse gases, vegetation, and creatures on land and in the oceans as a system that preserves and perpetuates the conditions that foster life.

Like any organism, Gaia can be pushed to its limits, says Lovelock, and warming the atmosphere is asking for such trouble. If Gaia could express a preference, Lovelock believes that "it would be for the cold of an ice age, not for today's com-

itself a greenhouse gas—should evaporate into the atmosphere and further increase the warmth. Similarly, oceans absorb less carbon dioxide the warmer they get, which leaves more CO_2 in the air to stimulate further warming. It's the balance between negative and positive feedbacks that

parative warmth." One example is that both drought and flood are a greater risk in warm climates than in cool ones, because water tends to evaporate more quickly the hotter it gets. Another example is that oceans warmer than 10°C (50°F) become stratified such that nutrients are quickly depleted in the warm sur-

James Lovelock. (Bruno Comby/Wikimedia Commons)

face layer. Lovelock believes that greenhouse gases could trigger an irreversible cascade of feedbacks that profoundly disrupt the self-regulating processes of Gaia. As for what to do about all this, Lovelock doesn't toe a standard environmentalist line. He's voiced support for increased use of nuclear power and natural gas, while criticizing the idea that wind and solar power alone can save the day.

It's been a long slog for Lovelock to convince disciplinary scientists of the merit in his all-embracing theory. Partial vindication came in 2001 when a group of leading Earth scientists issued the Amsterdam Declaration on Global Change. It asserts that "the Earth system behaves as a single, self-regulating system comprised of physical, chemical, biological and human components." Lovelock goes further by insisting this system has a goal: "to keep the Earth habitable for whatever are its inhabitants." Some researchers remain more comfortable with the idea that ecosystems and the environment coevolve without a particular goal. Among them are earth scientist Toby Tyrell (University of Southampton), author of the detailed 2013 analysis *On Gaia*. Tyrell finds the Gaia hypothesis fascinating yet ultimately flawed, but he adds that "we can at the same time appreciate Lovelock's originality and breadth of vision."

generally determines which way climate will turn at any given point. A big concern about the twenty-first century is that positive feedbacks, such as those now taking place in the Arctic, may overwhelm negative feedbacks, some of which (such as chemical weathering) take much longer to play out.

Whether they're looking a thousand or a billion years into the past, paleoclimatologists find that greenhouse gas concentrations are intimately linked with the rise and fall of global temperature. Ice cores and other records bear out this tight linkage over the last million years, and there's no reason to think it should be absent earlier on. There's a subtlety in this relationship, though—one often pointed out by global warming skeptics. Because the CO_2 changes are so closely coupled in time to the growth and decay of ice sheets, it's hard to tell exactly which came first and when. For example, there appear to be periods in which undersea volcanism produced vast amounts of CO_2, after which temperatures rose dramatically, and other periods where orbital cycles triggered ice-sheet growth and colder temperatures, with the process greatly enhanced by CO_2 decreases. Paleoclimatologists are still sorting out these chains of events. It appears that orbital changes play a major role in kicking off and ending ice ages, but during the growth and decay stages, there's a web of positive feedbacks at work among CO_2, ice, and other climate elements. For instance, cooler oceans absorb more carbon dioxide, which acts to support further cooling.

The important thing to keep in mind is this: CO_2 is a critical part of both global warming and cooling, whether the climate shift is kicked off by some other factor or by CO_2 itself—and it's the latter situation we find ourselves in now.

Hot in here: Earth's warm periods

What will the world look like if the more dire projections of climate change by the year 2100 come to pass? Computer models and basic physical theory tell us a lot about what to expect, but we can also gaze back into climate history for clues.

A good place to start is the **Mesozoic era**. It began after the world's greatest extinction episode, which occurred about 250 million years ago, and ended with another vast extinction 65 million years ago. In between are three periods well known to dinosaur lovers: the Triassic, Jurassic, and Cretaceous. All three intervals were notably warm, with no signs of any major glaciations or even much ice at all. This is in part because Earth's land masses were located away from the North Pole, mostly clustered in the warm supercontinent of Pangaea. However, parts of Pangaea extended to the South Pole, and the warmth continued even as Pangaea broke up and the Northern Hemisphere continents shifted closer to their present-day locations.

Two of the warmest intervals were the mid- to late Cretaceous and the early Eocene (part of the Tertiary epoch), which fall on either side of the asteroid-driven extinction that occurred 65 million years ago. In both periods, average global readings soared to 5°C (9°F) or more above present values. That's only a little beyond the IPCC's latest high-end projections of global temperature for the year 2100. Proxy data for these past periods indicate that carbon dioxide was several times more prevalent in the atmosphere than it is today. This enhances scientists' confidence that the extra CO_2 stoked the ancient warmth and that the greenhouse gases we're now adding could do something similar in our future.

Where did the extra CO_2 come from? The most likely candidate is an active period of **undersea volcanism** (see graphic below). Earth's tectonic plates clash most visibly along the world's volcanic coastlines, such as the Pacific's Ring of Fire. However, there's additional volcanic activity far beneath the oceans. Most major basins have a mid-ocean ridge, where tectonic plates separate and vast amounts of magma and carbon dioxide surge upward to fill the void. The magma eventually solidifies to form fresh crust, slowly elevating the center of the ridge as the older material spreads outward. It appears that, on average, the plates were separating up to 50% more quickly about 100 million years ago than they are now. This would have allowed more CO_2 to bubble upward and would have pushed the plates toward continents more vigorously, stimulating more activity (and more CO_2 emissions) at coastal volcanoes.

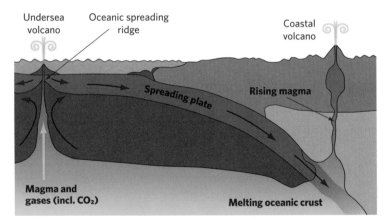

The process involved in undersea volcanism and the surface and atmospheric effects.

In a classic 1920 poem, Robert Frost observed that some people expect the world to go out in fire, while others think the end will be an icy one. Our planet edged toward those extremes at several different points in its history. Two examples brought vastly different consequences for ecosystems.

One of the coldest periods in early Earth history ran from about 750 to 580 million years ago (pictured). There's evidence that glaciers scoured most of Earth's large land masses, including some within 10° of the equator. The implication is that much or all of the planet was covered with ice for millions of years—a scenario dubbed Snowball Earth in 1992 by geologist Joseph Kirschvink of the California Institute of Technology. Since Earth's land was concentrated in the tropics, where precipitation is heaviest, chemical weathering (see p. 271) could have pulled large amounts of carbon dioxide from the air, cooling the global atmosphere. If the planet actually did freeze over entirely, the powerful positive feedback of a "white Earth" could have reflected most of the incoming sunlight and helped preserve the ice. How such a state would have ended isn't known, but a slow accumulation of carbon dioxide from volcanoes—on top of a cold-induced slowdown in chemical weathering—seems the most likely possibility. Whether or not ice actually covered the entire Earth, this period of widespread glaciation was followed by a sudden profusion of multicellular organisms. Some theorists speculate that a Snowball Earth and a subsequent

Another source of greenhouse gas might have provided even more of an influx at key times. Hundreds of billions of tons of carbon are believed to be locked in cold ocean seabeds in the form of **methane clathrates**, ice crystals that hold methane in their molecular structure (also known as methane hydrates). Methane clathrates remain stable if conditions are cold enough and/or if enough pressure is exerted on them (as is the case at the bottom of the sea). But a change in ocean temperature or circulation during a warm era might be enough to destabilize the methane and send it pouring into the atmosphere. This is a long-term concern for our future climate (see chapter 6), and it may have played a role in at least two massive warmings during already warm times. One was the temperature spike and mass extinction that kicked off the Mesozoic era 248 million years ago. The other is a distinct blip early in the already toasty Eocene epoch called the Paleocene–Eocene Thermal Maximum (PETM), which occurred about 55 million years ago. During the PETM, high-latitude air and ocean tem-

warm-up could have prodded the rapid evolution of primitive life into more complex forms.

At the other end of the spectrum, an especially intense warming around 250 million years ago spelled doom for most of the life forms present at the time. The Permian–Triassic extinction drew the curtain on more than 90% of marine species and more than two-thirds of land-based creatures. The die-off unfolded quickly in geological terms, over fewer than a million years. Greenhouse gases likely soared to many times their present-day amounts, and high-latitude oceans warmed to as much as 8°C (14°F) above present-day readings. The warmer oceans likely enhanced the separation of surface waters from cooler deep oceans, reducing the usual mixing between the layers that distributes oxygen and nourishes many marine organisms. Researchers haven't settled on a single cause for the extinction. A modeling study led by Jeffrey Kiehl (NCAR) suggested that enhanced volcanic activity over almost a million years could produce enough atmospheric warming through added carbon dioxide to slow the ocean currents that bring oxygen to deep-water marine life. Other possibilities include a giant meteorite (although there's no sign of an impact crater dating from that era), an outpouring of the greenhouse gas methane in the form of methane hydrates released from the ocean floor, a massive release of hydrogen sulfide produced as a result of oxygen depletion in the oceans, or some blend of these and/or other factors.

peratures soared by as much as 7°C (12.6°F) over a few thousand years and remained high for almost 100,000 years. Isotope studies by NASA indicate that a vast amount of greenhouse gas was pumped into the atmosphere during the PETM, though over a longer period and perhaps at a slower rate than our current emissions. Among the possible sources is a burst of volcanism in the North Atlantic, as suggested by recently analyzed igneous rock formations off the coast of Norway.

One thing that's fairly clear about past warm periods is that mid- and high-latitude areas were downright balmy as compared to today. Coral reefs grew as far poleward as 40°, close to the latitudes of present-day Philadelphia and Melbourne. Crocodiles, dinosaurs, and leafy trees flourished in polar regions, while other areas featured vast deserts covered with sand dunes. A 2006 study of sediments beneath the Arctic Ocean found that sea surface temperatures (SSTs) during the PETM soared to as high as 23°C (73°F), which is far beyond earlier estimates. In the tropics, SSTs may not have been much warmer than they are now, though some intriguing evidence of readings at or above 33°C (91°F) has been collected off the coasts of present-day South America and East Africa.

The oceans were at a high-water mark during these warm times. For millions of years, **sea level** was more than 200 m (660 ft) higher than it is now. This put vast stretches of the present-day continents under water, including the central third of North America and all of eastern Europe from Poland southward. The oceans were so high and vast for several reasons. One is that there wasn't any water locked up in long-term ice sheets. Another is that water, like any other fluid, expands as it warms up, just as the oceans are doing now (albeit on a much more modest scale). A third reason is the shape of the oceans themselves. The faster separation of tectonic plates noted above would have produced larger mid-ocean ridges as new crustal material surged upward. This, in turn, would have raised the height of the seafloor and the resulting sea level by surprisingly large amounts—perhaps enough to account for most of the difference between the Triassic–Eocene sea levels and today's.

From greenhouse to icehouse: The planet cools down

The most recent 1% of our atmosphere's history is mainly a story of cooling. Starting in the mid-Eocene, around 50–55 million years ago, sea levels began to drop, continents shifted ever closer to their current locations, and global temperatures began to fall. The cooling appears to have temporarily reversed between about 15 and 25 million years ago, then resumed with gusto. This long process wasn't linear—there were sharp warmings and coolings in between, each lasting many thousands of years—but overall, Earth's temperature took a tumble of perhaps 4°C (7.2°F) or more. This eventually pushed the planet into the glacial–interglacial sequence that began fewer than three million years ago and continues today.

What caused the cooldown? Once again, it appears that carbon dioxide was the most direct culprit. CO_2 in the atmosphere went from several times its modern-day amounts in the mid-Eocene to barely half of its present-day concentration during the last few ice ages. The question of what drove down CO_2 levels is far more challenging. Mid-ocean ridges were separating slower when the cooldown began, a sign that volcanic activity was pumping less CO_2 into the atmosphere. The biggest decreases in volcanism apparently occurred prior to 40 million years ago. That would correspond to the first pulse of global cooling, but the spreading rate had increased again by the time the second pulse began, leaving that one unexplained.

In their search for a solution to this riddle, one group of U.S. scientists cast their eyes toward the **Tibetan Plateau**. This gigantic feature began to take shape around 55 million years ago as the Indian subcontinent joined Asia, pushing the Himalayas and the plateau upward. The pace quickened about 30 million years ago, and today most of the

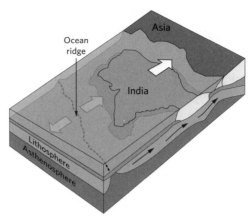

The Indian subcontinent about to crash into Asia 60 million years ago.

Alaska-sized Tibetan Plateau sits at elevations greater than 3700 m (12,000 ft). It's bordered by mountains that soar more than twice as high.

In the late 1980s, Maureen Raymo (Woods Hole Oceanographic Institution), Flip Froelich (now at Florida State University) and William Ruddiman (University of Virginia) proposed that the intense chemical weathering produced by the Tibetan Plateau could have pulled enough carbon dioxide from the air to trigger the global temperature drop that led to the most recent ice ages. Normally, a drop in global temperatures would be expected to act as a negative feedback, reducing precipitation and slowing down weathering. But in this case, we're faced with a change in the playing field (Earth's geography), and that could control the rate of weathering entirely on its own, overpowering any feedbacks. If ever there was a geological event conducive to weathering, it's the uplift of the Tibetan Plateau—perhaps the most massive above-ground feature in Earth's history. In the absence of enough data to confirm or disprove Raymo and colleagues, the debate over the advent of Earth's ice ages goes on.

There were other tectonic developments of interest during the long cooldown. **Antarctica** separated from South America between 20 and 25 million years ago. With air and water now able to circle around Antarctica unimpeded, the resulting tight circulation might have fostered the growth of the first major Antarctic ice sheet about 13 million years ago (though modeling studies have been inconclusive on this point).

Meanwhile, North and South America slowly approached each other, finally joining at the Isthmus of Panama nearly four million years ago. This

Core value: Getting information out of ice

Our understanding of the last few ice ages has been boosted immeasurably by the rich vein of data obtained through ice cores from Greenland, Antarctica, and several other glacier-studded areas. The task is laborious: each ice sheet is more than 3 km (1.9 mi) thick, but each cylindrical core is only about 10 cm (4 in.) in diameter, and each segment of the core must be drilled and removed a few meters at a time. Weather only permits a few short months of work each year. Even in midsummer, daytime temperatures can be colder than −20°C (−4°F) on top of the Greenland ice sheet; in central Antarctica, that's a mild summer day.

The first major ice cores were retrieved from both Antarctica and Greenland as part of the International Geophysical Year (IGY) of 1957–1958. In Greenland, the first cores were taken at camps far from the thickest part of the ice sheet. As interest in climate change intensified and researchers clamored for more of the ice, two parallel projects set out to gather definitive cores from the top of the sheet. By the early 1990s, these U.S. and European teams—separated by only about 32 km (20 mi)—had drilled from the summit of Greenland's ice sheet to bedrock, a depth of just over 3 km (9900 ft). The two cores they obtained each spanned a little more than 100,000 years of climate. In 2010 a new coring project hit bedrock in northern Greenland; it has led to data extending back almost 130,000 years, enough to capture the warmth of the last interglacial period.

Meanwhile, Russian scientists at Antarctica's Vostok Station, which was founded during the IGY, were engaged in their own decades-long drilling effort. By 1995, they'd reached a depth of around 3600 m (11,800 ft) before they halted to avoid contaminating a vast lake beneath the ice. The Vostok core is only about 20% longer than those from Greenland. However, since the highest reaches of the Antarctic ice sheet get so much less snowfall per year than Greenland's summit, each meter of the Vostok core details a much longer stretch of climate. The total Vostok record covers about 440,000 years.

closed the long-standing oceanic connection through which warm, salty water had flowed beneath trade winds from the tropical Atlantic into the Pacific. In a major readjustment, the Atlantic shifted into a mode much like the one we see today, with the tropical waters flowing northward along an intensified Gulf Stream and on into the far North Atlantic. This influx of warm, saline water into the Arctic would have cut down on sea-ice formation. Ironically, it could have also furnished the moisture needed to build ice sheets over land and thus hastened glaciation.

Scientists take ice core samples in cold, windy, and forbidding environments. (Ted Scambos and Rob Bauer, U.S. National Snow and Ice Data Center)

One of the most striking findings from both poles is that carbon dioxide and methane, two key greenhouse gases, rose and fell in near lockstep with the comings and goings of ice ages. Typically, the concentrations dropped by 20%–50% during a glaciation, then quickly recovered afterward.

The new champion of ice archives is the core completed in 2004 by the EU-sponsored European Project for Ice Coring in Antarctica (EPICA). Retrieved at a site atop Dome C, about 500 km (300 mi) from Vostok, the EPICA core has already yielded climate analyses extending back 800,000 years. Researchers with EPICA hope to use the core to look back nearly a million years—around the time when ice ages mysteriously transitioned to much longer intervals than before.

Thanks to some combination of the factors above—plus, perhaps, others yet to be discovered—Earth inexorably slid into what some experts call a "deteriorating" climate. Antarctic ice grew in a start-and-stop fashion. Glaciers began to appear about seven million years ago in the Andes, Alaska, and Greenland. Finally, **ice sheets** began spreading across North America and Eurasia about 2.7 million years ago. Since then, more than two dozen glaciations have ground their way across much of present-day Canada, northern Europe, and northwest Russia. In between each glaciation was a shorter, milder period, with ice extent and temperatures closer to those at present and even warmer in some cases.

At first, these **interglacial** periods occurred in between ice ages about every 41,000 years. This corresponds remarkably well to the 41,000-year cycle in the tilt of Earth's axis [see sidebar, "Round and round we go (and so does our climate)"]. When Earth's tilt is maximized, the extra solar energy received at high latitudes in summer melts ice sheets more easily (even though winters are getting less sun than usual). When the tilt is lower, summers are cooler, and ice sheets get a running start with unmelted snow from the previous year.

About a million years ago, however, the 41,000-year rhythm of ice ages ended. The ice ages got stronger and began to last much longer—about 100,000 years each, with interglacial periods on the order of 10,000–15,000 years long. What was behind this shift isn't yet clear. Earth's orbit also has a 100,000-year cycle in its off-centeredness, or eccentricity [see sidebar "Round and round we go (and so does our climate")], but most paleoclimatologists don't believe that this relatively weak cycle was enough in itself to wrench a change in the ice-age frequency. However, it could have interacted with some other cyclic process to provoke the timing.

Along with what produced the current tempo of ice ages, scientists are also keenly interested in what's been maintaining it. One of the more exotic hypotheses is that Earth's orbit may be passing through a disk of interplanetary dust about every 100,000 years. Another intriguing idea, advanced by Peter Huybers (Harvard University), is that the 41,000-year cycle in Earth's tilt is still the main driver but that ice sheets respond in an alternating fashion every second or third peak (about every 80,000 or 120,000 years), which would lead to an average frequency of about every 100,000 years. And a 2013 study in *Nature* led by Ayako Abe-Ouchi (University of Tokyo) elaborated on a long-standing hypothesis that the vast ice sheets may overextend themselves in North America, eventually becoming so large that a cyclic solar peak can melt more of the ice's surface than snowfall can replenish. There's much still to be learned, according to Huybers: "I think ice ages are really the outstanding mystery in Earth science," he says.

The period just before the last ice age began, from about 130,000 to 115,000 years ago, provides a sobering illustration of what climate change could bring to our planet over the next few centuries. That interglacial kicked off with a sharp warming most likely driven by orbital cycles. The 2013 IPCC report concluded that sea level during the interglacial likely topped out between 5 and 10 m (16–33 ft) above today's global average, even though greenhouse gases were less abundant than today. Computer modeling by Jonathan Overpeck (University of Arizona) and Bette Otto-Bliesner (NCAR) showed that major ice-sheet melting at both poles may have fed the sea level rise. The same modeling found that Arctic summer temperatures by 2100 will be close to those achieved in the last interglacial warming. Although it could take decades or even centuries for that warming to translate into a 6-m (20-ft) sea level rise—which would swamp major coastal areas around the world—Overpeck worries that the process of ice-sheet melt could prove irreversible sooner than we think.

Emerging from the ice

By the time of its peak about 20,000 years ago, the last ice age had carved out a very different-looking world than the one we know now. North America and Europe were half encased in ice, with global temperatures anywhere from 3° to 8°C (5°–14°F) colder than today's averages. South of these ice sheets, it was largely dry and windy, especially over continental areas. Tundra covered most of southern Europe, and spruce trees characteristic of high latitudes extended into the mid-Atlantic area of the present-day United States. One place where rainfall exceeded today's norm is the Great Basin: Nevada and Utah were freckled with dozens of large lakes that now exist only as salt flats. Sea levels were so low—more than 100 m (330 ft) below today's—that Australia extended northward into New Guinea, and Japan and Korea were separated only by shallow water.

The ice age may not have looked exactly like this through its entire 100,000-year lifespan. Until ice cores were plucked from Greenland and the Antarctic, many scientists had pictured ice-age climates as being relatively static. Instead, the cores revealed that Earth swung through a number of distinct warmings and coolings during the start of the last ice age, about 115,000 years ago, up to and beyond its conclusion around 15,000 years ago. As the clarity and completeness of ice-core data improved in the 1990s, the findings were accompanied by analyses of ocean sediments from several dispersed basins and stalagmites from caves in China. The new evidence made it clear that warm and cold swings during the ice age were far more widespread, and could unfold far more rapidly, than scientists had previously thought (see box opposite).

Two of the orbital **cycles** discussed in the sidebar "Round and round we go," involving Earth's tilt and its precession, synchronized to produce a strong peak in the summertime input of solar energy across the Northern Hemisphere about 15,000 years ago. Ice sheets began to retreat dramatically, but the path to a warmer climate wasn't a smooth one. As the melting

How fast can climate flip?

Evidence as far back as the 1970s hinted that climate during glacial periods might not have been as static as many believed. The pieces came together in the early 1990s, when scientists confirmed a sequence of major warmings and coolings that unfolded against the cold backdrop of the last ice age. Instead of furnishing a constant chill, it seems the 100,000-year ice age unfolded in a much more irregular fashion.

Gerald Bond and colleagues at the Lamont-Doherty Earth Observatory drew on data from sediments in the North Atlantic to trace the evolution of the ice age in eye-opening detail. The kind of upward and downward spikes they found don't seem to occur in warm regimes like our present-day, postglacial climate, but they serve as a reminder that climate can switch from one mode to another in the geologic equivalent of a heartbeat. The great ice sheets that coated much of Europe and North America never disappeared during these warmings and coolings, but conditions across many ocean and continental areas varied sharply as they came and went.

❋ **Going up.** On 25 different occasions during the last ice age, global air temperatures are believed to have climbed about half of the way back to their interglacial levels, then sank back to more typical ice-age readings. The warm-ups pushed average temperatures in Greenland up by as much as 16°C (29°F) in as little as 40 years, while the much slower return to glacial cold took about a thousand years. These warmings—which appear to have been concentrated in the Northern Hemisphere—are called Dansgaard-Oeschger (D-O) events, after Danish geophysicist Willi Dansgaard and Swiss geochemist Hans Oeschger.

intensified, water occasionally poured out of glacial lakes in spectacular bursts, some of which were large enough to influence ocean circulation and trigger rapid climate shifts.

Perhaps the most dramatic was the **Younger Dryas** period, a frigid encore that brought a return to cold nearly as extreme as the ice age itself. The period is named after a tenacious alpine flower—*Dryas octopetala*—that thrives in conditions too austere for most other plant life. The flower's pollen, preserved in Scandinavian bogs, provides evidence of extremely cold periods. The most recent (youngest) such period began nearly 13,000 years ago and persisted for some 1300 years. Ice cores from Greenland indicate that the Younger Dryas took as little as a decade both to set in and to

* **Going down.** Less frequently—at six points during the ice age—climate lurched in the other direction. Vast fields of icebergs poured from North America into the North Atlantic, disrupting the ocean circulation and cooling the climate for several hundred years. These so-called Heinrich events are named after German scientist Hartmut Heinrich, who discovered particles of Canadian soil scraped off by ice sheets and deposited in the North Atlantic by rafts of icebergs.

Together, the Heinrich and D–O events explain much of the variability in global climate that shows up in sediments and ice cores from the last 100,000 years. Scientists are still hunting for what might lie behind the timing and occurrence of both types of climate swings. For instance, each Heinrich event occurs after three to five D–O events, implying that the extended warm spells might have progressively destabilized the growing ice sheets and led to a Heinrich-style iceberg armada. Another interesting facet is that many of the D–O events are separated by around 1500 years, with a few spaced at about 3000 and 4500 years. "This suggests that the events are triggered by an underlying cycle . . . but that sometimes a beat or two is skipped," notes Stefan Rahmstorf of the Potsdam Institute for Climate Impact Research. Other paleoclimatologists believe these sub-beats may be little more than random variations. The D–O process gained public notice through the 2006 book *Unstoppable Global Warming: Every 1,500 Years*, in which skeptic-authors S. Fred Singer and Dennis Avery argue—contrary to IPCC and other consensus science—that we're now going through a natural warming event.

release its grip. During the interim, average temperatures across the British Isles sank to as low as −5°C (23°F), and snowfall increased across many mountainous parts of the world. Drought and cold swept across the Fertile Crescent of the Middle East. This may even be what prodded the ancient Natufian culture to pull back from wide-ranging hunting and gathering and shift toward what ranks as the world's earliest documented agriculture.

As for the cause of the Younger Dryas, most paleoclimatologists point to the massive draining of **Lake Agassiz**, a vast glacial remnant that covered parts of present-day Manitoba, Ontario, Minnesota, and North Dakota. Such an influx of fresh water could disrupt the thermohaline circulation that sends warm waters far into the North Atlantic (see chapter 7). However, there's spirited debate about whether the Agassiz melt flowed in the right direction to affect the North Atlantic. An even more perplexing piece of the puzzle comes from the Southern Hemisphere, where records from Antarctica and other areas show that the cooldown there began a thousand years earlier—closer to 14,000 years ago.

The Younger Dryas was followed by a much shorter and less intense cooling about 8200 years ago. Once those were out of the way, the next few thousand years were, by and large, a nourishing time for civilizations. Scientists who map out the climate for this period can draw on human history as well as other biological, physical, and chemical clues. (There's still a bit of a bias toward Europe, the Middle East, and North America, due in part to the concentrations of researchers and documented civilizations on those continents. And, as with any climate era, the proxies aren't perfect.)

Across northern Africa and southern Asia, the peak in summertime sunshine stimulated by orbital cycles during the postglacial era produced monsoons far more intense and widespread than today's. From about 14,000 to 6000 years ago, much of the Sahara, including the drought-prone Sahel, was covered in grasses and studded with lakes that played host to crocodiles and hippopotami. Traces of ancient streams have been detected by satellite beneath the sands of Sudan, while the Nile flowed at perhaps three times its present volume. Meanwhile, from about 9000 to 5000 years ago, most of North America and Europe were ice-free and summer temperatures there likely reached values similar to today, perhaps a touch warmer in some areas and at some times.

These mild times weren't destined to last forever. Starting about 5000 years ago, global temperatures began a gradual cooldown, interspersed with warmings lasting a few hundred years each. This pattern is consistent with ice-core records from toward the end of previous interglacial periods, although it's far from clear when the next ice age will actually arrive (see sidebar, "Sowing a savory climate"). One of the earliest and sharpest signs of this global cooling is the prolonged drought that crippled Near East civilizations about 4200 years ago. Orbital cycles had already weakened the summer sunlight that powered bountiful monsoons across Africa and Asia. At some point, the gradual drying killed off the grasses that covered much of the Sahara, which quickly pushed the region from savannah toward desert.

From AD 1000 to today (and beyond)

Because of its context-setting importance for what lies ahead, the last thousand years of climate have drawn intense scrutiny. The last millennium began near the peak of a 300-year stretch of widespread warmth called the **Medieval Warm Period** (also called the Medieval Climatic Optimum or Medieval Climate Anomaly), when temperatures in some parts of the world—especially North America—were close to modern-day levels. It may indeed have been an optimal climate for some Europeans, or at least a livable one, but people in parts of the Americas suffered from intense drought that ravaged entire cultures.

One example is the **Mayans**, whose empire extended from present-day Mexico across much of Central America. After nearly 2000 years of development, many Mayan cities were abandoned between about AD 750 and AD 950. Sediments in the nearby Caribbean reveal several strong multiyear droughts during this period. The Mayan collapse in eastern Mexico apparently coincided with that region's most intense and prolonged drought in over a millennium.

To the north, the region near the current Four Corners area settled by Pueblo peoples experienced periods of erratic rain and snow during the Medieval Warm Period. Tree-ring studies indicate an especially intense drought near the end of the 1200s, when the region's emblematic cliff dwellings were abandoned. Scholars in recent years have warned against pinning the Mayan and Pueblo declines solely on drought, pointing to signs of other factors such as migration, a high demand on trees and other resources, and power struggles within each culture and with neighboring peoples. However, climatic stresses can't have helped matters.

After the Medieval Warm Period faded, the long-term cooling resumed, intensifying during the 1400s and continuing into the mid-1800s. This was the Little Ice Age, which sent the global average temperature down by roughly 0.5°C (0.9°F) to its lowest century-scale levels in thousands of years. It wasn't a uniform cooling, though: some parts of the world appear to have seen little effect, while temperatures plummeted across much of North America and Eurasia. Periodic plagues and famines ravaged Europe, and glaciers descended from the Alps to engulf a number of villages. The chill wasn't uniform in time, either. As historian Brian Fagan notes, the Little Ice Age was marked by "an irregular seesaw of rapid climatic shifts, driven by complex and still little understood interactions between the atmosphere and the ocean."

One influence may have been a drop in **solar energy**. Isotopes of carbon in tree rings and beryllium in ice cores show a drop-off in solar radiation during much of the Little Ice Age. Moreover, sunspot observations that began around 1610 show a near-absence of reported sunspots between 1645 and 1715 (the so-called Maunder Minimum). However, recent studies have

Sowing a savory climate

Could we look to the arrival of a new ice age to counteract human-produced climate change? The core of the last three interglacials each lasted about 10,000 years. Given that the most recent ice age was largely finished some 10,000 years ago, you might think it's nearly time for a new glacial era to begin. However, the amount of carbon dioxide now in the air—much more than in the last million years of glacial cycles—is enough to prevent orbital cycles from triggering a new ice age for roughly 50,000 years, according to a range of model simulations discussed in the 2013 IPCC report.

If in fact the next ice age were already due, then human activity could be thanked (or blamed) for postponing its arrival. Veteran paleoclimatologist William Ruddiman argues that the development of agriculture boosted greenhouse gases and delayed a glaciation that should have already begun. Ruddiman introduced his theory with a provocative paper in 2003 entitled "The Anthropogenic Greenhouse Era Began Thousands of Years Ago." He's since elaborated it for a broader audience in the books *Plows, Plagues, and Petroleum* (2010) and *Earth Transformed* (2013).

Ruddiman's theory is based on ice-core records that show how greenhouse gases responded during previous interglacial periods. He noticed that in most interglacials, greenhouse gas levels tend to spike quickly and then gradually drop, setting the stage for the next icing. However, the most recent interglacial hasn't obeyed the pattern. After an initial spike and the beginnings of a drop, carbon dioxide amounts rose again by about 8% about 8000 years ago, then leveled off until the Industrial Revolution. Methane concentrations also increased, starting about 5000 years ago.

brought down the relative importance of this solar effect on the Little Ice Age. Debate continues, but one current view is that the solar slowdown acted mainly to help reinforce the direct effects of volcanoes, which appear to have erupted more frequently from the late 1200s onward than during the Medieval Warm Period. The 1815 eruption of Indonesia's **Mount Tambora**—the most violent ever recorded on Earth—led to a disastrously cold summer across much of the globe in 1816. That "year without a summer" brought crop failures to northern Europe as well as snows in Vermont as late as early June. (It also spawned a monster: while spending part of that dreary summer at Lake Geneva, a teenaged Mary Shelley started work on the novel that became *Frankenstein*.)

Ruddiman attributes these rises to the rapid and widespread growth of agriculture during these periods. The clearing of forests would have increased CO_2 levels, he says, and rice paddies and livestock would have emitted large quantities of methane. A comparison of ice cores from both Northern and Southern Hemispheres, as detailed in a 2013 *Science* analysis led by Logan Mitchell (Oregon State University), suggests that the methane surge is best explained by increased agriculture coupled with the effects of orbital changes on tropical wetlands, with neither factor able to fully explain the surge by itself.

To further bolster his case, Ruddiman points to the drop in agricultural activity and subsequent reforestation after plagues swept Europe in the 1300s and the Americas in the 1500s and 1600s. The timing, he points out, corresponds well with the reduction in both global temperature and carbon dioxide levels during the Little Ice Age.

Ruddiman's theory has met with enthusiastic agreement and stout resistance, both of which played out in a special 2011 issue of the journal *The Holocene*. Critics argue there wasn't enough farming thousands of years ago to influence CO_2 and methane, but Ruddiman cites new evidence that early farmers used far more land per capita. Others point out that an isotope associated with terrestrial carbon ($\partial^{13}CO_2$) has barely risen in the last 7000 years. This implies a cap on the CO_2 contribution from land processes, including agriculture. However, if huge amounts of carbon were taken up by boreal peatlands—as some analyses suggest—then other carbon could have poured out of newly cleared land, meaning that the $\partial^{13}CO_2$ limit isn't necessarily a deal breaker. Through it all, Ruddiman remains steadfast: "I really have not had a serious day of doubt."

After the mid-1800s, Earth's climate took a decided turn for the warmer. The overall climb in temperatures over the twentieth century, however, was punctuated by an apparent flattening from around the mid-1940s to the early 1980s (see graph, p. 10). At the time, it appeared to be global cooling, because the era's worldwide temperature analyses were skewed toward the Northern Hemisphere, where temperatures actually were dropping a bit. The causes of the northern cooldown are still being debated. For years it was thought to be related mainly to sun-reflecting pollution spewed out from North America and Europe in the industrial boom after World War II, though such effects are difficult to assess and quantify. Scientists have also been scrutinizing the role of the oceans. According to a pair of studies in 2008 and 2010 led by David

Painting the Little Ice Age

Some of northern Europe's greatest artists used oil and brush to set the mood that many associate with the Little Ice Age: cloudy, snowy, and dank. Pieter Bruegel the Elder may have used the frigid winter of 1565 as source material for the dull, greenish sky of *Hunters in the Snow*, part of his series of seasonal depictions. This was one of the first portrayals of a snowy landscape in European art, noted William Burroughs in the British journal *Weather*. Bruegel extended the wintry theme to other topics, including *The Adoration of the Magi in the Snow*. Many Dutch artists, notably Hendrick Avercamp, took to cold-weather depictions in the mid-1600s, another period of brutal chill across the region.

The northern Renaissance also spawned a new realism in sky portraiture. Back in the early 1400s, Flemish painter Jan van Eyck was one of the first to depict cloud types that a meteorologist today might recognize and label. Hans Neuberger quantified the treatment of clouds by U.S. and European painters in an unusual 1970 study that appeared in *Weather*. Sampling 41 museums in 9 countries, Neuberger examined more than 12,000 paintings produced between 1400 and 1967. He found that blue skies, which predominated up to 1550, gave way to low clouds in more than half of the post-1550 paintings. Neuberger didn't attempt to analyze how much of the trend was related to the Little Ice Age weather and how much to artistic fashion.

English landscape painters of the Little Ice Age held true to their island's cloudy climate. Every English sky examined by Neuberger had at least some cloudiness, and the sky was typically a pale blue at best. The English Romantic artist J. M. W. Turner specialized in foggy, misty tableaux as well as striking sunsets; the latter may have reflected the volcanic dust that added vivid hues to many sunsets in the early 1800s. Later in the century, the gigantic Krakatoa eruption of 1883 led to sunsets so striking they were noted in press reports in New York and London. According to astronomer Donald Olson of Texas State University, Krakatoa may also have inspired Edvard Munch's iconic masterpiece, *The Scream*. In describing what triggered the painting, Munch wrote of experiencing a "blood-red" sunset in present-day Oslo that resembled "a great unending scream piercing through

Thompson (Colorado State University), much of the midcentury hemispheric cooling can be traced back to two ocean-related factors, both of which pop into view after the ups and downs of El Niños, La Niñas, and volcanic eruptions are removed. One is an apparent drop in SSTs in 1945 that proved to be spurious, the result of a temporary increase in canvas-bucket measurements

Dutch painter Hendrick Avercamp (1585–1634) was a specialist in winter scenes, especially skaters on frozen rivers, canals, and flood waters. (Hendrick Avercamp [public domain], via Wikimedia Commons)

nature"—though Munch didn't give a date for this experience. Although a full decade separates the eruption from *The Scream*, Olson believes that Munch may have encountered a Krakatoa sunset and waited years to depict it.

The legendary frost fairs held on the River Thames in London during occasional freeze-ups were captured in a number of paintings, including *A Frost Fair on the Thames at Temple Stairs* (1684) by Dutch painter Abraham Hondius. However, these festivals weren't as frequent as one might assume. Outside of the especially frigid mid-1600s, the Thames froze at London only about once every 20 or 30 years from the 1400s until 1814, when the last freeze-up was recorded. Moreover, it wasn't the end of the Little Ice Age that ended the frost fairs. When London Bridge was replaced in the 1830s, it allowed the tide to sweep farther inland. This made it virtually impossible for the Thames to freeze at London, and it hasn't happened since.

at the end of World War II (see p. 243 for more on SST measurement techniques). The other is a major drop in Northern Hemisphere SSTs from 1968 to 1972, focused in the North Atlantic. This may be linked to a large influx of fresh water called the Great Salinity Anomaly that was observed around the same time, though the connections aren't yet fully mapped out.

Four ways to think of past and future climate

The answer to whether Earth is warming or cooling depends in large part on what time frame you're considering. Each graph below includes a very rough schematic of past (and plausible future) temperatures.

❋ **Over hundreds of years.** Around AD 1000, Earth was nearly as warm as it is now. After that, temperatures over much of the planet cooled by around 0.2°–0.5°C (0.4°–0.8°F), mostly during the period known as the Little Ice Age, from the 1400s to the mid-1800s. The global average has since rebounded by nearly 1°C (1.8°F) as human-produced greenhouse gases have increased. According to the 2013 IPCC report, further warming this century could run anywhere from 0.3° to 4.8°C (0.5°–8.6°F), depending on the rate of emissions. Further warming after 2100 could be substantial if greenhouse emissions continue to increase through this century, and the delayed response of ice sheets and deep oceans to warming could produce major sea level rise over several centuries.

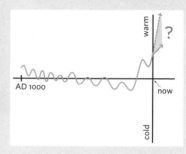

❋ **Over tens of thousands of years.** The most recent ice age began about 115,000 years ago and ended about 11,500 years ago. Then came a dramatic warm-up, which lasted until about 3000 BC. Since then, Earth's temperature has changed relatively little, with a very slight cooling interrupted by warmer periods and punctuated by the last century's sharp temperature rise. More than a thousand years from now, after humans have exhausted fossil fuels and the resulting greenhouse gases have left the atmosphere naturally (mostly through slow absorption by the

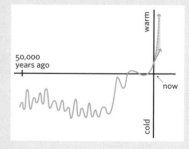

ocean), we may return to cooler times. If the length of the last interglacial is any guide, we're due for another ice age in the next few thousand years, although some scientists believe it's more likely to take 20,000 years or more. It's even possible that a mix of polar melting, land-use changes, and slowly waning greenhouse gas concentrations could postpone the next ice age for an undetermined period.

* **Over tens of millions of years.** A gradual, though sometimes erratic, cooling trend has been under way for at least 55 million years, perhaps due in part to carbon dioxide removed by weathering on top of the growing Himalayas. Northern Hemi-sphere glaciations began about 2.7 million years ago. Warm and cool periods come and go on

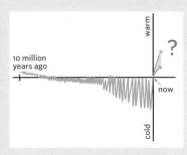

top of this overall cooling. Simple extrapolation would keep Earth on this extremely slow cooling trend, although there's no evidence to tell us how long that might last.

* **Over billions of years.** On this scale, the details are blurry, but the overall picture is clear. Changes in greenhouse gas levels have so far kept our climate relatively stable (not too hot or cold for life), even though the sun's output has risen by more than one-third since the solar system was formed around 4.6 billion years ago. The sun will con-tinue to heat up and eventually undergo changes to its size and structure, producing a climate on Earth hot enough to evaporate the oceans and make life impossi-ble. Eventually, the sun will shrink and start to fade, with the solar

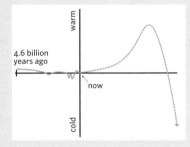

system reaching its cold "final configuration" in around 12 billion years.

Global warming resumed with gusto in the 1980s, and by the end of the twentieth century it was becoming clear that temperatures were at least on a par with those of the Medieval Warm Period. That recognition triggered one of the highest-profile climate change debates to unfold over the last few years, one that brought unprecedented attention to the fine points of paleoclimate analysis.

In the late 1990s, Michael Mann, then at the University of Virginia, joined Raymond Bradley (University of Massachusetts Amherst) and Malcolm Hughes (Laboratory of Tree-Ring Research) in the first attempt to reconstruct global temperature from the past millennium based on varied proxy data from around the world. Initial results were published in a 1998 *Nature* paper, and in a 1999 follow-up, the three scientists (often referred to by their initials, MBH) reported that the 1990s had likely been the warmest decade of the last 1000 years. Their work gained stature through its breadth and novel approach, as well as its timing. It arrived just after global temperatures hit a new high in 1998 and just as the IPCC was preparing its third major assessment. The key graphic produced by Mann and colleagues, nicknamed the "hockey stick" for its distinctive shape, was prominently featured in the IPCC's 2001 report (p.308).

Two Canadians questioned the MBH study: a former mining engineer and government analyst, Stephen McIntyre, and an economist, Ross McKitrick of the University of Guelph (the two are sometimes called M&M). Intrigued by—and somewhat suspicious of—the drama inherent in the hockey stick, McIntyre and McKitrick attempted to replicate the MBH findings. In a 2003 paper, M&M detailed a set of errors they found in the listing of proxy data for the 1999 *Nature* article. MBH responded with a 2004 correction in *Nature*, concluding that none of the errors affected their results.

After M&M found themselves still unable to replicate those results (in part, they claimed, because they lacked access to some of the relevant software), they took a different approach. They began to criticize the use of tree-ring proxies they felt had been chosen, grouped and analyzed in ways that—regardless of intent—artificially inflated the twentieth century's upward bend to the hockey stick. For example, the 1999 MBH paper incorporated tree-ring data from long-lived bristlecone pines in the western United States; some previous work had questioned the use of these trees as biological thermometers.

M&M also disputed MBH's use of collective proxy data in place of individual records, a strategy that's designed to minimize bias when a dense collection of proxies drawn from a relatively small part of the globe (mostly

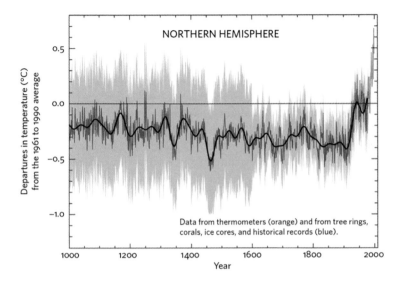

NORTHERN HEMISPHERE

Departures in temperature (°C) from the 1961 to 1990 average

Data from thermometers (orange) and from tree rings, corals, ice cores, and historical records (blue).

The "hockey stick" graph featured in the 2001 IPCC assessment, showing temperature (departures from the 1961–1990 average) in the Northern Hemisphere during the past 1000 years. (IPCC)

North America and Europe). Moreover, they showed how one method of using proxy groups to assess long-term trends, a technique called principal components analysis, had the potential to generate a hockey-stick curve even when there was no such trend in the underlying data. With these and other factors taken into account, M&M asserted, the Medieval Warm Period could indeed have been warmer than recent decades. Separately, climate researcher Hans von Storch (University of Hamburg) and colleagues used computer modeling for a 2004 *Science* paper to suggest that global temperature variations over the last millennium could have been more than twice as large as MBH had found.

The debate broke open in early 2005 as McIntyre and McKitrick outlined their arguments in *Geophysical Research Letters*. Their paper galvanized attention from the media and the U.S. Congress and set off a wave of concern, especially among those politicians and parts of the public inclined toward climate skepticism. Mann released all of the relevant data and software and a variety of other materials at the request of Congress (a request that itself raised concerns about political involvement in the scientific process). A new blog by McIntyre called Climate Audit quickly became a go-to site for climate skeptics.

A panel convened by the U.S. National Research Council looked into the controversy. The panel's 2006 report supported the type of multiproxy research pioneered by Mann and colleagues. It agreed that some statistical techniques could produce a hockey-stick shape where none existed, but even if such were the case, the report noted, additional evidence was at hand. Various other analyses and lines of evidence—including newer global reconstructions by other scientists—supported the rapid rise at the end of the hockey stick, the idea that recent warmth has indeed been unusual against the last four centuries. The panel did stress that, before the year 1600, proxy uncertainties are greater and confidence is lower. In the wake of the report, MBH and several colleagues extended their reconstruction back to the year AD 200, reaffirming that the last several decades stand out in their warmth even when tree-ring data are excluded. Meanwhile, McIntyre's blog offered an exhaustive running account of his quest to change the way proxies are used and proxy-based research is conducted. Sparks flew once again in 2009 after a trove of emails exchanged among climate scientists became public (see chapter 13).

While this grand battle raised issues of data and software access and transparency and showed the influence that investigators outside the climate science mainstream can wield, it also diverted attention from the unquestioned rise in both temperature and greenhouse emissions over the last several decades. Regardless of these and other dustups, one thing remained crystal clear: the greenhouse gas content of our atmosphere has entered territory never before explored in human history. Given what we know about these gases, the implications for our climate are profound.

CHAPTER TWELVE
Circuits of Change
MODELING THE FUTURE CLIMATE

In 1957, Roger Revelle and Hans Seuss of the Scripps Institution of Oceanography described the intersection of fossil fuels and global climate change in words that proved far more resonant than the authors might have imagined at the time. Their landmark study, published in the journal *Tellus*, included this oft-quoted passage:

> [H]uman beings are now carrying out a large-scale geophysical experiment of a kind that could not have happened in the past nor be reproduced in the future. Within a few centuries we are returning to the atmosphere and oceans the concentrated organic carbon stored in sedimentary rocks over hundreds of millions of years.

Like any good scientists, they were curious about the outcome: "This experiment, if adequately documented, may yield a far-reaching insight into the processes determining weather and climate."

It may seem callous to describe our current situation—glaciers melting, cities baking, species disappearing—as an interesting experiment. It's also quite accurate. Each year our emissions push the chemistry of the atmosphere a little further into territory never before explored in human history. If there were a million Earths in our solar system, each with a different concentration of greenhouse gases, we'd have a better idea of just what to expect in the future. Unfortunately, as Revelle and Seuss point out, we can't replicate the experiment that we're now conducting on planet Earth. The next best thing is to simulate it, and the only place that's feasible is inside a computer.

Reproducing the atmosphere with **computer models** has been a key activity in computer science ever since the field was born, just after World War II. At first the emphasis was on weather prediction. By the 1960s, forecasters were using 3-D models of atmospheric flow, or **general circulation models**, to gaze into the future of weather patterns several days out. Since then, breakthroughs in weather and computer science, together with Moore's Law (the maxim that computer processor speed doubles about every 18 months), have allowed for steady improvements in weather forecasting models and thus in forecast skill. Revelle and Seuss might have been happy to have a solid three-day forecast; now, 10-day outlooks are commonplace.

Models focusing on climate long lagged behind their counterparts in the weather field, and with good reason. The most obvious weather changes emerge from the interplay among a fairly limited set of ingredients in the atmosphere: pressure, temperature, moisture, wind, and clouds. Weather is also shaped by other factors in the surrounding environment, such as ocean temperature, vegetation, and sea ice. But these latter factors don't change much over the course of a few days, so a weather model can focus on tracking the atmosphere while keeping the rest of the environment constant.

Climate models don't have that luxury. As the weeks roll into months, vegetation thrives and decays. Sea ice comes and goes. Greenhouse gases accumulate. The ocean gradually absorbs heat from our warming atmosphere, sometimes releasing it in giant pulses during El Niño events. In the very long term, even the topography of Earth changes. All of these variations feed into the atmosphere and influence weather—often so subtly that the effect isn't obvious until it's been playing out for years. For all of these reasons, it's extremely difficult to go beyond a weather model's focus on the atmosphere and to depict Earth's whole environment accurately in a **global climate model**.

Difficult doesn't mean impossible, though. Climate modeling has undergone a rapid transformation in the last 30 years. As recently as the 1980s, the most sophisticated climate models portrayed the atmosphere and the land surface—and that was about it. Gradually, scientists have forged models that depict other parts of the Earth system. Many of these are incorporated in ever-growing global climate models. The best of the bunch now include land, ocean, sea ice, vegetation, and the atmosphere, all interacting in a fairly realistic way. Global models are far from ideal; as skeptics often point out, there are still major uncertainties about climate

Climate, weather, and chaos theory

One of the most common gripes about climate change projections is the idea that a computer model can tell us anything worthwhile about climate in the distant future. It's often phrased this way: "If they can't get the forecast right for next week, how can they predict the climate a hundred years from now?" Even the late Michael Crichton employed this tactic often (see chapter 13).

A weather forecast and a climate projection are two different beasts. Weather models can't see clearly beyond a few days because of inherent limits to the predictability of small-scale weather (i.e., what will happen in your neighborhood, as opposed to the globe as a whole). Weather observing stations are situated a few kilometers or miles apart at best, and in much of the world you can go 100 km (60 mi) between stations, not to mention the largely unsampled oceans. Small disturbances missed in the data can influence larger weather events over time—the "butterfly effect," as discovered in the 1960s by Edward Lorenz, the father of chaos theory. We'll never have enough weather stations to catch every one of these tiny weather makers. That's why we can't give a specific temperature forecast in February for the first day in August, although we can say with great confidence for most midlatitude cities that the month of August will be warmer than February.

Climate models aren't interested in the vagaries of individual weather events so much as the influence of long-term climate shapers such as greenhouse gases and polluting aerosols. Chaotic variations in weather normally average out over years and decades, so they don't corrupt a long-term climate projection the way they might a short-term weather forecast. The downside is that global models can't provide the precision in time and space that many policy makers and the public want to see, although progress is being made through techniques such as "downscaling" from global output to regional depictions.

processes that no model can resolve. However, the skill of recent global models in replicating twentieth-century climate, as we'll see below, is one sign that the models are doing many things right.

How computer modeling got started

The best analogy to a climate model may be a state-of-the-art computer game. Instead of cars tearing down a racecourse, or warlords fighting over territory, this game follows the heating, cooling, moistening and drying of

From great ideas to global views

The computer modelers now helping the world decide how to confront climate change owe a big debt to L. F. Richardson, a far-seeing British scientist of the 1920s. The story begins in World War I, when a team of meteorologists in Bergen, Norway, came up with the first three-dimensional theory of how weather worked. They named the boundaries that separated cold and warm air "fronts," after the battle fronts then raging in Europe. Meanwhile, Richardson, then an ambulance driver for the French army, was hatching a scheme to calculate the future of the atmosphere.

After the war, Richardson returned to England and set to work. Building on ideas that emerged from the Bergen group, he came up with seven equations based on physical principles from Isaac Newton, Robert Boyle, and Jacques Charles. If these equations could be solved, Richardson believed, one could not only describe the current weather but extend it into the future. He envisioned a "forecast factory," where hundreds of clerks would carry out the adding, subtracting, multiplying, and dividing needed to create a forecast by numbers. Richardson and his wife spent six weeks doing their own number-crunching in order to test his ideas on a single day's weather. The resulting outlook's skill was abysmal, but Richardson's test demonstrated to the world that one could—in principle, at least—carry out calculations and forecast the weather using more than intuition and rules of thumb that were standard in the 1920s.

Richardson's equations materialized again almost 30 years later in the world's first general-purpose electronic computer. The Electronic Numerical Integrator and Computer (ENIAC) was created in the United States at the close of World War II. In order to show the machine's prowess, its developers hunted for a science problem that could benefit from raw computing power. Weather predic-

the atmosphere. This isn't quite as sexy, but it's far more important to the future of the planet.

The action unfolds in a virtual space that's typically divided into a three-dimensional grid of rectangles. Imagine the steel latticework in a building under construction, but extended across the entire surface of the planet (see p. 302). At the center of each rectangle formed by the intersection of the imaginary beams is a **grid point**, where atmospheric conditions are tracked as the model's climate unfolds. In the atmospheric part of a typical global model, each grid point accounts for a horizontal area of roughly 1.25° longitude × 1.25° latitude. At midlatitudes (40°N), that reption fit the bill. Richardson's equations were updated and translated into machine language, and in Princeton, New Jersey, on March 5, 1950, ENIAC began cranking out the first computerized weather forecast. The computer took almost a week to complete its first 24-hour outlook, but ENIAC did do a better-than-expected job.

As weather modeling became established in the 1950s and 1960s, a small group of U.S. climate experts began to adapt some of the weather models for climate and to create new ones from scratch. One center of action was the Geophysical Fluid Dynamics Laboratory, based in Princeton (and now part of NOAA). Another

A function table from ENIAC, now housed in a museum. (Jud McCranie/Wikimedia Commons)

was the Courant Institute in New York, and a third was the National Center for Atmospheric Research, newly established in Boulder, Colorado. In the mid-1960s, NCAR's Warren Washington and Akira Kasahara built one of the first general circulation models that spanned the globe. (Many models at the time were two-dimensional or covered only the Northern Hemisphere.) Generations later, the seeds of that early model remain in NCAR's Community Earth System Model, used by university researchers around the world.

resents an area of around 140 km × 100 km (87 mi × 62 mi)—about the size of Connecticut or roughly half the size of Sicily. Some climate models now have a global atmospheric resolution as fine as 0.5° latitude and longitude, which corresponds to a midlatitude area of about 56 km × 43 km (35 mi × 27 mi), or roughly the size of a large city and its suburbs. Even then, each grid point represents the atmosphere over a fairly large area, much more so than for a weather model. Toward the poles, the width of each rectangle shrinks toward zero, which calls for special computational treatment. Some experimental weather and climate models are now replacing the 3-D global latticework with a hexagonal structure, resembling a soccer ball with thousands of tiny faces, that makes for smoother and more efficient computing. The typical global model also includes perhaps 25 or so vertical layers. Each layer spans anywhere from 100 to 500 m (330–1650 ft) close to Earth's surface, with the vertical distance between grid points increasing as you go up.

A climate model operates in a parallel universe of sorts, one in which time runs far more quickly than in real life. Each model advances in **time steps** that move the simulated climate forward by anywhere from five minutes to a half hour, depending on the model's complexity. At each time step, the model calls on formulas derived from the laws of physics to compute how all the atmospheric and environmental variables have changed since the last time step—the amount of sunlight hitting each spot on Earth, the convergence of winds, the formation and dissipation of clouds, and so forth. Each step only takes a few seconds to process, but it can still take weeks of dedicated time on a supercomputer to depict a century of climate in detail. The output from all this number crunching is saved in various formats along the way. After the processing is done, scientists can go back, sift through the results, calculate yearly averages and other statistics, and produce graphics.

In order to depict long-term climate change, scientists have to provide the model with input on how the environment is changing. One critical variable is fossil fuel emissions. Modelers rely on two major types of

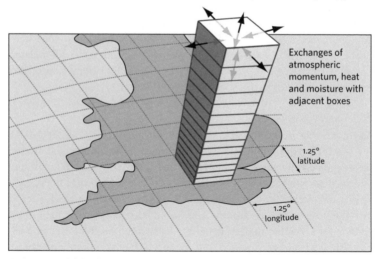

Exchanges of atmospheric momentum, heat and moisture with adjacent boxes

1.25° latitude

1.25° longitude

A climate model divides the atmosphere into 3-D chunks and measures the change in each one as weather unfolds over days, years, and decades of model time.

simulations to show how an increase in greenhouse gases affects climate. (In both cases, carbon dioxide is traditionally used as a stand-in for all human-produced greenhouse gases in order to save on computing time and expense, though in recent years the most detailed models have also incorporated methane and other key gases.)

* **Equilibrium** runs typically start off with an immediate, massive injection of carbon dioxide (for instance, enough to bring the airborne amount to twice or even four times the preindustrial level). The model's climate gradually responds to this extra CO_2 and eventually attains equilibrium at some new (and presumably warmer) state. The idea isn't to see how long the climate takes to respond—since we'd never add that much CO_2 instantaneously in the real world—but where the climate ends up.

* **Transient** runs more closely resemble reality. In these simulations, the CO_2 concentration in the model atmosphere is typically boosted in smaller increments that correspond to the actual amounts added by human activity—typically 1% per year, compounded year over year. In this case, scientists are interested as much in the process as the outcome. For instance, does the model climate warm smoothly, or does the temperature go up in fits and starts? The various global models have tended to show closer agreement for transient runs than for equilibrium runs—an encouraging sign, as the transient scenario is closer to how greenhouse gases increase in real life and is thus more relevant to climate policy over the twenty-first century.

Of course, you can't simply add CO_2 to a global model and leave everything else alone. As discussed in chapter 9, a large part of that CO_2 gets absorbed by vegetation and the oceans. There are also other processes that kick in as carbon dioxide warms Earth. Consider the sea ice that melts as the atmosphere warms up. As we saw in chapter 6, the loss of that ice means less reflection of sunlight and an amplifying, positive-feedback loop that leads to even more warming. The positive feedback from water vapor also has to be portrayed accurately: the warming produced by the extra CO_2 helps evaporate more moisture from the sea, and that water vapor raises global temperature further.

These environmental processes are quite different from the physics that drives day-to-day weather, and they can't be added to a climate model in an easy or simple way. The original global climate models of the 1960s and 1970s couldn't hope to account for these processes in a realistic fashion,

Hot spots for climate modeling

It takes vast amounts of computing power to run a comprehensive global model at a reasonable pace. Also needed are dozens of physical and computer scientists to ensure that a model's software is accurate and efficient. As such, the world's major climate models are produced by a fairly small number of institutions, most of them operated or funded by national governments. They include Australia's **Bureau of Meteorology Research Centre** and its **Commonwealth Scientific and Industrial Research Organisation**, both in the Melbourne area; the **Canadian Centre for Climate Modelling and Analysis**, based at the University of Victoria; Japan's **Frontier Research Center for Global Change** in Yokohama; the **Hadley Centre**, part of the Met Office, based in Exeter, United Kingdom; the **German Climate Computing Centre** (Deutsches Klimarechenzentrum) and the **Max Planck Institute for Meteorology**, both in Hamburg; and the **Laboratory of Dynamic Meteorology** in Paris, part of France's Centre National de la Recherche Scientifique. In the United States, there's NOAA's **Geophysical Fluid Dynamics Laboratory** in Princeton, New Jersey; NASA's **Goddard Institute for Space Studies** in New York City; several facilities of the **Department of Energy**; and the **National Center for Atmospheric Research** (NCAR) in Boulder, Colorado. NCAR's Community Earth System Model is one of the few open-source, high-end global models that anyone with sufficient computing power can download and use.

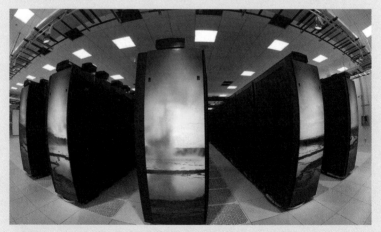

This IBM machine, dubbed Yellowstone, is located at the NCAR-Wyoming Supercomputing Center in Cheyenne. It can carry out more than one quadrillion calculations per second. (UCAR)

but as computing power has improved, the models have grown steadily in sophistication and detail. They have incorporated a growing number of submodels, dealing with each of the major climate-influencing elements. When an atmospheric model is yoked to one or more of these submodels, the new hybrid is called a **coupled model** (even when it has three or more components). If such a model also includes the flow of carbon into and out of the atmosphere, it's dubbed an **Earth system model**.

In a simple coupling, a submodel runs side by side with a global model and passes it information—a one-way dialogue, as it were. For instance, a global model might be fed information on airborne nitrogen levels from a chemistry model, but the resulting changes in climate wouldn't feed back into nitrogen production, as they would in real life. More realistic are **integrated** couplings, which allow two-way exchanges of information between model and submodel. However, it can take years to develop such linkages and verify their accuracy.

Climate modelers have typically avoided incorporating current weather conditions when projecting the climate decades into the future, since there's so much day-to-day variability in weather. However, bringing current ocean conditions into the model could help nail down how global temperature is likely to unfold over the next decade. The Met Office Hadley Centre used this strategy to issue a novel kind of experimental forecast unveiled in a 2007 paper in *Science*. By running a global model in two ways—either including or omitting the patchwork of present-day ocean temperatures—the group showed that incorporating current ocean conditions, and the "memory" they bring to the atmosphere, can add measurable benefit to a climate outlook. The model run captured realistic-looking peaks and valleys of global warm and cool spells caused by natural variability, but with an underlying rise that projected at least half of the period from 2009 to 2014 to be warmer than the record-setting year of 1998. As it turns out, temperatures didn't quite rise to this level, but the Met Office began issuing 5- and 10-year forecasts of global average temperature on a routine basis starting in 2009. Such forecasts are now showing at least a modest level of skillfulness.

Ocean conditions were also fed into 10- and 30-year model runs carried out in support of the 2013 IPCC report. Some of these runs were launched with the state of the ocean in the years 1960 or 1980, in order to see how well the modeled climate matched the observed conditions. Other 10- and 30-year runs peered into the future; these began with ocean conditions in 2005. With these simulations and related work to draw on, the IPCC's 2013 assessment was the first to provide near-term climate projections. The panel foresees a likely global temperature rise of 0.3°–0.7°C (0.5°–1.2°F) for the period 2016–2035 compared to 1986–2005, barring a major volcanic eruption or a sustained change in solar output.

Especially when the focus is on a weather or climate feature with restricted geographic extent, such as hurricanes, it makes sense to limit the model's extra-sharp precision (and the related computing burden) to the areas where it's most needed. This can be done using a **nested grid**, whereby the globe as a whole is simulated at a lower resolution and the areas of interest are covered at higher resolutions. Similarly, when studying how climate might evolve in a fixed area, such as parts of a continent, global models are often coupled to **regional climate models** that have extra detail on topography. The output from nested global models and regional models is especially useful in creating the climate assessments that are sorely needed by policymakers. However, bumping up a model's resolution isn't a simple matter: it involves a careful look at the physics of the model to ensure that processes that don't matter so much in a low-resolution simulation, such as showers and thunderstorms, are realistically depicted at the new, sharper resolution. Otherwise, improving a model's resolution can actually degrade its performance.

Whatever its design, every global model has to be calibrated and tested against some known climate regime in order to tell if it's functioning properly. The easiest way to do this is through a **control run** with greenhouse gases at their preindustrial level, plus a **validation run** with the gases increasing at their twentieth-century pace. By simulating the last century of climate, scientists can see if a model accurately depicts the average global temperature as well as regional and seasonal patterns, ocean circulations, and a host of other important climate features. Once it's passed this test, a model can go on to project the climate of the future—or the distant past, the goal of paleoclimate modeling.

Another safety check is that the world's leading models are regularly checked against each other through such activities as the Program for Climate Model Diagnosis and Intercomparison, conducted at the U.S. Lawrence Livermore National Laboratory. When the models agree, that bolsters confidence in the results; if one model deviates markedly from the others, it's a flag that more work may be needed. One of the program's main activities is running a mammoth Climate Model Intercomparison Project (CMIP), which is designed to structure and archive the modeling carried out at various centers around the world in support of each IPCC assessment. The 2013 IPCC report draws on the suite of modeling runs known as CMIP5.

The limitations

In both weather and climate models, it's not easy to calculate every iota of atmospheric change. **Clouds**, for example, are still a tremendous challenge. If they change in a systematic way, clouds can have major implications for how much radiation gets reflected and absorbed from both the sun and Earth. But clouds are far more varied and changeable in real life than any model can accurately represent, so modelers have to simplify them in their simulations. There's currently no gold standard for how to do this, and new techniques are being explored all the time. Typically, each grid point in a model will be assigned an amount or percentage of cloud cover on the basis of how much water vapor is present and whether the air is rising or sinking (rising air is conducive to cloud formation). The type of cloud and its reflective value are estimated from the height and temperature of the grid point.

To a greater or lesser degree, climate models also include **aerosols**—pollutants, dust, and other particles that can shield sunlight and serve as nuclei for cloud droplets and ice crystals. This allows scientists to see how changes in pollution might affect cloud formation and temperature, as in the case of the global dimming observed in the late twentieth century (see chapter 10). There's been substantial progress in modeling the direct effects of aerosols on sunlight and other radiation. All told, this direct influence appears to be a net cooling, even though some aerosols, such as soot, tend to absorb more sunlight than they reflect. Aerosols can also influence climate indirectly by shaping the formation of clouds. Although the cloud–aerosol interaction is tough to model, recent results point toward a slight overall cooling effect. Aerosols can also fall onto snow and ice and darken their surface, which hastens melting and slightly counterbalances the cooling effects noted above.

In order to test whether their cloud schemes are working well, scientists run their models for present-day climate and then compare the overall distributions of cloud cover to the real-world situation observed by satellite. (These datasets have their own limitations, though, since most satellites can't yet delineate multiple cloud layers very well.) The results are also compared to special cloud-resolving models that are too expensive to run routinely but that provide a valuable cross-check.

Oceans are also tricky. Real-life oceans feature a **mixed layer** on top, where the turbulence of waves and close contact with the atmosphere help keep the temperature fairly constant. The mixed layer may be as shallow as 10 m (33 ft) or as deep as 200 m (660 ft), depending on season, location, time of day, and weather conditions. Underneath the mixed layer is a zone of

Modeling for the masses: ClimatePrediction.net

At the turn of the twenty-first century, climate modeling was restricted to the supercomputers of a few universities and research labs. That all changed with ClimatePrediction.net, an online project that has set global models churning in more than 100,000 homes and offices worldwide. It's the brainchild of Myles Allen, a physicist and climate analyst at the University of Oxford. Allen was inspired in the late 1990s by the success of the SETI@home project, which has involved more than a million people around the world dedicating standby time on their personal computers (PCs) to assist in the search for extraterrestrial life.

The amount of computing time needed to analyze radio-telescope data for SETI is vast, but it can be parceled out in small doses. Each volunteer agrees to download a program that runs on her or his PC at times when the computer would otherwise be resting (the software never interferes with other tasks). These "screensaver scientists" get the satisfaction of contributing to a grand scientific endeavor with a minimal commitment of time, as well as the fun of joining a like-minded online community.

Right away, Allen saw the potential value to climate science of this distributed approach to computing. On an up-to-date home computer, a 50-year simulation of global climate could be run in six months, he figured. If thousands of volunteers were involved, the virtual horsepower would be far more than any research center could afford on its own. Moreover, this setup would be an ideal way to run ensembles—sets of closely related simulations, each with slightly different starting conditions. To drum up interest, Allen pitched the idea in a commentary for *Nature* in 1999. "Anyone respectable enough to sit on a peer review committee," he wrote, "will probably find the idea of getting schoolchildren to run full-scale climate models on their parents' PCs completely daft. But there will be others for whom the idea is as natural as Amazon.com." After several years of experiments and effort, Allen and colleague David Stainforth obtained financial backing from the United Kingdom's National Environmental Research

deeper, colder water that interacts with the mixed layer much more slowly. The earliest climate models depicted the mixed layer as a uniform slab but lacked a deep layer, a problem for long-term simulations. Fixed oceans were a frequent source of **climate drift**, which occurs when a model-depicted climate gradually veers into unrealistically warm or cold territory.

Today's global models are coupled with much more realistic ocean submodels that allow for the two-way exchange of heat, moisture, chemistry,

Council (NERC), settled on the name ClimatePrediction.net, and launched the experiment in 2003.

The greatest participation is in North America and Europe: more than half of the 90,000-plus computers involved as of 2010 were in the United States, United Kingdom, and Germany. But there's a healthy sprinkling of users all across the globe, with participants in more than 130 nations. The CP.net team works to engage volunteers with its extensive website, which includes background on climate science, techniques for comparing one volunteer's results against others and instant-messaging tools for the volunteer community.

The first science goal for ClimatePrediction.net was to examine sensitivity, the amount of global warming that a doubling of carbon dioxide would produce (see p. 311). The results grabbed headlines. Out of 2500 simulations using a version of the Met Office's Unified Model, most were in the ballpark of previous studies, but a few showed a climate sensitivity as high as 11.5°C (20.7°F). Allen believes the possibility of such high values was downplayed in previous research because traditional schemes for measuring sensitivity in any process (climatic or otherwise) don't portray the high part of the range well.

From its initial work, ClimatePrediction.net has moved on to investigate such hot topics as the impact of geoengineering on climate. With support from Microsoft Research, the project has completed more than half a million regional simulations for Europe, western North America, and southern Africa, working with partners that include Oregon State University. In 2010 it launched its first experiments with regional climate models, an effort dubbed weather@home. Results for the central U.S. heat wave and drought of summer 2012 suggested that anthropogenic greenhouse gas emissions contributed substantially to the high temperatures, but not to the lack of rainfall. Other efforts are in the works to apply the project's considerable modeling muscle toward assessing the role of human-produced climate change in extreme weather events.

and momentum among the surface, the mixed layer, and deeper levels, with 25 or more layers now depicted in some models. Still, the complicated circulations of the real ocean remain hard to portray. For instance, the North Atlantic's flow is often characterized as a simple loop, with warm water flowing north off the U.S. East Coast toward Europe, cold water streaming back south toward Africa, and warmer waters flowing west with the trade winds toward the Caribbean. In reality, however, the northward flow branches into several components as it approaches Greenland and the British Isles, and the water descends far undersea into a cold return flow that moves southward in several distinct layers. Where the contrasts in temperature and salinity are strong, small but powerful eddies can spin off. All this complexity is hard enough to measure, much less portray in a model.

The tropical Pacific remains one of the most difficult areas for climate models to depict in detail, not only because of ocean circulation but also because of the gigantic areas of rainfall spawned by the warm waters, especially in and near Indonesia. This warm pool periodically expands eastward or contracts westward during El Niño and La Niña (see p. 159). Most global models can now depict El Niño and La Niña, and some are able to represent the frequency of their comings and goings fairly well, but their geography and long-term behavior are still hard to pin down. Even in neutral conditions, it's a challenge for models to realistically capture the amount of heat and moisture pumped into the upper atmosphere by tropical showers and thunderstorms.

A sensitive topic: Predicting the warm-up

The question of what will happen in a greenhouse-warmed world is so essential to climate modeling that it's spawned an index that's widely used to assess model performance. This is the **sensitivity** of a global model, most often defined as the amount of temperature change produced in the model when carbon dioxide is doubled in an equilibrium-style run, as discussed above. The climate eventually settles down at a new global temperature, and the difference between that temperature and the old one is the sensitivity. It provides a convenient way to compare models and how readily they respond to greenhouse gas. This definition isn't cast in concrete: one might consider a tripling or quadrupling of CO_2 instead, and various starting or ending points can be used, such as the peak of the last ice age. However, the doubling of CO_2 since preindustrial times (generally taken as the mid-1800s) serves as a useful and time-tested convention familiar to all climate modelers.

At the current rate of increase, greenhouse gases will have roughly doubled in the atmosphere by the mid- to late twenty-first century (though we might be able to postpone the doubling, or perhaps even forestall it entirely, through a concerted effort to reduce greenhouse emissions). The doubling of CO_2 used in model sensitivity gives us a good idea of how much warming to expect once the atmosphere and oceans have adjusted to the extra greenhouse gas, which could take centuries longer than the doubling itself. Should there be big changes along the way—such as the partial or total loss of major ice sheets—then it could take many thousands of years more for global temperature to settle down. This potential longer-term rise is usually analyzed separately from the more conventional definition of sensitivity used below.

Even though sensitivity is expressed as a single number, it's actually the outgrowth of all the processes that unfold in a model. That's why equally sophisticated climate models can have different sensitivities. That said, what's striking about equilibrium climate sensitivity is how little the average value has changed as models have improved. Some of the earliest global climate models showed a doubled CO_2 sensitivity of around 2.5°C (4.5°F). As more models of varying complexity came online, the range of possibilities was estimated to be **1.5°–4.5°C (2.7°–8.1°F)**. This range—something of a best guess at the time—was first cited in a landmark 1979 report on climate change by the U.S. National Academy of Sciences. It was held constant by the IPCC in its first three assessments and was narrowed only slightly in the 2007 report (to 2.0°–4.5°C or 3.6°–8.1°F). Most of the major coupled models now show sensitivities clustered within a degree of **3.0°C (5.4°F)**, which is the best estimate pegged in the latest IPCC report. The fact that this value hasn't changed much after nearly 30 years of modeling is, arguably, testimony to how well modelers have captured the most important processes at work.

A skeptic might complain that the models have simply been tuned—their physics adjusted—to produce the result that people expect. Indeed, modelers have to guard against making so many adjustments to a model that it gets a "right" answer for the wrong reasons. But there's also a physical rationale behind the 3°C number. It's fairly easy to show that, if the atmosphere had no clouds or water vapor, the warming from doubled CO_2 would be somewhere around 1.0°–1.2°C (1.8°–2.2°F). It's believed that water vapor at least doubles the impact from the extra CO_2. Together with smaller effects from other greenhouse gases, this brings the total change into the 3°C ballpark.

Be that as it may, we can't be assured that larger (or smaller) changes are completely out of the question. The small but real risk of a dangerously high sensitivity has been borne out by a number of studies based on observations, as well as the massive experiment in distributed climate modeling known as ClimatePrediction.net (see sidebar "Modeling for the masses"). On the other hand, a flurry of studies in 2013 suggested that the likely range of sensitivity might actually be *lower* than long thought.

What exactly does probability mean?

There's an important difference between subjective and objective probabilities. In weather forecasting, a computer model might calculate a 60% chance of rain. That's an **objective** probability—it depends only on the physics or statistics in the model and the data from which it's working. However, if a human forecaster raised the value to 80% on the basis of information that the model didn't have, such as a glance out of the window that revealed gathering storm clouds, the new figure would be a **subjective** probability—that is, produced by a person, using her or his best judgment. Many objective probabilities can be drawn from past experience. For instance, if decades of climate in a given city show that it rains on average about one out of every four days in June, then a summer-solstice festival has a 25% chance of getting dampened in any given year. (Those odds may change as the event nears, much like the example above.)

Scientists can't use purely objective probability when it comes to climate change, because by definition the climate is entering new territory—there's no set of past human-induced warmings on which probabilities could be based. One way around this is for humans to apply expert judgment, as the forecaster above did, and assign probabilities on the basis of the evidence at hand. For example, in 2013 the IPCC stated that doubling carbon dioxide in the atmosphere would make an eventual warming of 1.5°–4.5°C "likely." Elsewhere in that report, the IPCC defines "likely" as a probability range of 66%–100%.

While these ranges can be useful to policymakers, they aren't based on hard-and-fast scientific experiments. Now that climate simulations can be carried out en masse, it's possible to derive probabilities as if those simulations were a set of climates similar to ours. That's how various groups are coming up with percentages and ranges like the ones discussed in chapter 14. These are more objective than before, but they're still based on models rather than real-world climate, given that we can never have an archive of thousands of global warmings against which an unfolding one could be compared.

In part, this reflected the pause in atmospheric warming that extended from the late 1990s into (at least) the early 2013s. When the hiatus was brought into models of the balance of incoming and outgoing energy, it tended to point toward a climate less sensitive to the relentless increase in greenhouse gases.

In the end, the IPCC report—while considering some of the new work on either side of the spectrum—chose to go with the long-accepted interval of 1.5°–4.5°C as the likely range of equilibrium climate sensitivity in its 2013 report. The panel noted, "Even though this assessed range is similar to previous reports . . . confidence today is much higher as a result of high quality and longer observational records with a clearer anthropogenic signal, better process understanding, more and better understood evidence from paleoclimate reconstructions, and better climate models with higher resolution that capture many more processes more realistically."

Clearly, there's still disagreement about just how sensitive climate might be, but recent work reiterates that the most likely outcome of a doubling of carbon dioxide is a global temperature increase close to the one scientists have been projecting for decades: somewhere not too far from 3°C (5.4°F). None of this rules out a much more or much less sensitive climate. It simply adds to the weight of evidence for an amount of warming that's certainly nothing to be cavalier about. A 3°C rise, after all, would bring Earth to its warmest point in millions of years. Plus, if greenhouse gas concentrations do reach the benchmark of doubled CO_2, they could easily go further still.

Listening to the models

It's up to policymakers and the public to decide how society ought to respond to the onslaught of data that climate models give us. Probabilities like the ones above can help immensely. It may seem confusing to think of a temperature range and a probability at the same time, but that's the direction modelers are heading, and it's the type of output that's most useful in making decisions. Once upon a time, weather forecasts would describe "possible showers" or "a chance of snow" without ever specifying a number. That made it hard to tell exactly how likely it was that your picnic would get rained out. It wasn't until computing became widespread in the 1960s that a new set of statistical weather models enabled forecasters to look at a wide range of outcomes and assign likelihoods. By the 1970s, "probabilistic" forecasts were the norm and people had grown accustomed to phrasings such as "a 30% chance of rain."

In much the same way, climate modeling is now affordable enough that some high-end models can be run a number of times. Each run might use a different rate of CO_2 increase, for example, to show the many ways the climate might unfold depending on how serious we get about reducing emissions. Alternately, the model might simulate the same CO_2 increase a number of times, but with starting conditions that vary slightly, in order to see how natural variations in climate affect the results. These large sets of simulations are called **ensembles**, and they've become increasingly

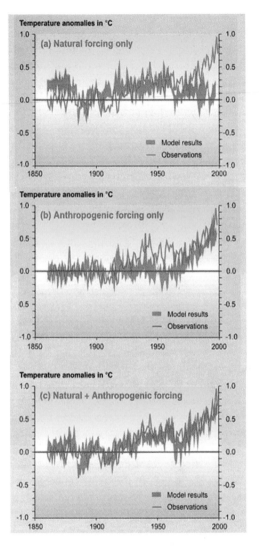

Temperature anomalies in °C

(a) Natural forcing only

Model results
Observations

Temperature anomalies in °C

(b) Anthropogenic forcing only

Model results
Observations

Temperature anomalies in °C

(c) Natural + Anthropogenic forcing

Model results
Observations

When models look only at natural climate forces (top) or human-produced ones, such as greenhouse gases and sulfate pollution (middle), they fail to accurately reproduce the last century's climate (the red line in each graph). A model that incorporates both kinds of forces (bottom) does a much better job. The gray shading shows the range of each model ensemble. (IPCC)

important in climate modeling. As in the examples cited above, some ensembles include more than 1000 simulations, which was unheard of as recently as the turn of the century. The sheer size of these ensembles allows scientists to calculate statistics and probabilities that a single model run can't provide. Apart from creating ensembles, scientists are hard at work making other improvements to their models. Among the current areas of keen interest are weaving in more interactive chemistry, adding more detailed land cover and ecosystems, and capturing the various effects of aerosols more precisely.

One might ask why we ought to believe anything that global climate models tell us. Some contrarians pose this question often, and it's healthy for everyone to recognize what climate models can and can't tell us. For instance, models will always struggle with uncertainty about the starting-point state of the atmosphere, which affects shorter-term weather forecasts. At the same time, our lack of foresight about greenhouse emissions becomes enormous when looking a century out. This leads to the interest in "sweet spots"—time frames close enough to be of interest for long-term business and government planning but far enough out so that models can provide insight that other techniques cannot. In a 2007 *Science* paper, Peter Cox and David Stephenson (University of Exeter) urged their fellow scientists to focus their research energies on the time frame from about 2040 to 2060. Along these lines, the 2013 IPCC report examined results for the period 2046–2065 as well as 2081–2100, with a separate chapter focusing on near-term projections (2016–2035).

One good reason to be confident in the general quality of today's climate models is their ability to simulate the past—especially the twentieth century, for which we have by far the most complete climate record. The three graphs opposite, first prepared for the 2001 IPCC report, show how one climate model captures the ups and downs in global temperature over the last century. It doesn't reproduce every year-to-year wiggle, but that's not the idea, just as a weather model doesn't aim to tell you the exact temperature in your backyard at 8:27 p.m. What this model does show accurately is the character of twentieth-century climate. It does so by including the gradual increase in greenhouse gases and sun-shielding pollutants, as well as the effects of natural features like volcanoes and the sun. When such a model goes on to tell us what might happen in the next century, we'd be foolhardy to ignore it.

Debates and Solutions

FROM SPATS AND SPIN TO PROTECTING THE PLANET

Protestors flock to central Copenhagen as the UN Conference of Parties opens, December 2009. (Robert Henson)

CHAPTER THIRTEEN

A Heated Debate

HOW SCIENTISTS, ACTIVISTS, SKEPTICS, AND INDUSTRY HAVE BATTLED FOR COLUMN INCHES AND THE PUBLIC MIND

In recent years, perhaps no realm of science has found itself under as much scrutiny as climate change. That's no surprise: at its heart, the issue counterposes the future of our planet's atmosphere and ecology against the most profitable industries on Earth—those that extract, refine, and sell fossil fuels. But there are many other reasons why global warming has been such a contentious topic for more than 20 years. The main cause, carbon dioxide, is literally invisible. The worst effects may occur many thousands of kilometers from where we live, or many decades from now, and unlike environmental problems that can be pinned on a few bad actors, virtually all of us produce greenhouse emissions through the course of our daily lives. All of these factors make it tempting to try to refute, dismiss, or ignore climate change rather than confront its profound implications head on.

Global warming politics didn't catch fire at first. For the most part, the topic remained in the scientific background until it became clear that the rise in greenhouse gases was real and serious. Even as research (and the gases themselves) continued to accumulate, it took other events to light the fuse in the late 1980s and make climate change a top-level global concern.

That's when industry saw the potential threat to its profits and began to act. Starting around 1990, many oil, coal, gas, and car companies and like-minded firms joined forces to sow seeds of doubt about climate change science. Working along similar lines as these companies (and, in some cases, getting funded by them directly or indirectly) was a tiny subset of contrarians—scientists of various stripes who bucked the research mainstream and downplayed the risk of global warming. Facilitated by a media eager to produce "balanced" coverage, this small group of scientists became familiar to anyone reading or hearing about climate change. At the other end of the spectrum, environmental activists argued passionately for the defense of the planet and its ecosystems, often with colorful protests tied to key diplomatic moments.

The landscape of climate change debate grew far more complex after the 1997 introduction of the Kyoto Protocol, the first global agreement to reduce greenhouse gas emissions. While the issue is often portrayed as a one-side-versus-the-other battle, there's plenty of nuance in the mix, even among those who agree on the big picture. Many corporations have taken up the banner of reducing emissions, in order to save on their energy costs as well as burnish their green credentials. Environmentalists have fiercely debated many angles among themselves, including the pace at which emissions ought to be reduced and the merits of nuclear power in a greenhouse-threatened world. A batch of emails leaked from top climate scientists in 2009 triggered controversy, after which international squabbling led to years of delay in forging the next global agreement on emissions reduction. All this has unfolded despite the fact that the world's leading scientific societies—including the national academies of more than 30 countries—stand firmly behind the recognition that humans are changing the climate on a global scale.

The early days

As we've seen, the greenhouse effect was discovered back in the nineteenth century, but serious scientific and media debate about climate change didn't take off until much later. One of the first news stories came in 1950, after global temperatures rose from around 1900 to the 1940s. *The Saturday Evening Post*, then one of the biggest U.S. magazines, asked a question one might hear today: "Was this past mild winter just part of a natural cycle?" Their article "Is the World Getting Warmer?" rounded up a variety of anecdotal evidence, including "tropical flying fish" sighted off the New Jersey coast.

As possible causes for the warm-up, the article cited solar variation and other natural factors. Greenhouse gases weren't even mentioned.

The scientific debate picked up in the 1960s, but at that time, there was plenty else to worry about—nuclear annihilation, for instance—so few people outside of scientific circles heard much about the risk of climate change. Things began to change in the 1970s, when the embryonic environmental movement called out air pollution as an example of humans' soiling of the planet. With early photos from outer space now highlighting Earth's stark aloneness, it was suddenly easier to believe that humans could affect the atmosphere on a global scale.

This *Saturday Evening Post* article from 1950 was one of the very first stories on climate change in the popular press. (*Saturday Evening Post*)

What grabbed most of the press in the 1970s, however, wasn't a global warm-up but a cool-down. Earth's temperature had been gradually slipping for some three decades, mainly in the Northern Hemisphere. A few maverick scientists speculated that dust and sun-blocking sulfate particles emitted from North America and Eurasia could be responsible for the cooling. A British documentary in 1974 called *The Weather Machine* warned that a single brutal winter could be enough to plaster northern latitudes with a "snow blitz" that the next summer couldn't entirely erase, thus leading to continent-encrusting ice sheets within decades. If nothing else, climate had started to seem more fluid and unstable than most people had ever thought possible.

Even as reporters chattered about cold, many scientists were concerned about the long-term outlook for warmth. In a 1972 *Nature* paper entitled "Man-made carbon dioxide and the 'greenhouse effect,'" J. S. Sawyer predicted a temperature rise of 0.6°C (1.0°F) for the rest of the twentieth century—a figure that was only slightly off the mark. A landmark 1975 paper in *Science* by Wallace Broecker (Lamont-Doherty Earth Observatory) asked if we were "on the brink of a pronounced global warming"—the first research paper to use that phrase in its title. Two studies late in the 1970s by the U.S. National Academy of Sciences confirmed that the ever-increasing levels of

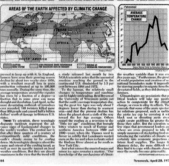

April 28, 1973: *Newsweek* reports on the fear of a forthcoming ice age. (*Newsweek*)

CO_2 in the air should lead to significant warming. Computer models were improving quickly, and they continued to indicate that warming was on the way. Finally, the atmosphere itself chimed in. By the late 1980s, global temperatures had begun an ascent that hasn't abated since, except for a sharp two-year drop after 1991's eruption of **Mount Pinatubo** (see p. 261) and a leveling off during much of the millennium's first two decades, albeit close to unprecedented highs.

The reports and findings continued to pile up through the 1980s, but with little fanfare at first outside of research labs and government hearings. Hypotheses of future global warming remained exotic enough to the public that many journalists kept the term "greenhouse effect" in quotes. The stunning discovery of the Antarctic **ozone hole** in 1985 was a turning point. Although it fostered long-lived confusion between ozone depletion and global warming (see sidebar, "Is the ozone hole linked to global warming?"), the finding was also a new sign of the atmosphere's fragility, borne out by vivid satellite images.

The other shoe dropped in the United States during its sizzling, drought-ridden **summer of 1988**. Huge tracts of forest burned in Yellowstone National Park, and parts of the Mississippi River were at record lows, bringing barge traffic to a virtual halt. On a record-hot June day in Washington, D.C., NASA scientist **James Hansen** delivered memorable testimony before Congress, presenting new model-based evidence and claiming to be "99% sure" that global warming was upon us, most likely induced by humans. Together, *The New York Times* and *Washington Post* ran more than 40 stories on climate change in 1988 after fewer than two dozen in the preceding

> " Climatological Cassandras are becoming increasingly apprehensive, for the weather aberrations they are studying may be the harbinger of another ice age." —Time *magazine, 1974*

Arguments and counterarguments

Fairly early on in the climate debate, skeptics developed a broad set of criticisms—many of which you still hear today—that were thrown at mainstream climate scientists and at the concept of global warming in general. Taken to the extreme, you could sum up the classic skeptical view like this:

> The atmosphere isn't warming, and if it is, then it's due to natural variation. Even if it's not due to natural variation, then the amount of warming is insignificant, and if it becomes significant, then the benefits will outweigh the problems. If by chance they don't outweigh the problems, technology will come to the rescue, and even if it doesn't, we shouldn't wreck the economy to fix the problem when many parts of the science are uncertain.

Probably no single skeptic would endorse the whole of that rather convoluted statement, yet each of the points within it has been argued vigorously over the years by various contrarians. Let's look briefly at each point in turn. For a set of succinct refutations of more than 100 arguments against human-produced climate change, see **www.skepticalscience.com**.

* **"The atmosphere isn't warming."** This one has been put safely to rest, although as recently as the 1990s some skeptics insisted there was no planet-wide warming at all, and the notion still crops up on the Internet. Fuelling this line of argument was the apparent lack of warming in upper-air temperatures as measured by satellites and radiosondes, but it's now clear that globally averaged upper-level temperatures are in fact warming at close to the same rate as the surface (see chapter 10). Although surface readings plateaued during the 2000s and early 2010s, the decadal global average remained higher than any in the history of weather observing, and new annual highs are almost certain to occur in the coming decades.
* **"The warming is due to natural variation."** This point is still argued often, even though the IPCC has concluded that the warming of the last century, especially since the 1970s, falls outside the bounds of natural variability (see chapter 1). In any event, the long-term ascent of global warming should outpace the peaks and valleys of natural cyclic processes such as the El Niño/Southern Oscillation or the Atlantic Multidecadal Oscillation in the long run, and variations in the sun don't appear to explain what's

been happening in the last few decades (see sidebar, "Is the Sun behind climate change?").

* **"The amount of warming is insignificant."** This claim mingles bona fide uncertainty about the future with a judgment call on how much warming should be labeled as significant. The genuine uncertainty is how much warming we can expect in the coming decades and centuries. As noted in chapter 12, the most widely accepted estimate for the eventual rise in global temperature from a doubling of atmospheric carbon dioxide is about 3.0°C (5.4°F) over preindustrial times. Though 3°C is a minor blip in day-to-day weather, it's a huge leap for global average temperature. The 2013 IPCC report finds that a warming of around 2.0°–3.7°C (3.6°–6.7°F) beyond the 1850–1900 global average is likely by century's end if CO_2 concentrations were to double by that point (RCP6.0; see p. 16 for more on RCPs). A number of skeptics claim that the low end of the IPCC temperature range is the most likely outcome, but in fact, we've already warmed about 0.9°C (1.62°F) since 1900. This puts us nearly halfway to the low end of the RCP6.0 range, with carbon dioxide close to 40% higher than its preindustrial value.

* **"The benefits will outweigh the problems."** Overall, some stimulation of plant growth can be expected, but it's not at all certain that the benefits will be prolonged or planet-wide (see chapter 9). Moreover, while CO_2 is already giving some forests a boost, the changing climate raises the risk of devastating fires and insect attacks. Against the pluses of CO_2 fertilization, and other benefits such as fewer cold-related illnesses and the possibility of sailing through the Arctic in midsummer, we have to balance the various negatives, including the risk of rising seas, widespread drought, and massive species loss.

* **"Technology will come to the rescue."** This isn't skepticism about global warming so much as an affirmation of human ingenuity. Some optimists believe that geoengineering might save us from the clutches of global warming (see chapter 16). Even if such an approach proves feasible, it would face an uphill trek to gain funding, international approval, and public confidence. Still, it's important to keep in mind the possibility and promise of technical innovation, while at the same time recognizing the reality of our present situation and the emissions trajectory we're on.

* **"We shouldn't wreck the economy."** For skeptics motivated more by economic than scientific considerations, this is the ultimate bottom line. If we don't know with absolute confidence how much it will warm and what the local and regional impacts will be, so the reasoning goes, perhaps we're better off not committing ourselves to costly reductions in greenhouse gas emissions. However, the eventual costs of environmental remedies have often proved much less than economic models indicated at first. Moreover, it's unclear how much further the scientific uncertainty around specific regional outcomes can be reduced. Perhaps more importantly, it would be foolish to assume that reducing emissions will cost more than coping with a changing climate—an expense that isn't always brought into economic analyses. In the following chapters, we'll discuss some technologies and approaches that could actually produce net savings in the long run, especially if markets are structured so that reducing carbon is rewarded.

* **Other arguments.** Many critiques of climate change that emerged long ago have popped up on talk shows, blog posts, and newspaper op-ed pages even after they've been debunked time and again. One example is the oft-cited argument, "If they can't predict the weather for next month, how can they predict the climate a hundred years from now?" Of course, these are two fundamentally different processes. A weather forecast tracks day-to-day changes at a given point. A climate projection looks at longer-term trends that in turn tell you about the type of weather we might expect. If you live in Germany or Minnesota and it's the first day of January, you can say with some confidence that the first day in July will almost certainly be warmer than today, even if you can't predict whether the actual high on July 1 will be 20°C (68°F) or 35°C (95°F).

Two other points of contention are the quality of the global models that project future warming and the data that tell us about past climate. The models certainly aren't perfect, but they've agreed for years that we can expect a significant warming. Likewise, shortfalls do exist in the records of past weather (which weren't really designed to detect climate shifts in the first place), but such issues aren't enough to rule out the overwhelming evidence of change already under way. It's hard to debate a world full of melting glaciers.

Is the sun behind climate change?

In an attempt to let carbon dioxide—and human actions—off the hook for climate change, it's often claimed that **solar variations** account for the last century's warmth. Over the very long term, variations in Earth's orbit that shape where and when sunlight reaches the planet are the main cause of ice ages (see chapter 11), but this doesn't apply to our current situation. It's true that the sun produced more sunspots in the late 1900s than it did in the early 1800s. However, the effects on Earth were mainly in the ultraviolet range of sunlight, which is only a tiny part of the solar spectrum. As a whole, the amount of solar energy reaching Earth changes very little over time. Across the 11-year solar cycle, it varies by only about 0.1%, and from the Little Ice Age chill of 1750 to the warmth of 2008, average solar output (factoring out the ups and downs of the 11-year cycle) climbed by no more than about 0.007%, according to the 2013 IPCC report. Interestingly, sunspot activity dropped to its lowest levels in a century in 2008–2009 as global temperatures remained close to record highs.

There's still a question mark or two when it comes to ultraviolet radiation, where the lion's share of solar variability occurs. It's possible that UV rays interact with ozone in the stratosphere to change circulation patterns, though more work is needed to clarify how this might occur. UV light also helps shield Earth from **cosmic rays** that bombard and ionize the atmosphere—a point much discussed by skeptics in recent years, thanks to work by Henrik Svensmark (Danish National Space Center) and others. This concept got major play in the 2003 book

four years, according to Katherine McComas and James Shanahan of Cornell University. *Time* magazine named "Endangered Earth" Planet of the Year, in place of its usual Man of the Year. Even conservative politicians took note. In August, U.S. presidential candidate George H. W. Bush declared, "Those who think we are powerless to do anything about the greenhouse effect forget about the 'White House effect.'" And although the meteorological drama of 1988 was focused on North America, the political waves reverberated far and wide. In September, British prime minister Margaret Thatcher warned the U.K. Royal Society that "we have unwittingly begun a massive experiment with the system of the planet itself." As Jeremy Leggett recalls in *The Carbon War*, "1988 was the year that broke the mold." Indeed, the events of that year were enough to convince Leggett, who was then teaching at Britain's Royal School of Mines, to join Greenpeace as a science advisor to its climate campaign.

The Chilling Stars, co-written by Nigel Calder and Svensmark, and on U.K. and Australian televisions in the 2007 documentary *The Great Global Warming Swindle*. The idea is that highly reflective low-level clouds might form more easily when tiny particles that serve as cloud nuclei are ionized, helping them to clump together more readily. If so, then an active sun would inhibit low-level clouds, thus allowing more sunlight to reach Earth and fostering warming. Laboratory and field studies have found some evidence for the clumping effect, but studies

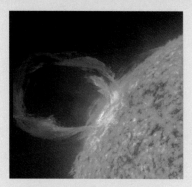

A solar eruptive prominence as seen in an image collected at extreme ultraviolet wavelengths on March 30, 2010. (NASA Solar Dynamics Observatory)

comparing satellite-derived cloud patterns to cosmic-ray counts have shown conflicting results, suggesting that any effect from cosmic rays is weak at best. Moreover, there's no clear evidence that more cosmic rays have actually made it into Earth's lower atmosphere over the last several decades. *The Great Global Warming Swindle* was criticized by many experts not only for downplaying these unknowns but also for using discredited data and inaccurate graphs.

From progress to roadblocks

In the aftermath of 1988's awareness-raising events, governments began to pour money into global warming research. The Intergovernmental Panel on Climate Change was established in 1989 as a means of channeling research from dozens of nations and many hundreds of scientists into an internationally recognized consensus. The IPCC produced its first report in 1990, underlining the risks of global warming, and activists did their best to alert journalists and the public to the problem.

By this time, the leading **environmental groups** in North America and Europe were well established, most with 15 years under their belts. No longer a fringe movement, their cause was now part of the fabric of public life. Years of activism had helped slow the growth of nuclear power to a crawl in many countries, with no small assistance from the Chernobyl debacle in 1986. Governments and politicians had become so attuned to environmental risk that it took fewer than three years—lightning speed by diplomatic standards—from the time the ozone hole over Antarctica was discovered in 1985 to the signing of the 1987 Montreal Protocol that's now guiding the planet toward eventual ozone recovery.

For a while, it looked as if the same success might be seen with global warming. In 1992, thousands of activists joined similar numbers of journalists and diplomats in Rio de Janeiro for the UN-sponsored **Earth Summit**. The meeting, and the global climate treaty that emerged, kicked off years of negotiations that led to the historic **Kyoto Protocol** in 1997. (For more on Kyoto, see the following chapter.)

As media attention grew and as the scientific evidence strengthened, people in many countries became more aware of the risks of climate change and the possible solutions. Yet something wasn't quite clicking. Although public support for climate change action was growing steadily, the support was "wide but shallow." In other words, many people were concerned, but not sufficiently concerned to force the issue up the political agenda, nor to take personal action to reduce their own greenhouse gas emissions.

The sheer scope of the problem was one factor. Fossil fuels are used in virtually every aspect of modern society, and climate change threatens to affect every country on Earth in one way or another. It's hard to motivate people to grapple with such an immense and seemingly intractable issue, and the many options for political and personal action could be too much to process. Moreover, even more than smog or acid rain, human-induced climate change is a classic "tragedy of the commons": the benefits of burning fossil fuels accrue to individuals, companies, and nations, while the costs accrue to the planet as a whole. And much of the benefit in avoiding a global meltdown remains distant in time and space for any single person.

> " Are existing environmental institutions up to the task of imagining the post–global warming world? Or do we now need a set of new institutions founded around a more expansive vision and set of values?"
>
> —*Michael Shellenberger and Ted Nordhaus,* The Death of Environmentalism

Another difficulty for climate campaigners was the fact that global warming hit it big just when many people were getting tired of fretting about the state of the world. From its earliest days, the environ-

A plenary session meets in the main hall of the Kyoto International Conference Center during the 1997 Kyoto Protocol negotiations. (Frank Leather/ United Nations)

mental movement had relied on stark, pseudo-apocalyptic imagery to motivate people. In her 1962 book *Silent Spring*, which set the template for environmental wake-up calls, Rachel Carson labeled pesticides and similar agents "the last and greatest danger to our civilization." In her footsteps came a series of similarly dire scenarios, from Paul Ehrlich's *The Population Bomb* to the notion of nuclear winter. Global warming lends itself especially well to this type of rhetoric. It's no exaggeration to talk about the risk of coastal settlements vanishing and the Arctic's summer ice pack disappearing. When activists do hit these points, however, it sometimes brings to mind other predictions of environmental doomsday that didn't come to pass (partly, of course, because society did respond to those earlier threats). Some researchers have found that fear-based rhetoric on climate change makes people defensive and resistant to recognizing the problem and can stoke the politicization of the issue.

Even the most painless ways to reduce global warming, such as improved **energy efficiency**, came with cultural baggage in some countries. Efficiency measures had swept the United States during the oil shocks of

Debates among the activists

Climate change is now a favored cause of most environmental groups, but key differences in strategy and ideology remain. One question is whether to embrace or reject certain **aspects of capitalism**. With consumerism driving the world economy, some eco groups are going with the flow—encouraging people to use their spending power to make climate-healthy choices, such as buying a hybrid car or using low-energy light bulbs. Other groups retain an abiding suspicion of the corporate world and the governments that support it, yet many of these groups are also working within the system to achieve results.

For environmentalists who accept the idea of green commerce, the struggle to isolate genuinely "ethical" choices can be tricky. For instance, Toyota makes the world's most popular hybrid—the Prius—yet it was one of the plaintiffs in a lawsuit that aimed to block California's strict emissions standards. Hence, some groups are encouraging consumers to push big companies in a climate-friendly direction through shareholder actions or socially responsible investing (see chapter 17).

Another difference between climate activists is **how to approach fossil fuels**. Some take a pragmatic approach, figuring that oil, gas, and coal aren't going away tomorrow. The U.S.-based Natural Resources Defense Council has thrown its considerable weight behind tighter restrictions on coal mining and coal-plant emissions. The NRDC notes the large number of coal plants already on order in China and elsewhere and stresses the need to develop cleaner technology. Some other groups take a more absolutist stand: for example, Greenpeace International is pushing for a global phase-out of coal as an energy source.

the 1970s—the top interstate speed limit was dropped to 55 mph (89 kph), for instance—but the practice of saving energy never lost its air of self-imposed deprivation. As oil prices plummeted in the go-go 1980s, efficiency quickly fell by the wayside and speed limits went back up. By the time global warming pushed energy efficiency back onto the national agenda, it was a loaded topic. Because so many activists had been proposing sensible energy-saving steps for years, it was easy for critics to paint them as opportunists, happy to use climate change or any other issue in order to advance their ulterior goals. To top it off, the 1990s produced a surge of people and groups doing their best to stop the public from getting too worried about global warming.

Looming in the background is **nuclear energy**, and the question of whether countries should turn to it to bridge the potential gap between fossil fuels and large-scale deployment of renewables. Some key environmental thinkers have lent support to the idea of using nuclear as a stopgap, including Gaia theorist James Lovelock and *Whole Earth Catalog* founder Stewart Brand. They were joined in 2013 by a quartet of eminent climate researchers (see p. 418). However, many of the most influential environmental groups, including Friends of the Earth and Greenpeace, remain adamantly opposed to nuclear power in any form.

A final point of difference is **picking goals**. A concrete target is the best way to motivate volunteers and supporters. The most commonly cited benchmark is to stabilize the climate at 2°C (3.6°F) above the preindustrial global temperature (see chapter 14). Long favored by the European Union, the 2°C target was formally adopted by delegates at the 2010 UN climate summit in Cancún. A more ambitious target is favored by the group **350.org**, which aims to bring the global concentration of carbon dioxide down to 350 ppm. To meet either goal, global emissions would have to be cut drastically over the next few decades. With this in mind, some environmental groups use other types of targets as well, including legislative ones. More recently, the idea of a global carbon budget has come into the limelight, spurred in part by 350.org founder Bill McKibben's "Do the Math" campaign. Shorter-term goals are still another approach, as employed by Britain's 10:10 campaign, which encourages individuals and organizations to aim for a 10% reduction in a year.

Skeptics and industry fight back

Especially in the United States, a group of scientists with skeptical views about climate change—perhaps no more than several dozen—have wielded far more influence than their numbers might indicate. Until recently, many, if not most, news articles about climate change included a comment from one or more of these contrarians. Their voices were backed up in many cases by the immense money and influence of the oil, coal, and auto industries, largely through conservative think tanks (see sidebar, "Pushing against climate change action"). This support enabled them to exert much sway on the U.S. Congress, the media, and, by extension, the global fight against climate change.

Pushing against climate change action

Like their peers, the few climate skeptics active in research are employed mainly by universities and private labs. Although a few have received grants from oil and coal companies, most rely largely on public funds to carry out their work. However, that work gets an extra dose of clout, especially in the United States, thanks to a number of think tanks and lobbying groups that cite their findings widely and use them in an attempt to convince legislators that climate change science is full of unknowns. Such centers are often influential, and many have been buoyed by funding from corporations with a lot to lose from carbon restrictions.

One highly visible group throughout the 1990s was the opaquely titled **Global Climate Coalition**, which formed in 1989 as the prospect of global diplomatic action on climate change appeared on the horizon. Based at the U.S. National Association of Manufacturers, the GCC included some of the biggest oil, coal, and auto companies in the world, including General Motors, Ford, BP, Royal Dutch Shell, and ExxonMobil. Along with lobbying at UN meetings, the coalition angled its way into becoming an oft-quoted presence in the media. They also financed Kyoto-related commercials warning that "Americans would pay the price" for the treaty.

The GCC began to fracture with the departures of BP in 1997, Shell in 1998, and Ford in 1999. By 2001, it was history, though arguably it had served its purpose and was no longer necessary. A 2001 memo written to ExxonMobil by U.S. undersecretary of state Paula Dobriansky, later obtained by Greenpeace, states that George W. Bush rejected Kyoto "partly based on input from you [the GCC]". In the group's own words, "The industry voice on climate change has served its purpose by contributing to a new national approach to global warming."

After the collapse of the GCC, most of the world's major oil companies shifted toward public acknowledgement of climate change, while **ExxonMobil**—the largest of them all—continued to cultivate doubt. From 2000 to 2003, according to *Mother Jones* magazine, the company poured more than $8 million into more than 40 organizations aligned with climate change skepticism. In a rare move, the United Kingdom's Royal Society wrote to ExxonMobil in 2006 asking it to stop funding skeptically aligned groups. The company announced in 2008 that it would stop contributing to such groups. In a 2013 speech to shareholders, ExxonMobil chief executive Rex Tillerson referred to climate change as "a serious issue" that presents "serious risk," though he also emphasized the uncertainty inherent in climate projections. Any such uncertainty, however, didn't stop ExxonMobil from inking a multi-billion-dollar agreement in 2011 with Russia's state-controlled oil firm Rosneft to embark on long-term drilling operations in the rapidly warming Arctic Ocean.

Working largely below the media radar is **Koch Industries**, a Kansas-based conglomerate with major oil holdings. Its founding brothers were ranked by *Forbes* in 2013 as two of the five richest Americans. Since the Koch Industries name doesn't appear on consumer products, the firm is virtually unknown to many Americans. However, *The Guardian* reported in 2013 that the Koch brothers had routed almost $120 million between 2002 and 2010 to a pair of trusts that, in turn, helped support dozens of skeptic-oriented think tanks, scientists, and policy forums. A company website, **KochFacts.com**, posted this in 2013: "Koch supports having an open, honest, and science-based debate about the extent to which human activity is responsible for climate change." The Charles Koch Foundation provided $150,000 to help launch the Berkeley Earth Surface Temperature project, which ended up confirming the twentieth-century warming that many contrarians had doubted (see chapter 10).

The global warming wing of the Washington-based **Competitive Enterprise Institute** became one of the leading public voices of climate skepticism as other entities pulled back or lost interest. Its leader, Myron Ebell, was censured by the British House of Commons "in the strongest possible terms" in 2004 after he told BBC's Radio 4 that Sir David King, the chief science advisor to prime minister Tony Blair, "knows nothing about climate science." On the 2006 release of the Al Gore documentary *An Inconvenient Truth*, the CEI issued a pair of glossy TV advertisements that noted how fossil fuels have made life more comfortable and convenient. They ended with the tag line "Carbon dioxide: They call it pollution. We call it life." As recently as 2013, CEI blogger Lawson Bader struck the now-familiar uncertainty chord: "So is the Earth warming? Probably. Does man have something to do with it? Probably. Will this be a bad or a good thing long term? Nobody knows."

Another group that found prominence through climate skepticism in the 2000s was the **Heartland Institute**, founded in 1984 to approach social and economic problems through the lens of the free market. Among its highest-profile activities was a set of skeptic-dominated meetings from 2008 to 2012 dubbed the International Conferences on Climate Change. Even among like-minded organizations, Heartland often went too far for comfort, particularly in a fire-breathing 2012 billboard campaign that featured a photo of Unabomber Ted Kaczynski with the tag line "I still believe in Global Warming. Do you?" The campaign was halted after just one day, and a number of funders and employees left Heartland in its wake.

Scientists who minimize the risks of climate change are often portrayed as mavericks or outsiders, and with good reason. Although a few are active in climate research, many aren't. Some of the most vocal have backgrounds in subjects like solid-state physics or mathematics, sometimes with impressive résumés in their fields but little if any experience specifically in climate change science. For example, physicist Frederick Seitz worked on atomic research in the 1940s, presided over the U.S. National Academy of Sciences in the 1960s, and then served as a consultant to the medical research program of the R. J. Reynolds tobacco company before becoming an outspoken climate change skeptic. Other dissenters are trained in atmospheric science but have published few peer-reviewed papers on climate change. Many are retirees, affording them the time and freedom to act as consultants, writers, and speakers without having to conduct scientific studies of their own. Of course, there have also been plenty of spokespeople for skeptical positions who aren't scientists at all (see box above), just as plenty of nonscientists speak out for climate change action.

Right from the beginning, uncertainty has been the overriding theme in the arguments of climate change contrarians. The core of greenhouse science—such as the consensus estimates on how much global temperature rise to expect from a doubling of CO_2—has held firm for decades. But climate change is such a multifaceted and complicated enterprise that it's easy enough to find minor weaknesses in one study or another. Furthermore, there are always exceptions that prove the rule, such as an expanding glacier or a region that's cooled in recent decades. When contrarians point to a single event or process as a disproof of global climate change, or when advocates tout a particular heat wave as iron-clad proof that humans are meddling with climate, they're often accused of **cherry-picking**: selecting a few bits of evidence that seem to prove their point while omitting counter examples. It's a classic rhetorical technique, one well known to skilled lawyers and politicians.

Starting in the early 1990s, skeptics seized such uncertainty and exceptions and—amplified by the public relations budgets of cor-

> " I am convinced that in fifteen to twenty years, we will look back on this period of global warming hysteria as we now look back on so many other popular, and trendy, scientific ideas."
> —*William Gray, Colorado State University, testifying before the U.S. Congress, September 2005*

porations heavily invested in fossil fuel use—gave the false impression that the entire edifice of knowledge about climate change might crumble at any moment (or even that the whole thing was a colossal scam, a claim voiced more than a few times). The tactic of stressing uncertainty wasn't a new one: it had been used to delay tobacco regulation for decades and to stall environmental action in other areas, as laid out in detail by science historians Naomi Oreskes and Erik Conway in their 2010 book *Merchants of Doubt*.

Skeptic attacks have put some climate scientists directly in the line of fire. In the 1970s, the late **Stephen Schneider** studied the powerful effects of polluting aerosols and, in his book *The Genesis Strategy*, stressed the need for society to prepare for intense climate shifts. With the media focused on global cooling during this period, Schneider was often quoted, but his book emphasized the risks of greenhouse warming as well as aerosol cooling. In 1988 Schneider wrote *Global Warming*, one of the first lay-oriented books on the topic. His visibility and prolific output soon made him a prime target for critics who pointed to Schneider's earlier work on aerosol cooling, took his quotes out of context, and derided his research on climate risks and assessments. Schneider later noted, "A scientist's likelihood of having her/his meaning turned on its head is pretty high—especially with highly politicized topics such as global warming."

The release of the IPCC's second assessment in 1995 drew one of the decade's sharpest rounds of skeptic vitriol. The report's beefed-up assertion that "the balance of evidence suggests a discernible human influence on global climate" was based in part on detecting climate change **fingerprints**, the 3-D temperature patterns one would expect from the observed increase in greenhouse gases (as opposed to different patterns of warming that might be produced by other factors, such as strengthening solar input). A procedural delay in incorporating review and revisions to one chapter following the IPCC's plenary meeting, and confusion involving text that had been leaked to the media, led critics to accuse the IPCC—and, directly or indirectly, convening lead author Benjamin Santer (Lawrence Livermore National Laboratory)—of "doctoring the documents" and "scientific cleansing." Seitz castigated the IPCC in a widely cited *Wall Street Journal* editorial, while Santer and colleagues emphasized that the edits were a required response to review comments, per IPCC rules, and that the chapter's conclusions were virtually unchanged. Eventually, the battle subsided, but it made a lasting impact on Santer. "Nothing in my university training prepared me for what I faced in the aftermath of that report," he later said. "You are prepared as a scientist to defend your research. But I was not prepared to defend my personal integrity. I never imagined I'd have to do that."

Bjørn Lomborg's skeptical environmentalism

Danish political scientist Bjørn Lomborg marshaled a slew of statistics and nearly 3000 footnotes to make his case that, overall, the environment is in better shape than we might think. In his influential 2001 book *The Skeptical Environmentalist*, Lomborg employed the climate-and-economy models used by the IPCC assessments to argue that major emissions reductions in the short term (à la Kyoto) are not only enormously costly but will have little impact on the longer-term climate outcome.

Lomborg's book got rave reviews in *The Economist*, *Rolling Stone*, and elsewhere, but it was panned in other publications and pilloried by some leading scientists. The dustup got to the point where the official Danish Committee on Scientific Dishonesty labeled Lomborg's book "objectively dishonest" (they later withdrew the finding). In his discussion of climate change, Lomborg glided past sea level rise with little discussion of the high-end possibilities. Moreover, Lomborg's economic focus failed to take into account the intrinsic, nonmonetary value of protecting particular species and ecosystems. Even so, the book's sunny-side-up view of economic and ecological progress and its critique of environmental doom and gloom drew many fans, especially from the skeptical side of the global warming aisle. Lomborg, in fact, ended his climate change discussion by claiming that society has the money to control greenhouse emissions if we deemed it a high enough priority. However, he argued that many other issues—such as preventable diseases—deserve to take precedence.

In his 2007 follow-up *Cool It: The Skeptical Environmentalist's Guide to Global Warming*, Lomborg staked out his turf even more firmly, remaining sanguine about such concerns as polar ice (in Antarctica, it's growing) and polar bears (their real nemesis is hunting, not warming). A 2010 documentary, also called

Two sides to *every* story

In spite of the flare-up over the second IPCC report, the media had largely lost interest in global warming by the mid-1990s. According to the Cornell study cited above, the number of climate change articles in *The New York Times* and *Washington Post* dropped from more than 70 in 1989 to fewer than 20 in 1994. In part, the drop-off was typical of how news stories come and go. Alarms couldn't ring forever, at least not without some major disaster to make climate chaos seem like an imminent threat. But the skeptics and lobbying groups undoubtedly played a role, having successfully convinced many journalists—and large swaths of the public—that global

Cool It, spotlighted Lomborg and his views. That same year, Lomborg—dubbed the "anti-Gore" by a *New York Times* writer—raised eyebrows with what seemed to be an about-face, declaring that the world should dedicate $100 billion per year to addressing climate change. That position emerged from the Copenhagen Consensus, a panel of economists organized by Lomborg. In 2004, the panel had ranked three climate measures (meeting the Kyoto Protocol goals, and assessing two types of carbon tax) at the bottom of a list of 16 actions on which governments might best spend $50 billion to advance global welfare. In 2009, Lomborg assigned a new question to the panel: given a set of

Bjørn Lomborg. (Emil Jupin/Wikimedia Commons)

potential responses to climate change, which ones would provide the best bang for the buck? Not surprisingly, taxes were the least favored approach, whereas geoengineering and research and development investment topped the list. Among Lomborg's recent positions was the idea of steering part of the money now spent on renewable energy subsidies toward research on improving the efficiency of green technologies and bringing down their cost.

warming was at best an unknown quantity and at worst "ideological propaganda . . . a global fraud" (in the words of *Daily Mail* journalist Melanie Phillips).

Even when climate change did appear in the news, it suffered from the media paradigm that seeks to give equal weight to both sides of every story, with an advocate from each side of the fence battling it out in a presumed fair fight. It's a time-honored form of reporting, honed to perfection in political coverage, but in the case of climate change it often conveys a misleading sense of symmetry, as if the skeptic camp represented half of the world's climate scientists rather than a small group of contrarians.

When global warming meets TV weather

For a variety of reasons, most TV forecasters are relatively quiet about global warming. With only two to four minutes available on a typical daily weather segment, there's little time to explain the greenhouse effect or other global warming science, and TV producers seldom turn to their resident weather experts for coverage that might tie global warming to local concerns and conditions. As consultant Matthew Felling told the website *Salon*, "The last thing any station wants is an activist weatherman."

There are far more weathercasters in the expansive U.S. television market than anywhere else—more than 1000 of them, with many hundreds trained in meteorology. Only some produce science or environmental stories on top of their regular weather-reporting duties, though many of them maintain extensive blogs and keep active on social media. The American Meteorological Society embarked on a campaign in 2003 to train weathercasters as "staff scientists," with an eye toward giving them a higher profile in environmental coverage at their stations. A number of broadcast meteorologists have taken on this mantle, creating thoughtful, locally relevant coverage of climate change. In South Carolina, weathercaster Jim Gandy found success with a recurring series of segments called "Climate Matters." A substantial fraction of weathercasters, however, fall in the skeptic camp. In a 2010 survey of more than 500 weathercasters across the United States, 26% agreed with the statement, "Global warming is a scam." That was the position of John Coleman, a weathercaster for more than 50 years and the founding president of The Weather Channel. Although he left the channel in 1983, only a year after its debut, Coleman traded on his pioneering role and an on-camera position in San Diego to air his skeptical views.

The Weather Channel itself, which reaches most American subscribers to cable TV, has proven to be a reliable source of news about global warming. In 2003 the network hired Heidi Cullen as its first-ever climate expert. Cullen—a former research scientist—produced numerous segments on climate change science and impacts before moving to the nonprofit Climate Central. The network continued to cover global warming research under its "Forecast: Earth" banner, and in 2013 it reintensified its climate coverage, adding a series titled "The Tipping Points." The Weather Channel's position statement on climate change, last updated in 2007, acknowledged "strong evidence that the majority of the warming over the past century is a result of human activities."

Despite its faults, the "dueling scientist" mode of coverage soon became the norm, especially in the United States. One study of seven leading U.S. papers showed that five top climate scientists from the United States and Sweden were quoted in a total of 33 articles in 1990 but in only 5 articles in 1994. The use of skeptics changed little over the period, so that what were once minority opinions were soon getting roughly equal time. In a later study for the journal *Global Environmental Change*, Maxwell Boykoff and Jules Boykoff detailed the sudden shift in U.S. reporting styles at four major newspapers. Most articles in 1988 and 1989 focused on the evidence that human-induced global change is a real concern. However, in 1990 these were eclipsed by "balanced" articles that gave more or less equal weight to natural variations and human factors as causes of climate change. This became the standard format into the 2000s, according to Boykoff and Boykoff. As they noted, "through the filter of balanced reporting, popular discourse has significantly diverged from the scientific discourse."

How did the skeptics gain such a high profile? One factor was the powerful public relations machine discussed above. With the blogosphere not yet a major force, and Facebook and Twitter yet to be born, several campaigns were designed to highlight uncertainties in the science through news releases, press conferences, and direct vehicles such as TV advertisements. In 1998, for example, just after the Kyoto Protocol was drafted, *The New York Times* learned of a proposal by an industry faction to spend $5 million campaigning against the treaty. "If we can show that science does not support the Kyoto treaty—which most true climate scientists believe to be the case—this puts the United States in a stronger moral position," noted the industry document, which called for "an action plan to inform the American public that science does not support the precipitous actions Kyoto would dictate." Victory will be achieved, it said, "when media 'understands' (recognizes) uncertainties in climate science."

It wasn't only lobbying groups and think tanks working to calm the public mind about climate change. After Kyoto, some government officials appeared to be out to stop—or at least slow—the research and views flowing from their agencies' own scientists to the media and the public. In June 2005, *The New York Times* reported that a political appointee of U.S. president George W. Bush had reviewed several climate reports, including the annual overview *Our Changing Planet*, with a heavy hand. For example, in the statement "The Earth is undergoing a period of rapid change," the term "is" had been changed to "may be." Within days, the Bush appointee left the White House and took a job with ExxonMobil. Similarly, several

Climate change and the church

Most environmental activists operate from a secular viewpoint, but that's not always the case. In the United States, there's a small but growing contingent of what one headline writer dubbed "earthy evangelists." They made the news in 2006, when nearly a hundred of them signed a statement in support of the fight against climate change. It was the first salvo in the Evangelical Climate Initiative, along with TV ads that included the tag line, "With God's help, we can stop global warming."

This was hardly the first faith-based action on global warming. Environmental groups with Jewish and other religious ties have entered the fray over the years (many of them as part of the U.S. National Religious Partnership for the Environment). On a personal level, the first director of the Intergovernmental Panel on Climate Change, John Houghton, invoked his Christian faith as a source of sustenance during his often-demanding IPCC work in the early 1990s. But for those accustomed to thinking of U.S. evangelicals as moving in lockstep with the nation's far-right wing, the 2006 statement was a startling move.

There are a lot of evangelical Christians in the United States—at least 30 million—and not all are on the same wavelength as the earthy evangelists. The activists spun off from the National Association of Evangelicals, which declined to endorse their project. A rival group quickly sprang up, the Interfaith Stewardship Alliance, featuring some of the nation's best-known conservative Christians, including James Dobson and Charles Colson. They wrote their own statement, claiming "global warming is not a consensus issue," and their positions aligned much more closely with traditional skeptic fare. A paper by one of their founders, E. Calvin Beisner, went so far as to draw an analogy between coal and Jesus: "Vegetation is sown a natural body. Then, raised from the dead as coal and burned to enhance and safeguard our lives, it becomes a spiritual body—carbon dioxide gas—that gives life to vegetation and, through that, to every other living thing." The group has since evolved into the Cornwall Alliance

researchers at NASA and NOAA claimed they were stymied by higher-ups in their attempts to speak with journalists on the latest findings in controversial areas, such as the effect of climate change on hurricanes. NASA's James Hansen told *The New York Times* and the BBC in 2006 that he was warned of "dire consequences" should he fail to clear all media requests for interviews with NASA headquarters. One of the political appointees charged with such clearance was a 24-year-old who resigned after it turned

for the Stewardship of Creation, whose 2009 "evangelical declaration on global warming" rails against mandatory reductions of greenhouse gases.

The earthy evangelicals and what might be called the "dominionists" come to their vastly different perspectives from a similar starting point. Both subscribe to the biblical view of humans as stewards of Earth, and both express concern over the fate of Earth's poorest residents, especially in the developing world. But where the activists point to climate change as "the latest evidence of our failure to exercise proper stewardship," Beisner says that "a truly biblical ethic of creation care simply cannot ignore the Biblical mandate for man to fill, subdue and rule the Earth."

While neither of these two groups has made a widespread imprint, climate change itself is now being addressed in multiple ways by broad sectors of the church. The National Association of Evangelicals released a 2012 report on the role of faith in addressing climate change's intersection with poverty, and several of the world's largest Christian relief and development organizations, such as World Vision, take the issue of global warming seriously. The ecumenical group Interfaith Power and Light has brought Christians, Jews, Muslims, and Buddhists together to discuss and act on climate change, and many individuals continue to speak out within and beyond the church. One of those is Katharine Hayhoe, director of Texas Tech University's Climate Science Center. Hayhoe and her husband Andrew Farley, a linguistics professor and minister, collaborated on the 2009 book *A Climate for Change: Global Warming Facts for Faith-Based Decisions*. In her work on regional climate assessment, Hayhoe connects the intangible, large-scale force of greenhouse gases to people's current and future real-world experience. When addressing people of faith, she makes similar connections from a Christian perspective, with an emphasis on concern for our fellow travelers on Earth. As she and Farley write, "Doing something, anything, about climate change is a step in the direction of caring for people."

out he'd falsely claimed a degree in journalism on his résumé. Both NASA and NOAA reiterated their public commitment to openness and transparency after these stories broke, and the U.S. National Science Board called for uniform federal guidelines to prevent "the intentional or unintentional suppression or distortion of research findings."

On the other side of the Pacific, a media storm erupted in 2006 with an investigative report from the Australian Broadcasting Corporation (ABC).

According to the ABC report, at least three scientists at Australia's Commonwealth Scientific and Industrial Research Organisation (CSIRO) were dissuaded from airing their views on climate policy in various reports. CSIRO discourages its scientists from commenting on policy, as is typical in the United States. However, at least one CSIRO scientist claimed that his comments on policy-relevant science—such as sea level rise, which could affect immigration from Pacific islands to Australia—were getting quashed as well. Among the casualties in this turmoil was a short-lived communications director at CSIRO who had joined the science agency after serving as a spokesperson for Australia's tobacco industry.

Even the U.K. government, a participant in the Kyoto Protocol, found itself under scrutiny. Sir David King, the United Kingdom's chief science advisor, stated in January 2004 that "climate change is the most severe problem we are facing today—more serious even than the threat of terrorism." A few weeks later, U.S. reporter Mike Martin discovered a floppy disk left in the press room at a Seattle science meeting attended by King. On the disk was a memo from Ivan Rogers, the chief private secretary to Tony Blair, asking King to avoid media interviews. The discovery led to an article by Martin in the journal *Science* and a follow-up from *The Independent* in London that was headlined, "Scientists 'gagged' by No 10 after warning of global warming threat." In the eyes of Martin—the only reporter who saw the actual wording—the memo was more of a suggestion than a gag order. But as U.S. journalism professor George Kennedy opined, "It doesn't seem unreasonable to read a 'request' made by one's superior as an 'order' to be followed on pain of consequences."

The tide turns?

Despite the naysayers and lobby groups, coverage of climate change was on the increase as the new millennium unfolded. In her study of three U.K. newspapers—*The Times*, *The Guardian*, and *The Independent*—communications scholar Anabela Carvalho found that the number of stories published on climate change per year jumped from about 50 per paper in 1996 to around 200–300 each by 2001. Carvalho noted that the articles were increasingly laced with a sense of urgency and more talk of extreme events, including such contemporary happenings as England's severe flooding in 2000 and the European heat wave of 2003. By 2006, front-page climate stories were a frequent sight in Britain, and David Cameron aimed to entice voters to the Conservative Party with the slogan "Vote Blue, Go Green."

By late 2008, the Climate Change Act—which obliged the United Kingdom to cut its emissions by 80% by 2050—had passed into law, a landmark in political response to global warming.

A shift toward mainstream acceptance of climate change was slower to arrive in the United States, where contrarians were highly visible in the editorials and opinion pages of right-wing papers such as *The Wall Street Journal*. But things eventually changed in a big way. One of the first concrete signs was a front-page story in *USA Today* on June 13, 2005, entitled, "The debate's over: Globe is warming." In the skeptic-friendly U.S. press, this was a banner development. Across the globe, climate change also started featuring more frequently on those most mainstream of media outlets: TV and film. BBC Television announced a series of 16 programs on climate impacts, themed to the massive Stop Climate Chaos campaign organized by a group of British nongovernmental organizations. Soon after, Al Gore's long-time mission to raise awareness of global warming suddenly went mainstream with the launch of *An Inconvenient Truth*, a feature-length documentary film (and accompanying book) which smashed per-screen box-office records on its launch in 2006. Magazines also took note, with titles from *Time* and *Time Out* to *Vanity Fair* releasing special environmental editions. By 2007, climate change was a prominent presence in most media—a situation bolstered in October, when the IPCC and Al Gore were co-awarded the Nobel Peace Prize.

While Gore himself was no longer a political figure by this point, other world leaders inclined to act on climate change were coming onto the scene. On his first day as Australian prime minister in December 2007, Kevin Rudd signed the Kyoto Protocol, leaving the United States as the only major industrialized country that hadn't ratified the treaty. A year later, Barack Obama swept into the White House. Although he didn't push to make the United States join Kyoto, Obama did promise in his campaign platform to make global warming a top priority. Given that U.S. politics had likely been the main obstacle to a truly global climate agreement (or, in the view of skeptics, the main bulwark against it), it seemed a new era just might be at hand.

> " Conspicuous by its absence has been any sense of urgency in the British media . . . the public has been left uninformed about a serious issue." —The British Medical Journal, *1996*

Climate change and the cinema

Movie theaters aren't exactly packed with films about climate change, but the topic has made a few appearances on screen, both major and minor. Filmmakers have used the backdrop of particular climates to great effect ever since the silent-film days, but the abstraction of climate *change*—the evolution of weather over time—is a much trickier concept to put into cinematic terms. Plus, as author and activist Bill McKibben has pointed out, the instigators of global warming are typically far removed from the consequences of their actions. In other words, you can't exactly resolve the plot with a thrilling chase.

Some of the first films to address the dystopian prospects of human-induced climate change were in the realm of science fiction. *Zombies of the Stratosphere* (1952) included a young Leonard Nimoy as part of a gang of Martians bent on exploding Earth from its orbit so that Mars can move sunward and benefit from a milder climate. The James Bond spoof *In Like Flint* (1965) featured hero Derek Flint, played by James Coburn, facing off against a sinister organization called Galaxy that plans to flood valleys and send icebergs into the Mediterranean. Then there's *Soylent Green* (1973), the first of the lot to present a scenario based in part on global warming science. It's set in an overcrowded New York City in the year 2022, with pollution and other environmental ills run amok. ("A heat wave all year long," grouses Charlton Heston. "Greenhouse effect," replies his partner, played by Edward G. Robinson.) With food scarce, residents turn to the concoction that provides the film's name—ostensibly a blend of soya and lentils (soylent), but actually something far more gruesome.

Waterworld (1995), with Kevin Costner as co-producer and star, was the first major U.S. film to draw on modern concerns about global warming. The most expensive movie ever made up to that point, at $175 million, it was considered a commercial and critical flop, even though it ended up making a profit overseas. In *Waterworld*, hardscrabble camps of people are left to fight and adapt long after the world has been entirely flooded by the melting of icecaps. (Such a sea level rise goes far beyond anything possible in the real world, even if every inch of Antarctic and Arctic ice melted.)

Climate change crept into other films of the period as well. In *The Arrival* (1996), aliens intent on occupying Earth pollute the atmosphere in order to boost the greenhouse effect and melt polar ice. *The March* (1990) depicted Africans migrating to a Europe stressed by global warming, and *Prey* (1997) featured a bioanthropologist who fears that a new über-species, nurtured by global warming, wants to kill off *Homo sapiens*.

Steven Spielberg set his futuristic *A.I. Artificial Intelligence* (2001) in a world where New York and other coastal cities have been abandoned due to rising sea levels. High water also figures in *Split Second* (1992), set in London circa 2008. Global warming coupled with "forty days and nights of torrential rain" (as noted in the prologue) has pushed the ocean into the city streets, where a policeman played by Rutger Hauer—who lives on "anxiety, coffee, and chocolate"—hunts for a mysterious, lethal creature.

It took until 2004 for Hollywood to produce a big-budget extravaganza with climate change at its center. The hero of *The Day After Tomorrow* is Jack Hall (Dennis Quaid), a paleoclimatologist who discovers that global warming has triggered a shutdown of North Atlantic currents. Within days, the Northern Hemisphere is plunged into an ice-age-scale deep freeze, thanks to cold air descending from the stratosphere (and disobeying the laws of physics, which dictate that the descending air would become warmer than the surface air it replaced). Before long, Hall is tramping through snowfields up the East Coast to rescue his son from the icebound New York Public Library.

Well before its release, *The Day After Tomorrow* got the attention of people on both sides of the global warming debate. The activist group **MoveOn.org** distributed fliers to people outside theaters showing the film, while skeptics bashed it as a propaganda tool. But whatever potential the movie had to sway real-world debate was compromised by its absurdly telescoped view of how fast climate change might unfold. Although the Atlantic currents that help keep Europe warm for its latitude have shut down before, and they could diminish in the future (see p. 162), the process should take years to decades to unfold, rather than mere days. Still, many saw the film's success as a sign of growing public interest in climate change.

Al Gore became an unlikely movie star in 2006 with the release of *An Inconvenient Truth*. The Oscar-winning documentary is a glossed-up version of Gore's stump speech, which he claims to have presented over a thousand times. *An Inconvenient Truth* rode the wave of two big trends—interest in global warming and a surge of hugely popular U.S. documentaries that included Michael Moore's *Fahrenheit 9/11*. Outside of a few nuggets on Gore's life story, the film is what it claims to be: the story of global warming, told in a style that's sober and thoughtful, yet visually rich and emotionally compelling, enhanced by Gore's 30-year storehouse of knowledge on the topic. Despite its success, the film (and Gore himself) became a lightning rod for Americans who rejected the scientific consensus on climate change.

Expanding on Gore's thesis, *The 11th Hour*, a powerful 2007 documentary narrated by Leonardo DiCaprio, linked climate change with other environmental woes. Among the many other nonfiction treatments of climate change that followed *An Inconvenient Truth* were *Climate Refugees* (2010), focused on people already being forced to migrate due to climate-related factors, and *The Island President* (2011), which spotlighted Maldives president and climate activist Mohamed Nasheed.

One of the sharpest takes on our global predicament emerged from the United Kingdom with 2009's *The Age of Stupid*. This hybrid creation involves a fictional archivist (Pete Postlethwaite) looking back dolefully from the warming-ravaged world of 2055. To find out what was and wasn't being done to fight climate change early in the twenty-first century, he reviews clips from the era—which are, in fact, current documentary-style profiles of people around the globe. *The Age of Stupid* was unveiled in a solar-powered tent at London's Leicester Square and simultaneously at 62 satellite-linked screens across the United Kingdom, making for the world's biggest and greenest film debut.

Global warming even penetrated the children's film market in 2006 with *Ice Age: The Meltdown*. This sequel to the animated feature *Ice Age*—featuring the tag line "the chill is gone"—follows a batch of prehistoric animals in a newly warmed world, with species going extinct and floods threatening. Though set in the great meltdown after the last ice age, the film subtly tips its hat to modern-day worries.

Copenhagen, Climategate, and beyond

In early 2009, with climate-friendly leadership now in place among most of the world's industrialized powers, hopes were high among activists and many leaders that an ambitious post-Kyoto agreement might take shape at the major UN summit in Copenhagen scheduled for that December. However, political realities soon dashed those hopes, and a newly revived sense of doubt was nourished by climate skeptics. The global economic meltdown that began in 2007 didn't help matters. Anything that might be construed as an extra cost, or an expense with benefits years away, was suddenly suspect. The U.S. cap-and-trade proposal was successfully recast as "cap and tax" by its opponents, which helped drag it into limbo in the U.S. Senate after it passed the House in 2009. With many Americans fearful that such laws might hit them in the wallet, U.S. support of climate change legislation—along with acceptance of the physical reality of climate change, much less the human element in it—dipped markedly. In a 2006 survey by the Pew Research Center for the People & the Press, a record 77% of Americans believed there was solid evidence that Earth was warming, regardless of the cause. By late 2009, that percentage had sunk to 57%, and it had only recovered to 67% as of late 2013.

The lack of a U.S. climate bill, together with other global obstacles, cast a pall over the UN's much-anticipated Copenhagen climate summit in December 2009. The meeting descended into chaos within days, with little real progress (see sidebar, "Copenhagen and its aftermath,"). Right from the start, the summit was handicapped with unfortunate symbolism. Not only did it unfold during a long stretch of cold and snow—hardly the setting that brings to mind a warmer world—but the sight of hundreds of world leaders travelling to Copenhagen in carbon-spewing private jets and limousines made the gathering an easy target for ridicule.

To make matters even worse, there was the unauthorized release in November 2009 of more than 1000 emails that had been exchanged among climate scientists over more than a decade. These messages were pulled from a computer at the University of East Anglia's Climatic Research Unit (CRU) and placed online for all to see. Whether the emails were actually stolen (no charges were filed) or merely leaked, the redistribution was no doubt meant to disparage the reputation of the scientists involved and, by extension, the validity of climate science. The media quickly dubbed the event Climategate—a widely used label that implied the emails contained scandalous content.

In fact, the vast bulk of the emails simply depicted the mundane process of climate researchers at work. Reporters and bloggers zeroed in on a tiny fraction of the emails that appeared to reveal something more sinister. Perhaps the most widely quoted snippet was from CRU's Phil Jones to Michael Mann (then at the University of Virginia), in which Jones cited a technique used by Mann that employed air temperature data in order to "hide the decline" in temperatures derived from tree rings over the latter part of the twentieth century. It isn't well understood why tree-ring data have diverged from air temperature in recent decades, but there's no doubt that the globe was warming during that time frame. Thus, the "hiding" actually produced a more accurate portrait of recent temperatures within the multi-century temperature record. The same email referred to this technique as a "trick," which some observers interpreted as a scheme to obscure global cooling. In context, it was clear that "trick" meant a satisfying approach to solving a problem, as opposed to a nefarious strategy.

A few other emails raised thornier issues, especially with regard to data and software access. Many climate scientists, including Mann and Jones, were facing an ever-growing tide of requests for their datasets and computer code. Some of these requests came from parties outside the climate-science mainstream, which meant that techniques might need to be documented and explained more thoroughly than they would be for peers—a task that took time away from research. Moreover, some records had been provided to CRU by national meteorological services without the right of redistribution. Because of these and other factors, responding to requests had become an increasingly burdensome task, and some of the released emails implied that scientists were trying to avoid at least a few of the requests.

A series of investigations from 2010 onward exonerated the researchers from allegations of scientific wrongdoing. However, several reports emphasized the need for greater transparency and responsiveness when it comes to making data and software available to the world at large. Those points were echoed in massive coverage by *The Guardian*, which entrained readers and scientists in an effort to produce a self-correcting, annotated online narrative of the Climategate saga. The debate over open access isn't limited to climate research: science as a whole is struggling with how best to document extremely complex software and massive amounts of data while making them more widely available in useful forms.

It's hard not to see the timing of Climategate, just one month before Copenhagen, as an attempt to torpedo the summit. However, the ill-fated meeting had so many other headwinds that the email release may not have been a make-or-break factor. The scene was different in the media, the blogosphere, and the court of public opinion, where the emails drew intense scrutiny and criticism—even from some of the greenest activists and writers around. Once the dust had settled, though, it became evident that many of the items that seemed outrageous at first weren't so shocking in context. And as many researchers took pains to emphasize, Climategate had no effect on the rock-solid scientific consensus, accumulated through the work of thousands of scientists, that greenhouse gases were warming Earth's climate and would continue to do so. In that sense, the whole episode was a colossal distraction—albeit a very effective one.

Another set of controversies that erupted around the same time involved errors in the 2007 IPCC assessment. One of the most prominent of these dustups, inevitably dubbed Glaciergate, centered on the report's claim that "the likelihood of [glaciers in the Himalayas] disappearing by the year 2035 and perhaps sooner is very high if the Earth keeps warming at the current rate." For this assertion, the IPCC cited a World Wildlife

Fund report from 2005, which in turn drew on a 1999 interview in *New Scientist* magazine with Indian glaciologist Syed Iqbal Hasnian. There was no peer-reviewed work behind the factoid, and it's generally accepted that Himalayan glaciers would take far longer to disappear than several decades (see chapter 6). Another part of the same IPCC paragraph contained a similar error involving the rate of loss of glacier-covered areas. These errors attracted little notice at first. Buried within the full IPCC assessment, they didn't make it into the summary for policymakers, which serves as the assessment's highest-profile document. But by early 2010, the errors had drawn flak from glaciologist Vijay Raina in an Indian government report, and the nation's environment minister was calling the IPCC "alarmist." Finally, the IPCC itself commented on the brouhaha: "In drafting the paragraph in question, the clear and well-established standards of evidence, required by the IPCC procedures, were not applied properly."

Although these and a few other unearthed errors involved only a minuscule part of the IPCC's exhaustive 2007 report, they did put a few dents in the panel's Nobel-backed reputation and raised questions about its review process and its leadership. A blue-ribbon group assembled in 2010 by the world's national science academies, acting through the InterAcademy Council, praised the IPCC's overall success while calling on the panel to "fundamentally reform its management structure" to better handle ever-more-complex assessments as well as ever-increasing public scrutiny. As the council noted, "The world has changed considerably since the creation of the IPCC." Many of these recommendations were put in place for the 2013 IPCC assessment (see chapter 15).

The various dramas swirling around climate change in 2009–2010 extended to the atmosphere itself. That winter, the U.S. East Coast and much of Europe were entrenched in a prolonged stretch of bitter cold and frequent snow, related to record-low values of the North Atlantic Oscillation (see sidebar, "The NAM, the NAO, and other climate cycles"). The onslaught was perfectly timed to build on skepticism fuelled by the revelations above. In February 2010, the family of U.S. senator James Inhofe—for many years the most outspoken climate skeptic in Congress—built an igloo in deep snow near the Capitol, with a sign reading "Al Gore's New Home." The igloo was nowhere in sight a few months later, when Washington endured its hottest summer ever observed. Even with the epic snow and cold taken into account, 2010 was one of the warmest years on record for the nation's mid-Atlantic coast.

Indeed, 2010 brought new monthly peaks in temperature for the globe as a whole, and the seasonally adjusted concentration of carbon dioxide topped 390 ppm for the first time. Nevertheless, the continued reality of a warming planet, and the overall solidity of climate science, couldn't seem to hold a candle to the skepticism fanned by politicians and commentators and bolstered by the emails and IPCC errors. From 2008 to 2010, public opinion polls in the United States, Britain, Germany, and Australia showed drops of 10%–20% in acceptance of, and concern about, human-produced climate change.

Political developments in 2010 foreshadowed the difficult path that climate-related legislation would face in the decade ahead. When David Cameron became U.K. prime minister that year, he immediately promised to lead "the greenest government ever." Yet the first British carbon tax to affect the economy as a whole, which went into effect in 2013, was followed by consumer utility hikes and a storm of public protest, which prompted Cameron to advocate a pullback of environment-related fees on utilities. Likewise, an Australian carbon tax made it into law in 2012 amid a series of political power struggles, but it was targeted for repeal only a year later. The U.S. Congress also swung sharply to the right in 2011, making it all but certain that any major American climate bill would be years away. (In spite of this, the Environmental Protection Agency continued moving forward on regulating greenhouse gases under the existing Clean Air Act, a path affirmed by the Supreme Court in 2007.) Developments were equally hamstrung on the international front, as the UN struggled to patch up disagreements that came to the fore in Copenhagen. Each of the subsequent annual climate summits through 2013 made only modest progress at best, with the future of any post-Kyoto agreement shrouded in uncertainty (see chapter 15).

As public opinion on climate change rebounded only gradually from its 2009–2010 dip, researchers began digging deeper into why so many people seemed so hesitant to accept the overwhelming consensus of climate scientists: that global warming since the 1950s was real and primarily human-caused. The apparent polarization of U.S. citizens took on more nuance with the advent of the influential "Six Americas" concept, which was based on extensive polling led by Anthony Leiserowitz (Yale University). The idea is that the U.S. public can be placed into six broad camps, ranging from "alarmed" to "dismissive," with the groups at either end most involved in the public debate. The impression of a split public fostered by those two outspoken groups obscures the widespread agreement found by Leiserowitz and colleagues that global warming is real, and the nation should act on it—although a majority of Americans still appear to doubt that humans are the primary cause. These concepts are exemplified in Texas, a politically deep-red state that was hit hard by record heat and drought in 2011. A Yale survey in 2013 found that 70% of Texans believe global warming is happening, and 60%–70% advocate action on individual, state, corporate, and federal levels. Yet nearly half (47%) believe that most scientists disagree on whether global warming is happening (never mind whether it's human-caused or not).

Part of Americans' growing concern about climate change may be related to some high-profile weather events, including the intense drought and record-smashing heat that struck much of the central United States in 2012 (see chapter 4). In particular, 2012's Hurricane Sandy jangled the nation's nerves. The storm was oddly timed and immensely large, and there was intense debate among scientists and commentators on whether large-scale weather conditions related to climate change might have fostered Sandy's extremely unusual path (see chapter 8). Yet many observers pointed to the fact that sea level rise alone would make any similar storm more damaging as time goes by. "As bad as Sandy was," cautioned New York mayor Michael Bloomberg, "future storms could be even worse."

Another devastating cyclone, Supertyphoon Haiyan, served as a climate change icon in 2013. Although the storm struck a normally typhoon-prone area in the Philippines, and most experts declined to pin the lion's share of its impact on climate change, Haiyan's stunning damage and its extreme intensity—including estimated winds as fierce as any in the 50-year satellite record—made their own impressions. Just a few days later, on the opening day of the UN climate meeting in Warsaw, lead Philippine negotiator Naderev "Yeb" Saño drew a standing ovation as he announced: "I will voluntarily refrain from eating food during this [conference of parties], until a meaningful outcome is in sight."

It seems a safe bet that economic concerns will continue to battle it out with extreme weather events for the hearts and minds of politicians and the public throughout the 2010s. With so many people worried about climate change, yet still unsure about the link to human activity, it's likely that the emphasis will increasingly fall on green actions that make both short- and long-term economic sense. Whether sufficient innovative policies and technologies can leap into place quickly enough to avert the worst scenarios of twenty-first-century global warming remains to be seen. In the next three chapters, we'll look at some of the leading political and technological solutions being proposed.

CHAPTER 14
The Predicament
WHAT WOULD IT TAKE TO FIX GLOBAL WARMING?

The global warming problem isn't going to be solved tomorrow, next week, or next year: we're in this one for the long haul, and there clearly isn't any single solution. A multifaceted approach could include governments agreeing to and enforcing targets, innovators developing low-carbon energy sources and improving efficiency, and individuals and organizations doing their best to reduce their own carbon footprints (and motivating others to do so). These three approaches are explored in the following chapters, but first, let's take a look at the problem—and what we need to do to solve it—as a whole.

Looking at the global-scale challenge before us, the most obvious worry is the sheer momentum that needs to be overcome. We've already added a great deal of greenhouse gas to the atmosphere, and we're adding more each year than the year before. Even if we can lower emissions, there's enough inertia in the physical drivers of climate to keep us rolling toward an ever-warmer future for many years to come. In short, then, if we're to reduce the chance of long-term climate trouble, we need to take real action on multiple levels sooner rather than later.

Understanding the goals

In terms of reducing emissions to mitigate climate change, at least three types of goals are commonly discussed by policymakers and activists working in the field of climate change.

* **Stabilizing emissions.** Making sure that each year we emit no more than the year before, and ideally less.
* **Stabilizing concentrations.** Reducing emissions enough so that the amount of greenhouse gas in the atmosphere levels off at a target and stays there (or falls back down).
* **Stabilizing temperature.** Keeping the atmosphere from warming beyond a certain point.

Obviously, these three types of goals overlap. Our emissions build up in the atmosphere, changing the concentration of greenhouse gases, which in turn alters the temperature. The graphic at right shows how these three quantities evolve along the four pathways outlined in the latest IPCC report.

Packed inside this complex-looking graphic are a few distinct messages. First off, you can see that simply leveling off emissions won't be enough to stabilize concentration and therefore temperature, because elevated levels of carbon dioxide persist in the atmosphere for many years, even centuries. After all, if water's flowing into a bathtub faster than it can drain out, you need to *reduce* the flow—not just keep it constant. And even though stabilizing global emissions would be an enormous short-term achievement, we actually need to go much further and make significant *cuts* in emissions (the "downhill" side of the four emission curves shown in the top graphic) in order to keep the world from getting significantly hotter. As we'll see later, some countries have already managed to stabilize or reduce their emissions, but globally, there's a long way to go.

Moreover, there's the **time-lag** factor to consider. If warmer temperatures bring lasting changes to the Earth system, such as the loss of major ice sheets or the release of trapped methane, then substantial climate change could be locked in for centuries to come. As the graphic shows, even if we do manage to make significant emissions cuts, it will take decades for the concentration to begin leveling off and more than a century for the temperature to stop rising—that is, unless we can implement major year-over-year reductions in global emissions starting in the 2020s and continuing thereafter (the RCP2.6 scenario).

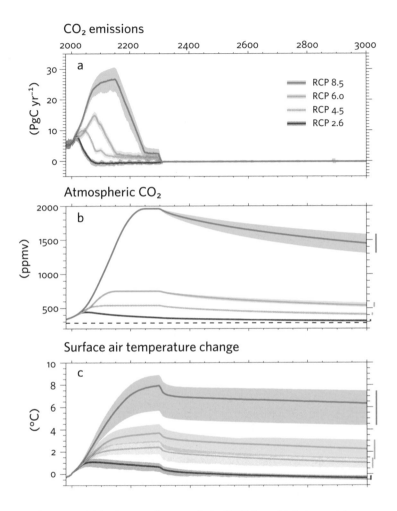

CO₂ emissions

Atmospheric CO₂

Surface air temperature change

Earth system models of intermediate complexity (EMICs) are detailed enough to capture important processes, yet simplified enough so that very long-term simulations are feasible. This graphic shows how EMIC runs carried out in support of the 2013 IPCC assessment depict three key variables over the next thousand years—carbon dioxide emissions (top), CO_2 concentrations (middle), and global surface air temperature (bottom)—for each of the four IPCC emission pathways, or RCPs. Temperature change is relative to the period 1986–2005. To determine the total temperature change since preindustrial times (1850–1900), add roughly 0.6°C (1.1°F). See p. 16 for RCP details. Shaded bands show the range for each model ensemble, with each solid line denoting an ensemble mean. The abrupt temperature drop in 2300 is a result of eliminating all CO_2 emissions and non-CO_2 forcing factors. (IPCC)

To help simplify this picture, many researchers and policy makers have begun focusing on the **cumulative carbon** that's put into the atmosphere, rather than the year-by-year emissions. There's quite a wide range of estimates on how much total carbon we can afford to emit, but surprisingly, the ultimate temperature jump doesn't appear to hinge on the timing of emissions—just on the total amount. As such, no matter what temperature rise we're prepared to risk, the longer we wait to start cutting emissions, the more stringent the later cuts will need to be.

In its 2013 assessment, the IPCC considered this type of global carbon budgeting in new detail. The panel tallied how much CO_2 we could burn and still keep the temperature rise at no more than 2°C over preindustrial levels (using the average temperature in the period 1861–1880 as the preindustrial benchmark). If we agree that a two-out-of-three chance (66%) of staying within the 2°C margin is an acceptable margin of risk, then humanity's total carbon emissions from CO_2—past, present, and future—must stay below about 1000 metric gigatons (and that's not even considering the added impact of other greenhouse gases, such as methane). However, as of 2011, we've already emitted roughly 515 metric gigatons of carbon in the form of CO_2. Thus, to retain that desired 66% chance of not exceeding the 2°C target, we'd need to keep our entire future emissions less than our total past ones—which would be truly a Herculean task, given the relentless upward trend so far.

Selecting a target

The IPCC's use of the 2°C benchmark raises a critical question: how much global warming would be truly dangerous? There's no single bright line that separates mild effects from serious threats, but researchers have done their best to identify points beyond which particular risks become more likely. Many studies suggest that a rise of 3°C (5.4°F) relative to preindustrial levels would be well past the edge of the comfort zone. It may be enough to trigger unstoppable melting of the Greenland ice sheet, for example, or a net drop in world food production. Any further warming could jeopardize parts of the Antarctic ice sheet and cause other dire consequences.

With these and many other factors in mind, the most commonly cited temperature target for climate stabilization is the familiar **2°C (3.6°F)** benchmark above preindustrial levels, or around 1.2°C (2.0°F) above the global temperature of the early 2010s. This value was agreed to by the European Union as far back as 1996. It's also shared by many researchers and activist groups. As German analyst Oliver Geden noted in 2013, "Despite the many uncertainties inherent in it, 2°C has been able to prevail as the global temperature threshold. It functions as the central point of reference in the climate debate, and as the one concrete objective on which key actors from policy, media, and research have been able to reach at least interim agreement."

Some parties have called for more ambitious goals. At the Copenhagen meeting in 2009, a group of more than 40 small island nations and like-minded activists pushed for a target of **1.5°C (2.7°F)**. The idea was to minimize the risk that rising sea levels could swamp low-lying countries. The accord that emerged from Copenhagen included the standard 2.0°C target, but it also called for an assessment (under way in the mid-2010s) to determine whether or not to strengthen the target to 1.5°C.

Even more stringent was the goal proposed by the grassroots effort **350. org**, which has organized more than 10,000 demonstrations and events since 2009. The group's name comes from its central goal of bringing global CO_2 concentrations down from their current values of around 400 ppm to **350 ppm**, which is deemed by the group (and by NASA's James Hansen, among others) to be the value consistent with a reasonable shot of keeping the eventual global temperature rise to 1.5°C. Reaching the 350 target is "a non-negotiable demand from the planet itself," says author and 350.org founder Bill McKibben. Even if it were ultimately unachievable, the 350-ppm goal serves as an icon of urgency and the need for action, as well as a reminder that we're already on a path toward much higher concentrations.

2°C—or beyond?

Given physical and political realities, the world may not be able to avoid 1.5°C at this point. After all, as we saw back in chapter 1, Earth has warmed by more than 0.8°C (1.4°F) thus far, and there's an additional 0.5°C (0.9°F) to come as the planet adjusts to emissions already in the atmosphere. Together, these bring us uncomfortably close to 1.5°C. Scenarios that would likely keep us below this target are so sparse (and perhaps so implausible now) that none have been examined thoroughly by multiple models, as the IPCC noted in its 2014 report on mitigation.

At least in theory, we might still have a chance of staying below 2°C, but the odds of reaching that goal are diminishing quickly. Global emissions have soared more than 50% since the **UN Framework Convention on Climate Change (UNFCCC)** process began in 1990, with increases of several percent notched in every year outside of the occasional minor drop during recessions (see below). Every so often there's a glimmer of hope, such as the news from the Netherlands Environmental Assessment Agency that global emissions in 2012 increased by a relatively modest 1.1%, even as the global economy grew by more than double that percentage. Experts took this as a sign that economic growth isn't necessarily yoked to emissions growth, as so many have argued in the past. Even so, the split between developed and developing economies that ultimately hobbled the Kyoto Protocol remains a huge obstacle, one that's proven extremely difficult to address (as evidenced by the schisms on display during the 2013 UN climate meetings in Warsaw). Given the labored pace of UN negotiations over the last few years, it's possible that any binding agreement to reduce emissions wouldn't take effect until well into the 2020s, if even then. Some analysts have already begun to explore the implications of revising or even abandoning the 2°C target should developments over the next few years prove it to be out of reach. Such a move would be tremendously risky, since it could lead to cynicism and fatalism about the global effort to keep greenhouse gases in check. This puts even more emphasis on the need for national, regional, and local initiatives—plus action by companies and individuals—to help stem the tide of emission growth.

All this effort could be for naught unless the act of burning carbon carries a price tag that's valid in some form or fashion around the entire world. Otherwise, there's a major risk of what's been dubbed **carbon leakage**. Should one nation reduce its emissions voluntarily, the supply of unburned fuel will rise and its price on the open market will go down, all else being equal. In turn, that would open the door for some other entity to purchase and burn the fuel that was conserved in the name of climate. For example, the United States moved from coal to natural gas in the early 2010s, which helped the nation to achieve unexpected emission reductions—but much of that coal ended up being burned elsewhere, as U.S. coal exports jumped to a record high.

A similarly thorny problem is **rebound effects**, which occur when the benefits of increased efficiency are directed into activities that burn more energy. Paradoxically, the result can sometimes be *more* carbon emission rather than less. For example, a family might decide to save money on utilities by insulating their home, but then spend that newfound cash on an extra plane trip each year—thus expanding their carbon footprint even more than they'd reduced it by sealing their drafty house.

This isn't to denigrate the importance of personal as well as national emission reductions, both of which are absolutely crucial in any effort to keep our planet from overheating. It's simply to recognize that our global economic system is hard-wired to consume whatever energy is available on the free market. If the cost to our global commons isn't somehow factored in, then achieving major emissions cuts will be a truly daunting effort. The true scope of this challenge was highlighted in 2012 through a hugely popular *Rolling Stone* article by Bill McKibben entitled "Do the Math," accompanied by a standing-room-only lecture tour. McKibben pointed out that, according to the Carbon Tracker Initiative, the world's proven reserves of fossil fuel in 2011—the oil, gas, and coal held by the planet's private and state-owned energy firms—represent a staggering 2795 billion metric tons of carbon. That's more than five times the total that can be burned from this point onward while still retaining a two-thirds chance of meeting the 2°C target, on the basis of the IPCC calculations noted above. It's difficult to imagine the world's big energy firms, which are among the largest and most profitable companies on the planet, voluntarily writing off more than 80% of their holdings (trillions of dollars' worth) and leaving them in the ground. Yet if even half of those proven reserves are burned, the chance of avoiding what the IPCC terms "dangerous interference with the climate system" will become slender indeed.

Overall, this looks like quite the discouraging picture. However, we have little choice but to face the predicament head on. That includes working in a variety of ways to ensure that any eventual multifaceted solution can benefit as much as possible from progress achieved on many fronts. As we'll see in the next two chapters, policy experts are looking at new ways to approach the carbon problem from an international standpoint, and plenty of new technologies are on tap to help reduce emissions substantially wherever there's an incentive for doing so. See the subsequent section (chapters 17–21) for tips on what you can do on a personal level.

When one considers greenhouse gases other than CO_2, the picture brightens just a bit. Most of those gases are less prevalent and shorter-lived than CO_2, but far more powerful. Together, they account for more than 30% of the global warming impact from gases we're adding to the atmosphere each year. The good news is that some of them could be reduced more easily and affordably than CO_2, and political agreements that can smooth the way are already in place, such as the Montreal Protocol and regional air pollution laws. In the EU, for instance, the nitrogen oxides emitted from

" The choice we face is between taking unimaginable risks with the planet and leaving vastly valuable fossil fuels in the ground." —*Mike Berners-Lee and Duncan Clark, The Burning Question*

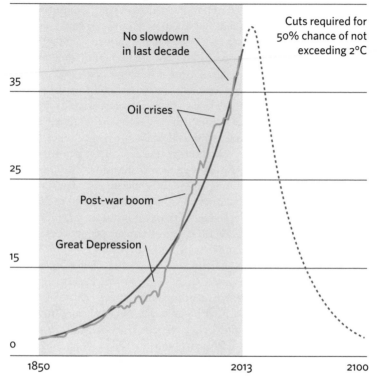

45 Billion metric tons of CO_2

No slowdown
in last decade

Cuts required for
50% chance of not
exceeding 2°C

35

Oil crises

25

Post-war boom

Great Depression

15

0

1850   2013   2100

CO_2 emissions since 1850 (orange), exponential growth (blue), and cuts to hit climate target (dashed). (Duncan Clark/*The Burning Question*)

diesel-powered cars and light trucks sold after 2009 were cut by roughly 20%; they're being reduced more than 50% beyond that new limit starting in 2014. There is some hope among policy experts and climate scientists that a two-step approach—cutting the emission rates of non-CO_2 greenhouse gases right away, plus reducing CO_2 over a somewhat longer time frame—might prove fruitful. New approaches to keeping carbon in Earth's ecosystem, including stronger **deforestation** limits, could also have a big impact relatively quickly.

Whether any particular temperature, emission, or concentration goal can be met depends on other factors as well, such as how fast **new technology** is developed and adopted and how seriously we take **energy efficiency**, which in turn depends partly on the political will to prioritize climate change. Another important factor is the state of the world **economy**. Global emissions of CO_2 from fossil fuels actually fell more than 4% between 1980 and 1983, a period of high oil prices and widespread recession, and they dropped by about 2% in 1992 and 1999, when the economies of eastern Europe and Russia were struggling. Despite the intensity of the worldwide downturn of 2009, global CO_2 emissions fell by only a little more than 1% that year, and in the following year (2010) they soared by more than 4%.

Naturally, all these factors are interrelated in complex ways. For example, high oil prices can simultaneously dampen economic growth, encourage efficiency, and stimulate investment in alternative energy sources—all of which are likely to reduce emissions. But if the economy suffers too much, governments may feel pressure from voters to prioritize short-term growth over long-term environmental issues. Despite all these complications, it seems at least theoretically possible that we could manage long-term net global emission cuts of a few percent within a decade or two, assuming there are incentives for reducing carbon use that carry weight in the global marketplace. And, though implementing such deep cuts would be liable to cause short-term fiscal pain (perhaps significant in some quarters), the long-term economic and environmental gains from energy efficiency and renewable energy could be enormous and widespread.

The wedge strategy

Since almost all human activity contributes to greenhouse gas emissions on some level, the task of reducing global emissions can seem somewhat overwhelming. What if we thought of it as a series of simultaneous smaller goals? That's the philosophy behind Stephen Pacala and Robert Socolow's "wedge" approach to climate protection. The two Princeton University scientists brought the wedge concept to a wide audience through a 2004 article in *Science*. The idea is to break down the enormous carbon reductions needed by midcentury into more manageable bits, or wedges, each of which can be assigned to a particular technology or strategy.

When it was introduced, the wedge concept triggered widespread interest and excitement, in part because Pacala and Socolow claimed that the needed emission cuts through 2050 could be handled entirely through existing technologies. The concept has proven hugely influential, and it's a handy way to compare and contrast various pieces of the overall emission-reduction puzzle.

The wedge concept originates from Pacala and Socolow's projection of historical CO_2 emissions into the future, starting from 2004, the year their paper was published (see diagram, p. 364). Let's assume it is 2004 right now, and let's also assume that emissions can be instantly stabilized—in other words, the yearly increases in CO_2 all go to zero right now—and remain that way until at least the 2050s. This is represented by the flat black line on the diagram. Pacala and Socolow argued that this would correspond to an eventual CO_2 concentration of about 500 ppm. Such a route falls somewhere between the two most optimistic emissions pathways (RCP2.6 and RCP4.5) in the 2013 IPCC report. This would provide us with a good chance of staying below 3°C by century's end, and at least a shot of remaining below 2°C, depending on how sensitive the climate is to CO_2 (see chapter 12). However, if emissions continue to increase as they have in the last several decades, at more than 1% per year—the red line on the diagram—then by 2054 we'd be adding twice as much CO_2 to the atmosphere each year (see the red "business as usual" line on the diagram). The result would be a warming far more severe than 2°C, in line with the most pessimistic of the IPCC pathways.

The triangle between the black and red lines shows the difference between the desired path of steady emissions and the dangerous uphill path. To get rid of the triangle, we'd need to come up with at least seven wedges, each of which would eliminate a billion metric tons (a metric gigaton) of annual carbon emission by 2054. Further wedges would probably be needed after 2054 to stay below the 2°C target.

Pacala and Socolow identified 15 examples of potential wedges (see descriptions beside graphic), each of which was already being implemented on a reasonably large scale somewhere in the world. They maintained that a 50-year goal was easier to grasp than a century-long one. Among other things, it's comparable to the length of a career or the lifetime of a coal-based power plant. As they put it, "Humanity can solve the carbon and climate problem in the first half of this century simply by scaling up what we already know how to do."

How do the wedges of 2004 hold up in the 2010s? Apart from occasional economy-produced dips, global emissions have been rising a bit faster than the 1.5% annual increase assumed in the Pacala and Socolow graph. In 2011, seven years after the breakthrough paper he coauthored, Socolow took a fresh look at the wedge strategy and concluded that, given the unabated emissions up to that point, we now needed nine rather than seven wedges. That's with the concentration goal now bumped up to 550 ppm, which adds perhaps 0.5°C to the eventual temperature peak and makes reaching the 2.0°C goal far more problematic. Taking into account the stark realities of our current path as well as the latest science, researchers led by Steven

Davis (University of California, Irvine) mapped out a far more ambitious variation in a 2013 paper called "Rethinking wedges." They propose a total of 19 wedges, each as large as those in the original 2004 set, that could bring CO_2 emissions close to zero within 50 years. Doing so, they hasten to add, would require "an integrated and aggressive set of policies and programs . . . to support energy technology innovation across all stages of research, development, demonstration, and commercialization. No matter the number required, wedges can still simplify and quantify the challenge. But the problem was never easy."

Indeed, while the wedge strategy is a helpful way to assess which directions we might want to go in, it certainly doesn't make the job a cakewalk, as Pacala and Socolow acknowledged from the start. "Each of these wedges represents a staggering amount of effort from the public and private sector," noted Joseph Romm in a 2008 *Nature* analysis. It's obvious that some proposed wedges are far more practical and/or politically palatable than others. At least one of the original wedges—the ramp-up of ethanol—has been fraught with problems that weren't fully foreseen a decade ago. Even if alternatives are deployed, there's no guarantee that they will supplant traditional fuels. And if the history of energy development is any guide, the economies of scale tend to push each sector toward a strongly preferred fuel type: gasoline for vehicles, coal for electric power plants, and so on. This would work against, say, the parallel large-scale growth of electricity generation from wind power, nuclear power, and cleaner types of coal. Moreover, key interest groups—whether it be fossil fuel companies or hard-green activists—are often hesitant to relax long-held positions that end up favoring only a small number of wedges rather than the full smorgasbord. There's also the obvious fact that we can't stabilize emissions today or tomorrow as if we were flipping a switch. Even using current technologies, such massive change would take years to implement. This is why many scenarios constructed by the IPCC and others include a ramp-up phase of continued growth, followed by substantial cuts in emissions—ultimately bringing us well below today's emission levels—in contrast to the flat-line "stabilization trajectory" shown on the above graph. One other concern: bolstering the clean-energy wedges may not reduce the global use of dirty energy if there aren't economic incentives that transcend national boundaries, as noted above.

Looming in the background, and not considered directly in the wedges, is the inexorable rise in global **population**, which threatens to swamp the emission cuts made in efficiency and technology. Because the idea of stabilizing global population has proven to be such a political minefield, it's

The wedges (expressed in terms of the year 2054 versus 2004):

▲▲ Doubling vehicle fuel economy and **cutting the distance driven per car in half** (one wedge each). Fuel economy is on the rise in many wealthy nations, including the United States, but auto use is also increasing rapidly in the developed world.

▲ Installing lights and appliances with state-of-the-art energy efficiency in all new and existing residential and commercial buildings, thus reducing emissions by about 25%. There's been incremental progress in recent years, especially with the advent of compact fluorescent (CFL) and LED lighting.

▲ Improving the efficiency of coal-fired power plants from 40% to 60%, plus cutting in half the energy lost when fossil fuels are extracted, processed, and delivered to those plants. Progress has been relatively slow in this area.

▲▲ A 50-fold expansion in wind energy, replacing coal. As of 2012, the global total of installed wind-power capacity was more than five times the 2004 value. A second wedge could be obtained by adding another four million turbines to generate hydrogen for fuel-cell vehicles.

▲ A 700-fold expansion of photovoltaic (PV) solar energy, again replacing coal. This would require enough panels to blanket every inch of an area the size of New Jersey. However, many of the panels could be mounted on roofs and walls. The total capacity of solar PV panels installed around the world grew roughly 20-fold from 2004 to 2012, with no end in sight to the rapid growth.

▲ A 50-fold expansion in ethanol, displacing petroleum. Even with plants that yield far more energy per acre than the corn and sugar cane used now for ethanol, these fuel crops would still require a region about the size of India. Biofuels production more than tripled from 2004

to 2010, but the trend then flat-lined over the following two years.

▲ A halting of current deforestation, coupled with plantations on non-forested land, together covering a total area about 40% the size of Australia by 2050. There's been major progress on deforestation, due in part to agreements made at the 2009 Copenhagen summit.

▲ Employing conservation tillage on all cropland, in which farmers avoid plowing and thus reduce the amount of CO_2 escaping from tilled soil. It's now used on only a small fraction of cropland globally, but interest in these techniques is spreading.

▲ Tripling the amount of energy now generated by nuclear sources, adding about 700 one-gigawatt plants as well as maintaining or replacing all nuclear plants now in use. Nuclear power has been on the decline globally in recent years, with global electricity production from nuclear plants dropping more than 3% from 2009 to 2011. It plummeted another 7% in 2012, largely due to the Fukushima disaster and the power-down of other nuclear plants in Japan.

▲ Quadrupling the use of natural gas in power plants, replacing an equal number of coal-fired plants. Electricity production from natural gas grew by more than 10% from 2004 to 2010, though it dipped slightly in 2011.

▲▲▲ Sequestration—capturing carbon emitted by large fossil fuel power plants and storing it underground—with one wedge each coming from standard coal or natural-gas plants, synfuel plants (which generate synthetic fuel from coal), and hydrogen plants that draw on coal or natural gas. Although research continues, even experimental deployments have been few and far between.

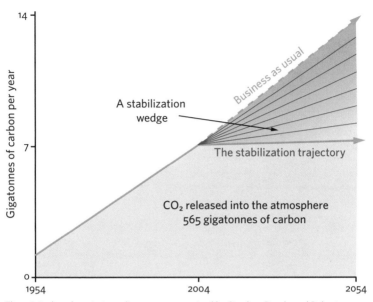

The original wedge-strategy diagram, as conceived by Stephen Pacala and Robert Socolow.

sometimes omitted from climate change discussions. But there's no getting around the fact that more people on the planet will—all else being equal—lead to more greenhouse emissions.

Despite all these various qualms, the wedge concept remains useful as a rough framework for examining the total emissions picture. Each wedge is technically feasible—it's a matter of society choosing which ones to emphasize. There's been real progress since 2004 on some of the wedges, including wind and solar energy (see chapter 16). Government subsidies or other incentives can further shape this course, and it's possible that different countries will opt for different technologies, thus providing a well-rounded portfolio on the global scale. The wedges also allow one to work backward and set even shorter-term goals, taking into account that some wedges might grow more rapidly at first while others would take a while to get going.

We'll be considering the above options in chapters 16–21, after we explore what governments are doing—and not doing—to speed us along the path to a low-carbon future.

CHAPTER FIFTEEN
Political Solutions
KYOTO AND BEYOND

In one way or another, climate change must be dealt with on a global level, and this poses an unparalleled challenge in international diplomacy. For one thing, a whole host of intractable issues, from excessive consumption to economic inequity, tend to glom onto climate change, as if the latter weren't a daunting enough problem in itself. Over more than 20 years of addressing climate change, the United Nations has made fitful headway. Nearly all countries are on record as taking the problem seriously, and the UN's 1997 Kyoto Protocol—though weaker than many had hoped—was ratified by countries that together represent the majority of people in the developed world (though only a third of global emissions). However, the effort to forge a binding post-Kyoto agreement experienced a major set-back in 2009 with the near-collapse of UN negotiations in Copenhagen, and progress in the early 2010s toward a new plan for global action was painfully slow.

For reasons we'll discuss below, it could be very difficult to bring about major emissions reductions for the planet as a whole without a strong agreement that involves virtually all nations. However, many governments aren't waiting for the next Kyoto Protocol. Some countries that opted out of Kyoto are exploring other mechanisms for grappling with the carbon problem. For instance, a number of large U.S. states and cities volunteered for reductions that match or exceed Kyoto goals, and a whole smorgasbord of new ideas will be on the table over the next few years as diplomats attempt to hammer out a global agreement to succeed or replace Kyoto.

The road to Kyoto

In some ways, global warming couldn't have arrived on the diplomatic radar screen at a better point than in 1988. A year earlier the UN had forged the **Montreal Protocol**, which called for the gradual phase-out of the industrial chemicals (chlorofluorocarbons) implicated in the Antarctic ozone hole (an issue separate from global warming, as explained on p. 32). The Montreal Protocol came about only after a thorough assessment of the science behind ozone depletion. Previous work had already shown the connection to chlorofluorocarbons, but it was the assessment that gave an international imprimatur to the findings. It showed that researchers from around the world agreed with the science, which in turn provided the green light for global action. The protocol even had a happy side effect for climate: because CFCs are potent greenhouse gases, their phase-out has been one of the biggest emission-reduction successes to date. Had nations not acted to cut CFCs and other ozone-depleting chemicals from the 1970s onward, the total greenhouse impact of these gases by 2010 would have nearly rivaled that of carbon dioxide, according to an analysis led by Mario Molina (University of California, San Diego). With this in mind, similar gases might be added to the protocol—especially hydrofluorocarbons (HFCs), which are ozone-friendlier than CFCs but just as dangerous from a greenhouse standpoint.

The Montreal Protocol's success made it natural for the UN to ask for another high-level scientific assessment when global climate change reared its head. In response, the World Meteorological Organization and the UN Environment Programme established a group called the **Intergovernmental Panel on Climate Change (IPCC)**. Few, if any, people at the time could have guessed how influential, and at times controversial, the IPCC would turn out to be. Its findings have been featured in countless news stories and cited in thousands of reports. The IPCC's five major assessments—released in 1990, 1995, 2001, 2007, and 2013–2014—have become the definitive reference tools on the state of climate science, and the panel shared a Nobel Prize with Al Gore in 2007 for its cumulative ef-

> **I told the world I thought that Kyoto was a lousy deal for America."**
>
> *—U.S. president George W. Bush, March 2006*

forts (see overleaf for more about how the IPCC works). Several nations have taken a cue from the IPCC and produced their own comprehensive assessments.

The IPCC's first assessment in 1990 underscored the seriousness of the climate change threat, so there was plenty of momentum driving world leaders to the 1992 UN **Earth Summit** in Rio de Janeiro. Global warming was only one of a host of environmental issues on the table at this Woodstock of green politics, but it played a starring role. The UN Framework Convention on Climate Change (**UNFCCC**) was introduced at the summit and signed by 166 nations that summer (and, by 2013, all UN member states except for the new South Sudan). Because the United States and several other countries were steadfast in their opposition to binding emissions reductions, the UNFCCC itself doesn't include any targets or timelines, much to the dismay of climate activists. Yet it was, and is still, a landmark agreement. Among its key points are the following:

* **The ultimate objective** is to stabilize climate "at a level that would prevent dangerous anthropogenic interference with the climate system" and in a time frame "sufficient to allow ecosystems to adapt naturally to climate change, to ensure that food production is not threatened and to enable economic development to proceed in a sustainable manner."
* **International equity** is given a nod with a reference to "common but differentiated responsibilities and respective capabilities," a theme that ended up in the Kyoto Protocol.
* **Uncertainty in climate science**—at that point already one of the favorite criticisms from skeptics—was acknowledged but put in its place: "Where there are threats of serious or irreversible damage, lack of full scientific certainty should not be used as a reason for postponing such measures."
* **Countries agreed to inventory their greenhouse emissions** and publish yearly summaries.

With the UNFCCC in place, the world's industrialized countries plus "economies in transition" (mostly in central and eastern Europe)—together referred to as **Annex I**—agreed to reduce their emissions to 1990 levels by the year 2000. This was fairly easy for Russia and the former Soviet republics because of their lackluster economic performance and the resulting drop in energy use, but many other countries missed the goal. Britain managed it only because of the phase-out of subsidies for coal-fired power

Inside the IPCC

The IPCC has only a few permanent staff, but it's a far larger enterprise than the term "panel" might suggest. Indeed, it's one of the biggest science-related endeavors in history. That said, the IPCC doesn't conduct any science of its own. Its role is to evaluate studies carried out by thousands of researchers around the world, then to synthesize the results in a form that helps policymakers decide how to respond to climate change. Each assessment is a bit different, but typically each of several IPCC working groups generates an exhaustive report as well as a summary for policymakers. All of the reports are available online. For their 2001, 2007, and 2013–2014 reports, the three working groups dealt with: **the basis in physical science** (how climate change works); **impacts, adaptation, and vulnerability** (options for dealing with climate change); and **mitigation** (options for reducing emissions or removing them from the atmosphere in order to minimize climate change). Each of the three working group reports includes the summary for policymakers, typically released a few weeks ahead; a synthesis report concludes the process by incorporating findings from all three groups.

Every IPCC assessment involves several hundred researchers from many dozens of countries, generally nominated by their governments or by a nongovernmental organization. Each working group—and typically each chapter topic within a working group—is headed by a pair of scientists, one each from a developed and a developing country. By and large, these scientists volunteer their time to be involved with the IPCC with the blessing of their employers. They survey peer-reviewed science studies and other pertinent materials; meet with peers to gather input as needed; and draft, revise, and finalize reports. Several hundred other experts then review each report. Finally, each document is scrutinized by technical reviewers within each government and accepted at a plenary meeting. The policymaker summaries take shape on a parallel track; they're approved by a panel of governments on a word-by-word basis.

With so much riding on its conclusions, it's no wonder that the IPCC has drawn scrutiny from those aiming to discount the risk of climate change. The second assessment provoked an attack from the Global Climate Coalition (see chapter 10) and other skeptics, who criticized the sequence of events through which the wording of one of the summaries for policymakers was prepared (including the fateful statement "the balance of evidence suggests that there is a discernible human influence on global climate"). The process was revised for the next assessment in response to this and other concerns, although no IPCC rules had been violated and the panel's earlier conclusions were unchanged. After the third assessment came out, skeptics' focus turned to its "hockey stick" graphic depicting climate over the last thousand years, including the sharp upturn of the twentieth century (the "head" of the stick—see chapter 11). Controversy aside, the IPCC's pronouncements on the state of global climate resonate worldwide. As noted earlier in this book, the IPCC has grown increasingly emphatic about its conclusions on human-induced climate change.

Susan Solomon with a draft of the 2007 IPCC assessment. (Robert Henson)

Although the IPCC doesn't commission research, the world's major climate-modeling activities are timed to coordinate with IPCC schedules, with the results compiled and coordinated through the Coupled Model Intercomparison Project (CMIP). The CMIP5 suite of models that supported the 2013 assessment included the type of century-long simulations featured in past reports, as well as a set of "near-term" analyses that each spanned 10–30 years. The near-term modeling, which was jump-started using actual ocean and sea ice conditions, was designed to provide more detail on the time frames of keenest interest for policymakers and other users.

Putting together an IPCC report is unlike any other job in the science world. Susan Solomon, the scientist who unraveled the role of polar stratospheric clouds in creating the ozone hole (see chapter 2) while at NOAA, served as co-chair of Working Group I for the 2007 assessment. "It's a very intense activity," says Solomon, now at the Massachusetts Institute of Technology. She estimates she went through more than 17,000 comments from more than 500 reviewers on the first draft of her group's report. "I've learned a tremendous amount about climate, but it is demanding, both personally and professionally," said Solomon shortly after the work concluded. More than anything, she stressed the community aspect of the panel: "It's very important for people to understand that the IPCC is not one scientist's voice."

plants and a resulting shift to natural gas (much as the growth of fracking has helped the United States trim its own emissions in recent years). As a whole, the Annex I emissions were 6% above 1990 levels by 2002.

It was already apparent by the mid-1990s that some type of **mandatory reductions** were needed to make real progress on emissions. Setting targets, though, was far easier said than done. At a 1995 meeting in Berlin, it was agreed that industrialized countries (the ones that had caused the lion's share of the greenhouse problem thus far) would bear the full brunt of the first round of agreed-upon emissions cuts. This was in keeping with the principles of the UNFCCC. The idea of mandated cuts eventually gained support from U.S. president Bill Clinton, albeit with weaker cuts than those promoted by the EU. But in 1997 the U.S. Congress voted 95–0 against any treaty that did not specify "meaningful" emission cuts for developing as well as developed countries.

The standoff roiled the diplomatic world right up to the fateful meeting in Kyoto, Japan, late in 1997, where targets would be set. In the end—after virtually nonstop negotiations and some last-minute assistance from U.S. vice president Al Gore—the parties agreed to a Kyoto Protocol that would exempt developing countries and produce an average 5.2% cut among Annex I countries by 2008–2012 compared to 1990 levels. This time span was dubbed the **first commitment period**, in anticipation of a second phase that would extend from 2013 to 2020.

The United States and Australia signed the treaty, but neither country proceeded to ratify it right away. (Australia eventually did, as we'll see.) However, the Kyoto Protocol was crafted so that it could become international law even without U.S. participation—but just barely. The rules required ratification by at least 55 Annex I countries that together account for at least 55% of the total Annex I emissions in 1990. Thus, if every other major developed nation apart from the United States and Australia ratified the protocol, it could still take effect. Most of the industrialized world quickly came on board, but it wasn't clear whether or not Russia would. Years of suspense passed before Russia, after stating its opposition to Kyoto in 2003, ratified the protocol in late 2004. The protocol became international law 90 days later, on 16 February 2005. Australia belatedly joined the treaty in December 2007, with incoming prime minister Kevin Rudd signing the ratification document on his first day in office—which happened to coincide with the kickoff of that year's UN summit in Bali (see below). The only major country to withdraw from Kyoto during the first commitment period was Canada, which dropped out in 2012.

How Kyoto works

The meat of the Kyoto Protocol (which is technically an amendment to the UNFCCC) was the requirement for developed nations to cut their emissions of six greenhouse gases: CO_2, **nitrogen oxides**, **methane**, **sulfur hexafluoride**, **hydrofluorocarbons**, and **perfluorocarbons**. Different countries had different reductions targets and, for various reasons, some countries were actually allowed an *increase* in their emissions. The targets were as follows:

* **Bulgaria, Czech Republic, Estonia, Latvia, Liechtenstein, Lithuania, Monaco, Romania, Slovakia, Slovenia, and Switzerland:** 8% reduction;
* **United States:** 7% reduction;
* **Canada, Hungary, Japan, and Poland:** 6% reduction;
* **Croatia:** 5% reduction;
* **New Zealand, Russia, and Ukraine:** no change;
* **Norway:** 1% increase;
* **Australia:** 8% increase;
* **Iceland:** 10% increase; and
* **all other countries:** no mandated targets.

These targets applied to emissions during the interval 2008–2012, which was the close of the first commitment period, and were expressed in CO_2 equivalent, which takes into account the varying power of each of the six gases (see p. 44). With a few exceptions, the targets were relative to 1990. Those for the United States were hypothetical, since it didn't ratify the treaty.

As you might guess, the targets that emerged in the Kyoto Protocol were political creatures. Australia is a good case in point: how did this prosperous country win the right to increase its emissions? As did the United States, Australia came up with voluntary reduction plans in the 1990s, hoping to avoid mandated cuts, but these failed to do the trick. Its negotiators then argued that Australia was a special case because it was a net energy producer with a growing population and a huge, transport-dependent land area. They also pointed to reductions in Australian deforestation after 1990. Kyoto diplomats were keen to get all of the industrialized world on board, so in the end they acceded to Australian wishes, but the country still opted out of the treaty for nearly a decade.

Another wrinkle is the so-called "EU bubble." The European Union was assigned an 8% target but was allowed the freedom to reallocate targets country by country within its overall goal, given that some EU economies would find the 8% limit easier to meet than others would. The resulting targets ranged from reductions as big as 29% for Denmark and Germany to increases as high as 19% and 20% for Portugal and Luxembourg, respectively.

The emissions market

Because Kyoto didn't become international law until 2005, many countries found themselves playing catch-up in order to try and meet their targets. They were assisted by several market-based mechanisms woven into the protocol, designed to help countries meet their targets at the lowest possible cost. The basic idea is that countries or companies finding it tough to cut their emissions directly could meet their targets by purchasing reductions achieved more cheaply elsewhere. A secondary goal was to promote the export of clean technologies to emerging economies. The mechanisms included the following:

* **Emissions trading** is just what its name implies. Annex I countries that exceed their emissions targets can buy allowances from another Annex I country that's doing better than its goal. This idea was inspired by the successful U.S. drive to reduce sulfur dioxide pollution by allowing corporations to buy and sell emission credits from each other under a federally imposed cap on total emissions. Such systems are often labeled "cap and trade" because they operate by capping the total emissions from a group of companies or countries (which assures the overall target will be met) and by allowing the various participants to trade emission credits (which allows market forces to help prioritize the lowest-cost reduction strategies). The systems are launched with a set of permits that are allocated to companies that fall under the plan and/or auctioned off to the highest bidder(s). Ironically, it was the United States that lobbied hardest for a cap-and-trade component in the early days of Kyoto negotiations, while Europe balked. But the United States ended up out of the Kyoto picture, while the EU not only participated in the Kyoto trading scheme, which began in 2008, but also created an internal trading system of its own (see below).
* **The Clean Development Mechanism (CDM)**, first proposed by Brazil, allows developed countries to get credit for bankrolling projects such as reforestation or wind farms that reduce emissions

in developing countries. The scheme was slow to get off the ground, with many investors waiting to see if Kyoto would become official. By 2011, though, more than 2400 CDM projects had been registered, with the majority of the emissions savings taking place in China and India. Together, these projects were expected to keep a total of more than 1.8 billion metric tons of CO_2 equivalent out of the atmosphere between 2001 and 2012, with an addition one billion each in 2013–2015 and 2016–2020. These are substantial amounts, on the order of a few percent of the entire CO_2 emitted globally from human activity per year. However, the CDM is not without its detractors, some of whom object on philosophical and environmental grounds while others criticize the methodology (see "The pros and cons of clean development"). For example, more than a dozen large factories, mainly in China and India, have received hundreds of millions of dollars for CDM projects aimed at destroying trifluoromethane (HFC-23), which is a by-product of making the refrigerant HCFC-22. Critics argue that the CDM itself has motivated companies to generate vast amounts of HFC-23 in order to gain the credits for destroying it (an action that's typically more profitable than manufacturing the HFCF-22 itself). Although the UN banned new factories in 2007 from gaining credits for HFC-23 destruction, preexisting factories represented almost 20% of all CDM credits as of 2012, making them the largest single category. Such credits were banned from Europe's emissions trading system in 2013. U.S. researcher Michael Ware argued in a 2007 *Nature* article that CDMs would be far more effective if they were devoted purely to CO_2 than to all six Kyoto gases.

※ **Joint Implementation Projects** allow an Annex I country to earn emission credits by subsidizing an emissions-reduction project in another Annex I country. As with the CDM, these projects officially began in 2005 and took a while to gather steam. As of late 2010, there were more than 200 projects under way in 14 host countries. The favored locations have been Russia and the former Eastern Bloc, where there are plenty of inefficient technologies ripe for upgrading.

Because of the complexity of assessing and verifying emissions for each nation, the results for the initial Kyoto commitment period (2008–2012) weren't available when this book went to press. See graphic on p. 378 for the numbers through 2011. It seemed likely that Annex I countries would manage to meet their collective target of a 5.2% cut over 1990 levels, even

The pros and cons of clean development

The Clean Development Mechanism seems like a win–win part of the Kyoto Protocol. It allows nations that come up against their emissions targets to pour money into emission-reducing projects in developing countries. Yet not everyone likes the CDM. Some oppose it on principle, claiming it's simply a tool that legitimizes the polluting legacy of Annex I nations by allowing them to buy their way out of any commitment to actually reduce emissions. Others are more concerned with how the CDM is implemented. The parties who arrange a CDM project have to demonstrate that it produces a reduction in developing-country emissions that wouldn't have happened otherwise. However, many of the initial projects got started before they came under the CDM umbrella, thus complicating the task of deciding what would have happened without the CDM.

Some activist groups, including the WWF, have lobbied to see certain projects ruled off-limits for CDM, such as those involving coal, large-scale hydropower, and forest-based carbon sinks. A particular sore spot is the Plantar project in southeastern Brazil, funded in part by the World Bank. In 2012, it became the first reforestation project to win certified emission reduction credits from the United Nations. Grassland at the Plantar site is being replaced by a eucalyptus plantation spanning about 230 km² (89 sq mi). The trees are being harvested and converted to charcoal that goes to local pig-iron smelters, providing them with a less carbon-intensive form of fuel than coal. However, activists say that such monoculture plantations have been a problem in Brazil for decades, displacing local residents as well as indigenous ecosystems. Another CDM applicant is the Barro Blanco Dam in western Panama, which is being built with loans from state-owned banks in Germany and the Netherlands. Though the resulting reservoir is projected to cover only about 2.5 km² (1 sq mi), it will convert a free-flowing river into a managed waterway. The neighboring Ngäbe-Buglé peoples, con-

before considering any help from the mechanisms above. However, this collective success hides a great deal of variation among countries, as can be seen in the graphic. As of 2011, several of the wealthiest Annex I members, including Australia, Canada, New Zealand, and Spain, had racked up double-digit emission *increases*. Meanwhile, the United Kingdom's emissions were down by more than 25%, well below its Kyoto target of a 12.5% decline—though even here, things look far less positive if you consider that the United Kingdom has massively increased its reliance on imported goods from countries with rising emissions (see chapter 3).

cerned about potential impacts to agriculture, fishing, and water supply, have partnered with activist groups to mount vigorous protests. They note that the project gained approval without formal Ngäbe-Buglé consent.

In a more distributed approach, the world's largest off-grid energy project took root in Bangladesh through CDM support. Since 1999 the Bangladeshi bank Grameen Bank (winner of the 2006 Nobel Peace Prize, with founder Muhammad Yunus) has provided small, no-collateral loans to many thousands of householders—most of them women—which allow them to install solar photovoltaic systems. Fewer than 30% of homes in Bangladesh are on the national electric grid. The others typically use kerosene for lighting and cooking and perhaps a

Bangladeshi woman with one of the thousands of solar panels bringing electricity to homes nationwide. (Sarah Butler-Sloss/www.ashden.org)

lead–acid battery, recharged at a regional center, to power a TV. Adding a solar system not only reduces greenhouse emissions but relieves women from the drudgery of maintaining kerosene lamps and provides cleaner indoor air and better light for reading. In 2012, the project reached the milestone of serving one million homes, with a second million expected by later in the decade.

What enabled Annex I to meet its Kyoto goal were the "economies in transition" that emerged from the fall of the Soviet Union. These nations experienced big drops in emissions during the 1990s simply because of the economic pain they endured in the initial post-Soviet years. As a result, many of them, including Russia, are now emitting anywhere from 30% to almost 60% less than they did in 1990. This comfortable margin has been derisively termed "hot air," since it arose from the peculiarities of the Kyoto target-setting process and the downfall of the Eastern Bloc economies rather than from any concerted effort to cut emissions. Were it not

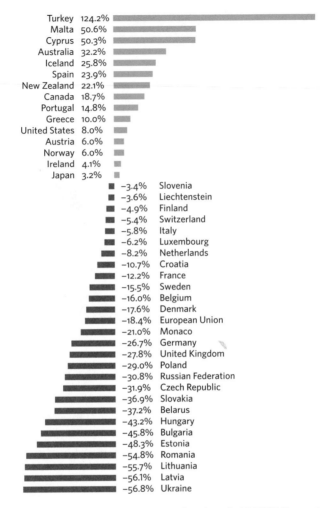

Turkey	124.2%
Malta	50.6%
Cyprus	50.3%
Australia	32.2%
Iceland	25.8%
Spain	23.9%
New Zealand	22.1%
Canada	18.7%
Portugal	14.8%
Greece	10.0%
United States	8.0%
Austria	6.0%
Norway	6.0%
Ireland	4.1%
Japan	3.2%

−3.4%	Slovenia
−3.6%	Liechtenstein
−4.9%	Finland
−5.4%	Switzerland
−5.8%	Italy
−6.2%	Luxembourg
−8.2%	Netherlands
−10.7%	Croatia
−12.2%	France
−15.5%	Sweden
−16.0%	Belgium
−17.6%	Denmark
−18.4%	European Union
−21.0%	Monaco
−26.7%	Germany
−27.8%	United Kingdom
−29.0%	Poland
−30.8%	Russian Federation
−31.9%	Czech Republic
−36.9%	Slovakia
−37.2%	Belarus
−43.2%	Hungary
−45.8%	Bulgaria
−48.3%	Estonia
−54.8%	Romania
−55.7%	Lithuania
−56.1%	Latvia
−56.8%	Ukraine

Changes to total greenhouse gas emissions 1990–2011. Data from the UNFCCC. Does not include land-use changes such as deforestation.

for these countries, the Annex I group would have missed its Kyoto target by a long shot—ending up with overall emission gains instead of cuts. In fact, some Annex I nations were on track to miss their goals despite the option of taking advantage of the CDM and/or buying emissions reductions through the European trading system.

Even if all countries were set to meet their 2012 Kyoto targets through bona fide emission reductions, it wouldn't have made a great deal of difference to the atmosphere. Together, all of the Annex I parties to the protocol represented only about 30% of the world's 1990 greenhouse gas emissions, and only about 20% of 2005's global emissions. This is largely due to the absence of the United States from the protocol and the lack of targets for China and other large, fast-developing countries. Even if the Kyoto-prescribed reductions were maintained for a full century—rather than until 2012—they would only bring down the year 2100 temperature increase by a few percent, according to a 1998 analysis by Tom Wigley of the National Center for Atmospheric Research. Critics cited the weakness of even the potential best-case outcome in order to bash the protocol altogether. But as Wigley noted about his findings, "This does not mean that the actions implied by the protocol are unnecessary." Indeed, the larger goal of Kyoto was to establish a template through which the world could work for even greater reductions in subsequent agreements.

Beyond Kyoto: What's next?

There's no expiration date woven into the Kyoto Protocol itself: participants are free to choose a new commitment period with new targets. In fact, that's just what they ended up doing in the early 2010s. The idea of a second phase to Kyoto was in the air for years, but it was typically overshadowed by hopes for an entirely new agreement that might win broader participation from key emitters (see below). However, by the time delegates met in 2011 in Durban, South Africa, it was clear that no fresh treaty would be feasible in the near future. With time running short, delegates at Durban launched the official process of creating a second commitment period for Kyoto, which now extends from 2013 to 2020. While this action helped save some face, the second phase of Kyoto falls short of even the modest ambitions of the first phase. The new targets lack the legally binding status of the initial ones, and a number of Kyoto participants opted not to join the second commitment period at all, including Japan, Russia, and New Zealand. As of early 2014, those nations that had signed on represented a mere 15% of global emissions, which is hardly enough to move the climate needle in a big way.

Ambitions were much higher for a post-Kyoto agreement, but that project has faced an uphill road for years, perhaps exacerbated by "climate fatigue" as well as the grinding aftermath of the global economic meltdown.

A road map first emerged from the UN's 2007 climate summit in Bali, and two years of subsequent planning culminated in the mammoth meeting that took place in Copenhagen in December 2009 (see "Copenhagen and its aftermath.") However, the results were a stunning disappointment to almost all parties involved. The resulting Copenhagen Accord, drafted by a handful of leaders at the last minute, was merely "noted" by participants rather than ratified by them. Nations were asked to submit their own voluntary targets for 2020, which were later added to the agreement—but unlike Kyoto, the accord was not legally binding. Weak as the deal was, it did manage to codify unprecedented pledges from China (a 40%–45% reduction in carbon intensity) and the United States (a 17% reduction over 2005 emission levels by 2020). In addition, participants agreed to beef up efforts to protect rainforests and to establish a Green Climate Fund, with $100 billion in support pledged by the year 2020, that would help developing countries mitigate and adapt to climate change.

The hangover from Copenhagen's dashed hopes left momentum sagging for years afterward. Even the Green Climate Fund became more controversial over time, with few of the pledged contributions coming forth (only $75 million as of mid-2013) and nations tussling over what type of financial instruments should be favored (large capital funds versus microfinance, for example). One concrete step toward the next global agreement took place at the 2012 Durban meeting, where delegates agree to a deadline of 2015 for crafting a treaty that would come into force in 2020. Participants at the 2013 meeting in Warsaw continued to aim for a 2015 agreement, although there was also fresh debate over how and whether the developed world should reimburse vulnerable developing nations for "loss and damage" from impacts relatable to climate change. As this book went to press, it was still unclear whether the persistent tensions and conflicts in the UN process would be resolved in time to make an ambitious global treaty possible.

If U.S. participation is critical to the next binding global treaty, that agreement may be a long while in coming. During his first year in office (2009), Barack Obama launched a number of energy-efficiency measures, some of them tied to massive recession-recovery bills. However, in the aftermath of Climategate (see chapter 13), the U.S. Congress found itself returning to skepticism on the value of cap and trade and other policy mechanisms for reducing carbon emission. Obama himself acknowledged the near impossibility of passing cap-and-trade legislation for at least several years. As late as 2013, several leading U.S. senators on both sides of the aisle were openly pessimistic that Congress might ratify any new global agreement on the UN's timeline above—even if such an agreement were to include binding targets on high-emitting China and India, as the United States was insisting.

All of the above adds to the urgency of emission reduction efforts being undertaken on local and regional scales, as discussed below, even though the lack of a binding global agreement will leave the door open for market-driven shifts that could blunt the usefulness of emission cuts within a given country (see chapter 14). The troubled outlook for climate diplomacy also highlights the need for serious work on adaptation (see box, p. 322), given that some degree of climate change is inevitable. For many years, adaptation was considered to be a flag of surrender to climate change, with policymakers encouraged to focus purely on reducing emissions, but governments are increasingly recognizing that there's no sense in remaining unprepared for impacts that are increasingly likely.

What might an eventual successor to Kyoto look like? There's been no shortage of suggestions over the years. A 2004 report from the U.S. Pew Center on Global Climate Change (now the Center for Climate and Energy Solutions) summarized more than 40 ideas, with names that ranged from the grandiose (Orchestra of Treaties and Climate Marshall Plan) through the humble (Broad but Shallow Beginning and Soft Landing in Emissions Growth) to the droll (Keep it Simple, Stupid).

The main points in question include the following:

* **What's the best time frame to consider?** Some plans focus on the 2020s, while others extend all the way out to 2100.
* **What type of commitments should be specified?** There is a whole array of possibilities, from emission targets by nation or region to non-emission approaches such as technology standards or financial transfers.
* **How does the world make sure that commitments are enforced?** It's an issue that many say Kyoto hasn't fully addressed, as Canada's unexpected exit just before the end of the first commitment period made clear.
* **How should the burden of climate protection be shared among developed and developing countries?** This remains a key sticking point, as was the case from the very beginning. The issue has grown only more complex over time. For example, U.S. CO_2 emissions were more than 50% greater than China's when Kyoto was first crafted in 1997—but by 2013, China's CO_2 emissions were almost twice those of the United States. At that point, the per capita emissions rate for China was still half that of the United States, but on par with that of the European Union.

* **Should there be a single global plan or an array of decentralized alliances?** From the ashes of Copenhagen emerged a stronger interest in new kinds of partnerships outside the Kyoto-style global model, such as the impromptu group of nations that pulled together the summit's last-minute accord.

Developing countries are often highly skeptical about international groupings outside of the UN system, as they've traditionally been excluded from them. Certainly, the European Union has made real progress in emissions reductions in a variety of ways, thanks in large part to its rule-making power. However, regional alliances that don't have such teeth may not be able to accomplish so much. In 2005, the United States teamed up with Australia, China, India, Japan, and South Korea (with Canada joining in 2007) to form the Asia-Pacific Partnership on Clean Development and Climate. The group was hailed by U.S. president George W. Bush as a "results-oriented partnership" preferable to the Kyoto process. The group's members were heavy hitters indeed—accounting for more than half the world's carbon emissions, GDP, and population—but the partnership itself focused on voluntary, non-binding actions such as "aspirational" emission goals and the sharing of energy technologies. This strategy was widely slammed as "Kyoto-lite." Even prominent conservative John McCain blasted the formation of the group, calling it "nothing more than a nice little public relations ploy." The partnership quietly folded in 2011.

One of the persistent challenges facing any post-Kyoto agreement is to find a path that allows developing countries some measure of economic growth without cutting into rich nation's economies so much that they refuse to participate. Among the most intriguing and influential plans offered to date along these lines is the contraction and convergence (C&C) model developed by the Global Commons Institute, a British group headed by Aubrey Meyer. It was introduced by the Indian government in 1995 and adopted by the Africa Group of Nations in 1997 during the run-up to Kyoto. The plan has also received votes of support from the European Parliament and several U.K. and German advisory groups. The two principles at the heart of C&C are:

* **contraction**, the need to reduce overall global emissions in order to reach a target concentration of CO_2; and
* **convergence**, the idea that global per capita emissions, which now vary greatly from country to country, should converge toward a common amount in a process more or less parallel to contraction.

In short, C&C calculates how much carbon the world can emit in order to reach its target, then apportions that total between nations in a way that moves quickly toward an equal allocation for each person around the world. It's an elegant concept that moves the process toward a climate-protected future in a way that's widely recognized as fair.

There's no disguising the fact that C&C and many other suggested plans call for a massive explicit and/or implicit transfer of wealth from rich to poor countries, at least in the beginning, in order to get results. One might argue that such a transfer is unavoidable if the developed world is to take full responsibility for its outsized cumulative contribution to the greenhouse effect. However, the fierce political resistance in some quarters to such an approach is likely to continue, especially now that emissions have soared in some of the largest developing economies. It's also been argued that emission cuts won't happen in a big way—especially in fast-growing countries like China and India—until technological breakthroughs make low-carbon energy sources so cheap that countries can reduce emissions while improving their standard of living. Otherwise, when individuals and countries juxtapose the promised long-term benefits to climate against their own near-term economic future, the climate can easily lose out.

There's growing acknowledgement, at least among those who favor some type of global action on climate, that a price of some sort must be put on carbon. As noted above and discussed below, **cap-and-trade policies** put a cap on the total *amount* of carbon and other greenhouse gases used by the participating parties. This aims to provide some reassurance that a particular net emissions goal will be met, but it leaves the *cost* of using fossil fuels uncertain. An alternate approach is a **carbon tax**, which provides certainty on the per-unit cost to fossil fuel users but doesn't put limits on how much carbon can be emitted. Either of these approaches would have the most impact on climate if it were universal; otherwise, nonparticipants could easily take advantage of the situation by being able to burn fuel without constraint, and the carbon-leakage problem noted above could grow apace.

William Nordhaus (Yale University), a globally acknowledged expert who has studied the economics of climate change for decades, argues that either taxation or a cap-and-trade approach could work, as long as one or the other (or a hybrid) is implemented hand in hand with increased public awareness and intensified research into low-carbon technologies. However, he warns in his 2013 book *The Carbon Casino* that "an effective policy must necessarily be global in scope."

Finland and Sweden pioneered the practice of taxing carbon on a national level in the early 1990s. Only a handful of countries have followed

Copenhagen and its aftermath

The 2009 Conference of Parties to the UNFCCC—more commonly referred to as the Copenhagen Summit—was the world's biggest-ever climate change summit. It was long expected to be the venue where a post-Kyoto treaty would be hammered out. The United States arrived at the summit backed by a new president and new Congress, both far more amenable to climate change policy than their predecessors.

Despite the hoopla, Copenhagen ended up as a bitter disappointment for activists and policymakers alike. The dank, chilly weather reflected the evolving mood inside the cavernous Bella Center, as delegates were torn between the demands of poorer countries, including African nations and island states, and the mutual reluctance of the United States and China to commit to emission cuts that would go beyond Kyoto levels, which neither nation had agreed to in the first place. A round of spirited protests midway through the meeting was quashed by Danish police, sending hundreds into detention. Afterward, the conference went into virtual lockdown for its final week. Most of the 22,000 registrants from nongovernmental organizations were shut out in the final two days to make way for heads of state and their vast security teams.

On the final day, negotiations culminated in an offsite meeting among the leaders of Brazil, China, India, South Africa, and the United States. From behind those closed doors emerged the bare bones of the Copenhagen Accord. Instead of a step beyond the Kyoto Protocol, the document was seen by many as a step backward. Other countries were left out of the drafting process, and Kyoto-style binding targets were omitted. Dissatisfaction among delegates was so widespread that, rather than formally adopting the document, the assembled parties managed only to "take note" of it.

Although a few participants and observers sounded notes of optimism—"We have come a long way," said British prime minister Gordon Brown—these voices competed with a chorus of critics. The accord was "nothing short of climate change skepticism in action," said Sudan's Lumumba Di-Aping, chief negotiator

in their footsteps, though several U.S. states and Canadian provinces have done so. In 2007 the city of Boulder, Colorado, became the nation's first to carbon-tax itself, thanks to the approval of voters who went on to pass a five-year extension in 2012. Of course, taxes of any type aren't the easiest thing to sell, and if people see their energy costs rising and attribute this to a recently instituted carbon tax, the protests can be sharp—even if green

for the G77 group of developing countries. Even the conference's own news service acknowledged that "enthusiasm for the Copenhagen Accord is scarce." Dubbing Copenhagen a "fiasco" and a "foundering forum," the *Financial Times* claimed that the big winners in the process were climate skeptics, China, and the Danish economy. The losers, according to the *FT*: carbon traders, renewable energy companies, and the planet.

(Robert Henson)

One of the few bright outcomes of the Copenhagen process was the creation of a "green climate fund" as well as new mechanisms for transferring technology and reducing deforestation. The Kyoto Protocol had assigned credits for renewing destroyed forests and planting new ones, but it had no mechanism for preventing deforestation in the first place. Without a global agreement, any nation's preservation work might simply push deforestation to the next country. But in Copenhagen, at the urging of a coalition of activist groups—as well as such luminaries as Jane Goodall and Richard Branson—billions of dollars were pledged to help cash-poor, rainforest-rich nations gain more value from preserving their forests than from exploiting them. The agreement emerged after years of research in the area known as REDD (reducing emissions from deforestation and degradation). Newly precise monitoring by satellite will help keep track of progress.

measures aren't really to blame. In 2013 U.K. prime minister David Cameron, who had run on environmental credentials just years earlier, started rolling back climate-related fees on energy generation and use amid rising concern about spiraling bills, despite the fact that price hikes had largely been caused by increases in the cost of natural gas. Australia's carbon tax, introduced in 2012, set off fireworks as well. In the following year, Tony

Abbott was elected as prime minister after having made the repealing of the tax a key campaign pledge.

One way to help make a carbon tax more palatable is to ensure that the costs and benefits are closely related and easy for everyone to grasp. Roger Pielke, Jr. (University of Colorado Boulder) has suggested that a new global agreement might start by setting a relatively low, but slowly rising, global tax on carbon, levied at the point where the fuel is produced, with the revenues flowing directly into intensified research and development on decarbonized energy sources. That way, the link between sacrifice and gain would be clearer and shorter term. Many other policy specialists have advocated for a carbon tax, including several former administrators of the U.S. Environmental Protection Agency. Scientist and activist James Hansen also believes a global carbon tax is the way to go, but he calls for a fairly large one—to help drive immediate emission cuts—that's revenue-neutral, with benefits returning to citizens. The latter idea, often dubbed a "feebate" or fee-and-dividend plan, was pioneered by Amory Lovins of the Rocky Mountain Institute. According to Hansen, "As long as fossil fuels are the cheapest form of energy, their use will continue and even increase."

The next few years should continue to make for fascinating and hugely important politics, as various plans vie for consideration and leaders from around the world jostle for position. In the words of the C&C position statement—as true now as when C&C was first proposed in the 1990s—"the global community continues to generate dangerous climate change faster than it organizes to avoid it. The international diplomatic challenge is to reverse this."

National, regional, and local schemes

Governments of all sizes are taking a variety of actions independent of Kyoto and its potential successors. Some of these actions involve setting reduction targets based on much longer time frames than Kyoto specifies, while others focus on specific sectors or market mechanisms.

A big part of the EU's strategy for climate protection is its **emissions trading scheme (ETS)**, launched in 2005. It's similar in spirit to, but separate from, the Kyoto emissions trading. When the EU's scheme began, each nation assigned each of its largest industries the right to emit a certain amount of carbon. The scheme covers a variety of activities, including oil refining, cement production, iron and steel manufacture, glass and ceramics works, and paper and pulp production, as well as fossil fuel–based power generators of more than 20 megawatts. Together, these companies were emitting more than two billion metric tons per year—almost half of the EU's total CO_2 emissions—at the program's outset. In order to meet its own targets, each of these big EU companies could either cut its emissions or "pay to pollute" by purchasing credits from another company that's done better than targeted. The credits could even be purchased from a country outside Annex I, via the Kyoto mechanisms discussed previously.

Largely because the initial allocations assigned by EU governments to their biggest companies were so generous, there were far more sellers than buyers for carbon credits in the first ETS phase (2005–2007). The price plummeted from a peak in 2006 of nearly €30 per emissions allowance (one metric ton of CO_2) to only a few cents per allowance by late 2007. Prices recovered as the second phase (2008–2012) began; it included an expansion from CO_2 to all six of the greenhouse gases covered by Kyoto, as well as clean-development and joint-implementation credits from the Kyoto process above. However, with the European economy in a prolonged slump, emission targets were becoming easier to reach, and permits soon began to accumulate in the marketplace. The price of ETS credits steadily fell through the second phase, dropping below €10 per allowance by 2012. The third phase of the ETS (2013–2020) included still-tighter emissions targets, this time across the entire union rather than country-by-country. Still, by the end of 2013, the price per allowance had fallen below €5. To help combat the slide over the next few years, the EU is gradually increasing the fraction of new permits that will be auctioned (where companies bid on the right to emit) rather than allocated without charge.

It's hard to know exactly how much good the EU's scheme has accomplished, but some studies suggest that the first phase of emission trading helped reduce the region's carbon emissions by several percent over a business-as-usual approach. As noted in a 2012 review of the system by the Environmental Defense Fund, "low market prices for emissions allowances may indicate that firms are achieving emission reductions at lower costs than predicted." Elsewhere, New Zealand, Switzerland, Tokyo, California, and a group of northeastern U.S. states had adopted emission trading schemes by 2010. Among the nations on record as planning to launch various types of emission marketplaces before the end of the decade are Australia, South Korea, India, Thailand, and Vietnam.

Aside from these efforts, countries have tackled emission reduction with a wide range of policy tools that, for any one country, could include a blend of taxes, support for low-carbon fuel sources, campaigns to raise public awareness about energy efficiency, and environmental aid to developing countries. The **United Kingdom** took a leap forward with its Climate

Change Act 2008, which mandates an 80% reduction from 1990 levels of the six Kyoto greenhouse gases by 2050—among the most stringent goals set by any government. The United Kingdom has also put into place a range of directives dealing with everything from cogeneration to energy efficiency in buildings (see chapter 16).

The most stunning example of national energy transformation may be **Portugal**, where CO_2 emissions had soared by more than 40% from 1990 to 2004—the most of any Annex I country except for Turkey. In 2005, Portuguese prime minister José Sócrates led the push to revolutionize the country's use of energy, including a massive increase in wind and solar farms and a carefully designed mix of government-owned transmission lines and privatized energy producers. With only modest increases in utility costs to consumers, Portugal now generates roughly half of its electricity from renewable sources. The nation's CO_2 emissions dropped nearly 14% from 2005 to 2008, although an intense drought in 2012 put a crimp in hydropower production and caused a short-term spike in emissions.

The **United States**, by contrast, has put only a smattering of climate-protection policies in place, and some American laws seemed to encourage more greenhouse gas rather than less. For instance, hulking sport utility vehicles (SUVs) were required to meet only minimal fuel-economy standards for years, and the passenger-car standard in 2010 was the same as in 1985. The rules are now tightening, thanks to legislation passed in 2007 for passenger vehicles and 2010 for trucks and buses. The final standards, issued in 2012, mandate a steady increase in fuel economy for cars and light trucks (including SUVs), with a combined average of 54 miles per gallon by 2025—more than double the average that prevailed before the new law. That still leaves the United States outpaced by China, Japan, and the EU, which are all on track to reach similar benchmarks by 2020. However, given the size of the U.S. motoring public and the weak rules that had previously been in place, the new standards are an accomplishment worth noting.

Despite the lack of major climate legislation, U.S. output of carbon dioxide has dropped in recent years by margins capable of surprising even a jaded observer. After peaking in 2005, the nation's CO_2 emissions fell more than 12% by 2012. There appear to be several factors in the mix, including the 2008–2009 economic downturn, a trend toward more energy-efficient homes and commercial buildings, and an unexpected, sustained reduction in the number of highway miles driven after 2007. The single biggest player from 2011 onward may be the nation's epic shift away from coal and toward natural gas, a transformation made possible by hydraulic fracturing, or **fracking** (see chapter 16).

One major player looming on the U.S. emission-reduction scene is the Environmental Protection Agency's regulation of greenhouse gases under the Clean Air Act, a stance that was given the green light by the Supreme Court in a landmark 2007 ruling. The EPA had already made big gains through its Energy Star efficiency labeling system (see chapter 18), but the Supreme Court decision paved the way for a variety of additional emission-paring actions that likely wouldn't have made it through the U.S. Congress. Many of these rules won't take effect for years, but they could soon have a significant impact on U.S. greenhouse gas output—particularly from both new and existing coal-fired power plants. Barack Obama directed the EPA to issue the restrictions on coal as part of a climate action plan put forth in 2013. With coal still a big part of the nation's energy grid, any EPA regulations are sure to face resistance from utilities and coal-state legislators, and perhaps from the courts, that could take years to play out.

Otherwise, much of the U.S. action has been happening on the **state** level, where there's more of a chance for a few motivated legislators and policy analysts to make a real difference. **California**, as usual, is a bellwether. Its greenhouse emissions are comparable to those of Australia or Brazil, so its actions could have a real impact in themselves, as well as through their influence on other U.S. states. In 2005, California's governor and erstwhile movie star Arnold Schwarzenegger signed an executive order calling for massive emission reductions in the long term—a reduction to 1990 emission levels by 2020 and a further reduction of 80% by 2050. State voters soundly rejected a move to repeal the law, and in 2012 the state embarked on a cap-and-trade system for power plants and refineries. Meanwhile, a group of nine northeastern states extending from Maryland to Maine formed the Regional Greenhouse Gas Initiative, an emissions trading scheme focused on large power plants. Originally, the goal was to cut those plants' CO_2 emissions 10% below 1990 levels by the period 2015–2020; this target was strengthened in 2013 to an impressive 45% cut by 2020. A similar regional partnership, but focused on western North America, started in 2007. Though it involved seven western U.S. states and four Canadian provinces at its peak, all of the U.S. participants except California had peeled off by 2011.

Across the world, **cities and towns** are also getting involved. More than 1000 U.S. mayors, representing more than a quarter of Americans, have signed pledges to meet Kyoto targets. In the United Kingdom, Newcastle—

a cradle of coal mining for centuries—made a splash in 2002 when it set out to become the world's first carbon-neutral city. It's still aiming for that goal, along with a growing number of other municipalities.

At the same time, a few brand-new settlements are being planned with energy efficiency topmost in mind. One of the leaders in this movement is **Masdar City**, a project seemingly at odds with the oil-fueled extravagance of its home turf of Abu Dhabi. The original 2006 plan was highly ambitious—a carbon-neutral, car-free landscape powered by wind and solar farms and housing up to 50,000 residents. The project was delayed by the financial downturn, and the plans have been tweaked somewhat, but the UAE-based development firm behind Masdar City is still aiming for build-out by 2025. Some elements are already in place, including parts of the Masdar Institute of Science and Technology campus. Another high-profile eco-city, **Dongtan**, was intended to take shape on Chongming, China's third-largest island, just northeast of Shanghai. The population—originally projected to reach 10,000 by 2010 and 500,000 by the year 2050—would drive hybrid cars or pedal bikes along a network of paths. Renewable power for homes and businesses would flow from a centralized energy center, drawing in part on a wind farm located on the island's west edge. These grand plans haven't materialized, though. The project's government co-ordinator was jailed on corruption charges, and as of 2014 there were no Dongtan residents and little sign of progress except for wind turbines and a bridge to the mainland.

There are other eco-cities on the drawing board, especially in China, but such showpieces are facing other headwinds in the form of disfavor among environmentalists. Although WWF is a supporter of Masdar City, some activists have derided purpose-built green towns as exclusive communities irrelevant to the challenge of making everyday cities more sustainable.

Given the scope of the U.S. economy, private endeavors can make a big difference. Teaming with the Environmental Defense Fund, **Walmart**—the world's largest company—aimed in the early 2010s to more than offset the growth of its emissions by cutting 20 million tons of emissions from its global chain of suppliers by 2015. On the down side, Walmart's carbon footprint remains gigantic even if it's been trimmed a bit. Although a 2013 EPA analysis placed Walmart as the nation's fifth-largest user of renewable energy, with more than 750 million kilowatt-hours (kWh) in a year's time, that represented a paltry 4% of the company's total energy usage. The top firm on the list, Intel, drew on 100% renewables for the 3.1 billion kWh it used. (The EPA rankings include renewable-energy offsets purchased by companies.)

How much will it cost?

No analysis of climate change politics would be complete without considering economic costs and benefits. As major players appeared to finally coalesce around the need for action on climate change, the presumed economic impact of taking action became a hot topic. Conservative commentators and politicians, especially those in the United States, tended to stress that emissions cuts could bring major economic downswings. Green-minded writers and legislators, on the other hand, typically suggested that the cost of *not* acting could be even higher.

The terms of the debate shifted sharply in 2006 with the release of the *Stern Review on the Economics of Climate Change*. Commissioned by Gordon Brown, then the head of the U.K. treasury, this massive study headed by Nicholas Stern (formerly the chief economist for the World Bank) reverberated in financial and political circles around the globe. Like a mini IPCC report, Stern's study drew on the expertise of a wide range of physical and social scientists to craft a serious, comprehensive look at the potential fiscal toll of global warming.

Stern's conclusions were stark. Our failure to fully account for the cost of greenhouse emissions, he wrote, "is the greatest and widest-ranging market failure ever seen." He estimated that climate change could sap anywhere from 5% to 20% from the global economy by 2100, and that "the benefits of strong, early action considerably outweigh the cost."

While the *Stern Review* drew massive publicity and galvanized leaders around the world, it wasn't universally praised by economists, many of whom declared it was based on overly pessimistic assumptions. A major sticking point was how Stern handled the discount rate—the gradual decline in value of an economic unit over time. In traditional analysis, a unit of spending today is valued more highly than a unit spent tomorrow, because the present is more tangible and the unit will be worth less in the presumably richer future. But climate change is a far more complex matter because of its global reach and the long-term effects of current emissions. In that light, Stern chose not to "discount" tomorrow the way most economists would: "if a future generation will be present, we suppose that it has the same claim on our ethical attention as the current one."

While he praised the Stern report for its attempt to balance economics with environmentalism, William Nordhaus argued in a 2007 analysis that its various assumptions made it a political document at heart: "The *Stern Review* is a president's or a prime minister's dream come true. It provides decisive answers instead of the dreaded conjectures, contingencies, and qualifications."

The New Great Game: Who owns the Arctic?

Throughout the 1800s, Britain and Russia vied for power in central Asia in a century-long contest dubbed the Great Game. Today, the geopolitics around the fast-thawing Arctic region have been described as the "New Great Game," as the erosion of the region's summertime ice pack opens the door to high-latitude wheeling and dealing. Later this century, shipping routes that now link Europe, Asia, and North America through the Arctic could be open months longer than they are now, and huge vaults of oil and gas that are believed to lie beneath the sea may become accessible for the first time.

The five countries with Arctic coastlines—Canada, Denmark, Russia, Sweden, and the United States (via Alaska)—long paid little mind to each other's business in a region seemingly devoid of usefulness. Intersecting borders across the Arctic were once drawn informally or not at all. Now they're a topic of intense debate, as countries angle for the best shipping routes and undersea resources.

The competition heated up in 2007, as did the Arctic itself. Russia planted a flag on the seafloor beneath the North Pole, which it claims is an extension of the Lomonosov ridge running northward from Russia's Arctic coast. Canada and Denmark, meanwhile, are attempting to claim the ridge—which may hold billions of dollars' worth of fossil fuel—as their own, while the United States asserts that the ridge extends beyond all three nations' borders. These and other disputes will be resolved through the Law of the Sea, the United Nations convention ratified by more than 150 nations, including all of those bordering the Arctic except the United States. Although there's bipartisan U.S. support for the Law of the Sea, a small group of conservative legislators has long blocked a vote on ratification. At the same time, U.S. geologists have sailed across the Arctic

Stern's report also drew fire for its assumptions about weather extremes. In the journal *Global Environment Change*, for example, Roger Pielke, Jr., criticized the report for extrapolating a 2%-per-year rise in disaster-related costs observed in recent decades over and above changes in wealth, inflation, and demography. As it happens, 2004 and 2005 brought a number of expensive U.S. hurricanes, and Pielke argues that this happenstance skewed the trend used by Stern. (See chapter 8 for more on the complex and still-debated science on hurricanes, tornadoes, and climate change.) At the same time, Pielke has granted that the *Stern Review* "helped to redirect attention away from debates over science and toward debates over the costs and benefits of alternative courses of action."

each year to measure the outermost extent of the continental shelf claimed by the United States. The Law of the Sea dictates that each nation has an "exclusive economic zone"—including the right to drill for oil and gas—extending 200 nautical miles (about 370 km) poleward from its Arctic coastline. That leaves a sizeable area under contention, from near the North Pole westward toward the Siberian coast. As it happens, the unprecedented melt of summertime ice in the Arctic has opened much of this area, with prevailing currents pushing much of the remaining ice toward Greenland.

Entrepreneurs and companies have also gotten involved in the New Great Game. For example, Pat Bode of Denver snapped up the port of Churchill, Manitoba, from the Canadian government for a token US$7 in 1998, as part of an US$11 million purchase of train lines running into Canada's heartland. Though it's only fitfully busy now, Churchill's port—which lies on the west side of Hudson Bay—could become a major transport hub if climate change thaws the region as much as projected. In the increasingly ice-free waters off the Alaskan coast, oil drilling resumed in 2012 after a two-decade hiatus. However, Shell's first foray into the region was plagued by malfunctions that year, and drilling across the region was again halted indefinitely.

Though the Arctic's development remains tentative, a future with less ice can only make it easier for nations and companies to exploit the region. In his book *The World in 2050*, Laurence Smith draws parallels between this century's Arctic and the wide-open, resource-rich western United States. As he puts it, "Imagine the Arctic in 2050 as a frigid version of Nevada—an empty landscape dotted with gleaming boom towns."

Long before Stern's report, the insurance industry realized that climate change threw an uncertain element into the detailed calculations it uses to gauge risk. As early as 1989, the Lloyd's of London insurance market began incurring massive extra losses. The head of the American reinsurance association said in 1993 that "changes in the number, the frequency and the severity of natural catastrophes are threatening to bankrupt the industry." According to industry giant Swiss Re, the global total of insured losses from natural disasters hit US$120 million in 2005 (as expressed in 2009 dollars). That was nearly twice the constant-dollar record set only a year earlier. Part of this rise was due to steep rises in property prices in hurricane-prone regions such as the U.S. Gulf Coast, and the years 2006–2009 saw numbers falling back to levels seen earlier in the decade. Moreover, research by Pielke and colleagues finds that non-atmospheric factors (increased wealth, for instance, or population growth focused on coastlines) can play an outsized role in the increased costs from several different types of natural disasters.

Even so, Swiss Re cited global warming as an important co-factor in the long-term rise in damages from natural disasters, and the years 2011 and 2012 ranked second and third only behind 2005. Both Swiss Re and Munich Re have been among the strongest corporate voices calling for climate protection, issuing reports and raising public awareness. They're also using their leverage as institutional investors (in the United Kingdom, they own around a quarter of the stock market) to persuade other companies to take climate change on board.

By comparison, U.S. insurance companies have been rather mute on the issue, despite suffering massive financial hits. Even in 2013, the Institute for Business and Home Safety, a major industry group, had no sections about climate change on its website. A former president of the group told climate reporter Ross Gelbspan in 2003 that U.S. insurers are "burying their heads," dropping customers and abandoning high-risk areas. Indeed, hundreds of thousands of Floridians saw their home insurance cancelled in 2005 and 2006, after the state's string of hurricanes. Over two million people along the U.S. Gulf and Atlantic coasts turned to "insurers of last resort" established by state governments—with the public often paying the tab if major disaster strikes. Most home insurance policies don't cover flooding, even when produced by hurricanes, and the U.S. National Flood Insurance Program, which covers more than $1 trillion in property, slashed subsidies in 2013 to help alleviate a growing debt. That left many homeowners in high-risk areas, especially along hurricane-prone coasts, suddenly facing premiums several times higher than before.

The risk of international conflict

Given how difficult it's been to achieve global agreement on even minimal steps to fight climate change, it's not hard to imagine international tensions bubbling over at some point, especially as the climate itself heats up. Defense planners have begun paying more attention to how such conflicts might unfold over the coming decades. Thus far, their analyses have focused on "what if" scenarios in which the effects of global warming mingle with other environmental and geopolitical pressures in dangerous ways.

Such research has played out in the American military sphere since at least 1978, when the report *Climate Change to the Year 2000* was issued by the National Defense University amid a period of intense cold snaps in North America and Europe. The report included expert opinion from 24 climatologists in seven nations on how Northern Hemisphere temperatures might evolve by the close of the century. The actual readings ended up near the warm end of the experts' predictions.

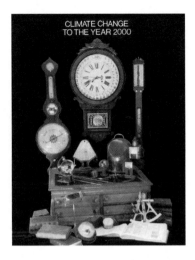

One landmark defense-oriented assessment of climate change was a widely publicized report prepared for the Pentagon in 2003. It considered a specific scenario for abrupt climate change in which regional and global conflicts over land and water use grow through the century. If the planet's ability to support its growing population were to drop abruptly, "humanity would revert to its norm of constant battles for diminishing resources," the report warned. The message was underscored in 2007 with another Pentagon-commissioned report, *National Security and Climate Change*, in which a group of retired generals and admirals examined climate change research and commented on its implications. "Projected climate change poses a serious threat to America's national security," concluded the team, which envisioned conflicts brewing even in now-stable parts of the world. Starting in 2008, U.S. military planners were required to consider the effects of climate change on future missions, and the U.S. Department of Defense issued its first Climate Change Adaptation Roadmap in 2012.

The United Kingdom's defense ministry has also been taking global warming into account, as with its 2009 *Climate Change Strategy* report. Already, the ministry has reported occasional disruptions in its activities because of fire and extreme heat, and some of its coastal facilities are vulnerable to erosion. Much of the 2009 U.K. report deals with reducing the military's own greenhouse emissions. It's been estimated that U.K. defense activities are responsible for 1% of the nation's carbon dioxide emissions. Likewise, U.S. forces reportedly account for 1.5% of American oil use. The U.S. military is now taking major steps toward bolstering its energy efficiency through solar farms, electrified vehicles, and other advances. By 2025, the nation's armed services aim to be drawing roughly 25% of their power from renewable sources.

While some observers interpret military concern about climate change as a sign that the issue has gone truly mainstream, others worry that the notion of global warming itself could become "militarized." In this view, climate refugees and water wars might simply take the place of more traditional conflicts as a rationale for military activity—or by the same token, an emerging climate-related crisis could open the door for intensified efforts in border control, surveillance, and the like.

Scholars have also continued to tussle over the long-standing question of how the environment intersects with social factors to help breed warfare. Drought, in particular, is a perennial marker of conflict. Unequal access to water has long been associated with unrest in the Middle East, and the collapse of several ancient cultures has been linked to drought. But the connection isn't a simple one. While early historians often saw drought as a primary trigger of societal ills, many cultures have been able to adapt to even severe water pressures. It appears that drought isn't a prime cause of war so much as a critical backdrop that can help lead to conflict if it persists long enough and if it syncs with an unstable political structure. The horrific violence in Darfur, Sudan, offers one example. It followed decades of strained relations between nomadic herders and farmers, coinciding with a sustained drop in rainfall from the 1970s onward. However, the dramatic onset of warfare in 2003 doesn't appear to be directly linked to short-term dryness.

As the world's subtropics heat up and dry out, many important regions—including parts of the southwest United States and northern Mexico, and the nations of Europe and Africa ringing the Mediterranean—are projected to see an intensifying risk of major drought, which would only feed other sources of societal stress and international conflict. These same parts of the world could also be well suited for hosting mammoth solar farms to send energy across international lines, making for another potential wrinkle in the geopolitics of tomorrow.

CHAPTER SIXTEEN

Technological Solutions

ENERGY, ENGINEERING, AND EFFICIENCY

Human ingenuity got us into our greenhouse mess, and we'll need to call on it again in order to find our way out. Even if we can develop affordable ways of storing carbon underground or removing it from the air—both of which are far from certain—the world faces a titanic challenge in reducing its voracious appetite for oil, coal, and gas, especially with many developing countries on a fast growth track.

The good news is that plenty of alternatives exist that produce little or no greenhouse output. However, to make a real dent in carbon emissions, they'll need to be deployed and coordinated on a far larger scale than they are now. And even if that can be accomplished, it doesn't ensure that the remaining fossil fuels will stay in the ground—which is one reason that technologies for capturing CO_2 from power stations and storing it underground could end up being crucial.

Of course, it's not just where we get our energy that counts. Equally important is how *efficiently* we use it. Hence, this chapter also covers technologies and policies designed to raise efficiency and reduce energy waste. Finally, we look at geoengineering plans—schemes to pull greenhouse gases directly from the atmosphere or to reduce the amount of sunlight reaching Earth. No matter what technologies come down the pipeline, the economics and politics discussed in the previous chapter will play a huge role in shaping their cost effectiveness and thus how quickly they're adopted.

The future of fossil fuels

For the time being, fossil fuels remain at the core of the global energy supply. Of course, these fuels are all finite resources, which, if we continue to use them, will cease to be available at some stage [see sidebar, "When will the oil (and gas) run out?"]. But even if fossil fuel discoveries came to a sudden halt, that wouldn't be enough to stop climate change. There's certainly enough remaining oil and gas to push us deep into the greenhouse danger zone—especially when you include carbon-intensive "unconventional" oil reserves—and we'll consume extra amounts of energy retrieving some of those hard-to-get supplies. As of 2013, the vast oil sands of western Canada were providing the United States with more than twice as much crude oil as was Saudi Arabia. This oil is either stripped from the land or cooked out of the soil, and the process takes more energy than regular oil production. Recent studies have pared down the "well-to-wheels" estimate of greenhouse gases emitted during the entire life cycle of oil sands, but even some industry voices acknowledge that emissions using current oil-sand technology are a few percent more than for a typical barrel of U.S. oil. (The widespread use of strip mining in oil-sand production and the resulting effect on landscapes is an environmental discussion all its own.)

Even if oil and gas were to run out, there would still be enormous reserves of **coal** to consider. In 2012, according to the International Energy Agency, around 41% of the world's grid-based electricity came from coal-fired power plants (versus about 22% from natural gas, 20% renewables, 12% nuclear, and 5% oil). Coal is also responsible for more than a quarter of global CO_2 emissions. What's scary from a climate perspective is that there are virtually no brakes on greenhouse emissions from coal. The world's most extensive coal reserves are in countries that fall outside the Kyoto Protocol's emissions mandates. This includes China, India, and the United States, which happen to be the world's three most populous countries as well as the three nations with the greatest coal reserves (more than half of the global total).

There's a lot of momentum driving the world's reliance on coal. The typical coal-fired power plant is in operation for 60 years or more, and hundreds of them are on the drawing board across the globe. Once built, these could generate almost as much power than all of the coal-fired plants now in use globally. Thus, even with improved efficiency, we can expect substantially more carbon dioxide to emerge from the world's coal plants in 2030 than from the smaller number now in place unless some of that CO_2 can be captured and buried (see below) or unless the rate of construction slows down, which is something many activists are calling for.

China and India are currently building the most new plants by far. A 2012 report by the World Resources Institute found 455 coal plants being planned in India and 353 in China. While some of these include emission-reducing technologies, the sheer volume points toward a major spike in future emissions. As of 2013, China was already using about as much coal as the rest of the world combined, and the new plants on order would boost its consumption to even greater levels. By 2020, it's possible that coal will outpace oil as the world's most heavily consumed fossil fuel. However, some interesting crosscurrents have emerged lately, including the global boom in natural gas production [see sidebar "When will the oil (and gas) run out?"].

China drew global attention in 2013 when it announced that no new coal plants would be built in three industrial districts near the cities of Beijing, Guangzhou, and Shanghai. The move was driven by concerns about air quality more than climate change, and the decision only affects a small part of the nation's total coal use. Still, some experts have begun to speculate that China might begin moving away from coal and toward renewables, nuclear power, and natural gas sooner than expected.

Roadblocks to coal are also popping up in other parts of the world. As early as 2007, a state agency in Kansas denied permits for a pair of new plants on the basis of their anticipated greenhouse emissions. This marked the first time climate change had been invoked to block a U.S. coal plant. A much bigger concern for U.S. coal is the prospect that the Environmental Protection Agency will issue rules by the mid-2010s effectively forcing all new plants to capture and bury their CO_2 emissions (see below). The agency argues that, with carbon and air-quality restrictions likely to become more stringent over time, it's actually in the best long-term interest of the coal industry to embark sooner rather than later on sequestering CO_2. (The rules are expected to apply only to new U.S. coal plants, not existing ones.) Similarly, a 2009 ruling mandated that any new coal-fired power plants in the United Kingdom would require CO_2 sequestration (see below), and in 2013 the nation announced that it would no longer provide funding for conventional coal plants in other countries. A number

When will the oil (and gas) run out?

One key question in predicting the energy mix of the future is the amount of oil left in the ground. Not long ago, the concept of "**peak oil**"—the point at which we've extracted half of the world's accessible oil reserves—was gaining prominence. The claim wasn't that oil would suddenly become scarce, but that production would start to drop, demand would outstrip supply, and prices would skyrocket, causing massive economic dislocations.

Lately, the concept of peak oil has come upon hard times. Estimates of accessible oil have jumped dramatically over the last few years with the advent of **hydraulic fracturing** (or **fracking**), as well as oil-sand extraction and other so-called unconventional ways of getting oil out of the ground. After plateauing in 2005, global oil production began climbing again in the early 2010s. Most of that worldwide growth can be traced to the unexpected surge in the United States, where fracking has led to oil booms in Texas and North Dakota.

Putting such wild cards aside for a moment, the concept behind peak oil remains valid. In 1956, M. King Hubbert, a geologist at Shell, told colleagues that American oil production would peak in 1971. Few people took his warning seriously—after all, production was climbing year upon year through the 1960s—but sure enough, the peak arrived in 1970, just one year from Hubbert's forecast. In that year, more than 9.6 million barrels per day of crude oil were produced. And as of 2013, U.S. oil production has yet to return anywhere close to those 1970s levels. However, the trend has taken an unexpected detour from the bell curve predicted by Hubbert. Fracking has sent U.S. production on the sharpest multiyear climb in its history: from about five million barrels per day in 2008 to more than seven million barrels per day by 2013.

Just as the boom itself came with little warning, there's no way to know when or how it will end. Optimistic projections have surged lately, as new fields are discovered across the world and drilling technology continues to advance. Yet fracking may not gain the same foothold in countries outside North America that lack the needed legal or physical infrastructure or where environmental concerns rank high. Even in the United States, at least two states have refused to allow fracking at all. And at some point—perhaps decades away, possibly sooner—even these unconventional oil resources would be expected to follow the same Hubbert-style pattern that traditional U.S. oil did. In its 2013 World Energy Outlook, the International Energy Agency (IEA) projects that total U.S. oil production (including fracking) will push past its 1970 peak within a few years, but then they expect it to begin dropping again, perhaps before the year 2030. How long fracking-based oil remains profitable depends in large part on the global balance of energy supply and demand, including factors such as potential carbon restrictions as well as the growth of conservation and renewables. According to the IEA outlook, as much as half of the global growth in unconventional oil resources will merely compensate for the gradual decline in conventional supplies.

Fracking has given an even bigger boost to **natural gas**. It's being produced in record amounts in the United States and worldwide, and a major shift from coal toward gas helped the United States trim its emissions of carbon dioxide in the early 2010s (see chapter 15). The IEA foresees continued growth in world natural-gas production for decades, though there's again uncertainty in several areas: how long unconventional techniques will bear fruit, which countries will prove most hospitable to fracking, and to what extent global trade in natural gas might increase. Natural gas is more costly and complicated than oil to transport by sea, as liquefied natural gas (LNG) is fiercely explosive and few coastal cities are eager to play host to LNG tankers. Even so, natural gas may have an important role to play in displacing coal for electricity generation, according to the IPCC's 2014 report on mitigation. Many of the more stringent emission-reduction scenarios look to natural gas as a bridge fuel over the next several decades, with the assumption that its use will fall back below current levels by 2050.

For now, the oil and gas boom is helping fossil fuels to remain king of the energy castle. They've represented around 80% of total world energy use for decades, despite the rapid growth in renewables. The 2013 IEA projection shows overall energy use rising so dramatically that even a continued surge in renewables won't throw fossil fuel use off its century-plus growth path anytime soon. However, any future shift in oil or gas supplies could have massive effects on the economics of other energy sources. Oil is used above all as a portable fuel, while much gas is burnt for domestic heating and cooking. Turning to alternatives—from hydrogen cars to electric heating—would require mammoth increases in electricity generation. For this, the world would need to call on a mix of coal, natural gas, renewables, and/or nuclear power. The relative investment in each of these options could be the single greatest factor in our success or failure in combating climate change.

of existing plants in the United Kingdom are being shuttered, in large part because of tightened European Union rules on non-CO_2 pollutants.

How much can coal plants themselves be cleaned up? Incremental improvements in technology over the years have raised the efficiency with which coal-fired plants can pulverize and burn their fuel by a percent or so each year—no small achievement. Yet as of 2013, only about a third of the energy in the coal burned at a typical plant across the world was translated into electricity, according to the World Coal Association. The most modern coal-fired plants have efficiencies closer to 45%.

Another approach to cleaning up coal is **gasification**. Instead of burning the coal directly, a gasification plant heats the fuel to a high temperature with just a dash of oxygen. The resulting gases are either burned or further refined into **synfuels**—synthetic versions of oil-based fuels such as gasoline or diesel. This isn't a new idea; Germany was driven to synfuels when its oil supplies ran low during World War II. Going forward, it's easy to see how the coal industry might turn to this approach, assuming that renewables taken on an increasing share of electricity generation while demand continues to grow for fuels to transport people and goods. On the face of it, turning coal into a gas or liquid might seem like an inefficient way to use an already problematic fuel. However, there are some potential benefits to the process. Chief among them, in terms of greenhouse gases, is that CO_2 can be easily separated out of the stew of gases after combustion. There's even the option of underground gasification, another old idea now being re-explored at various sites around the world. In this case, the coal isn't mined at all—instead, it's heated and burned in place below the surface, where the resulting carbon dioxide could theoretically be stored.

Carbon capture and storage

The benefit of isolating CO_2 created in a coal gasification plant—or any other power station—is the chance to capture the gas and keep it from entering the atmosphere. This process is known as **carbon capture and storage (CCS)**, or more generally as **sequestration**. There was no economic or regulatory motive for most of the world to explore sequestration until the Kyoto Protocol kicked in. Now there are dozens of pilot projects scattered around the world, especially in Europe and North America. Most involve large power plants, as these are the biggest point sources of greenhouse gases.

The push for CCS slowed notably in the late 2000s and early 2010s for a variety of reasons, including the high start-up cost for such projects, the global economic downturn, and the continued lack of a price on carbon, especially in the leading coal-producing nations. Moreover, many activists view CCS as an extension of the throwaway culture that's led to climate trouble in the first place. Still, CSS is likely to become an increasingly important tool over time—not least because it's hard to imagine a situation where all the existing and planned fossil fuel power plants get switched off or canceled.

The idea behind CCS is to put the captured CO_2 (which could represent as much as 90% of a power plant's carbon footprint) in some sort of underground location where it will remain stable for a long time. Some proposals initially explored injecting CO_2 into the sea—at shallow or intermediate depths, where it would dissolve into the water, or at the ocean bottom, where the high pressure would keep it liquid for long periods before it eventually dissolved. But those concepts have proven unable to overcome a serious environmental objection: we know that CO_2 is making the oceans more acidic (see chapter 7). A more promising solution is to put the carbon underground. If it's pumped to depths of more than 800 m (2600 ft) below ground, CO_2 takes on a liquid-like form that greatly reduces the chance of it rising to the surface. Given the right geological formations, the carbon could remain underground almost indefinitely. There's also the possibility of storing carbon as a solid by combining the CO_2 with magnesium and calcium-based minerals—not unlike the process of chemical weathering, by which Earth itself can pull CO_2 from the atmosphere (see chapter 11). A group based at the University of Newcastle, Australia, is testing how mineralizing carbon might allow for the creation of carbonate "bricks" that could be used in construction, thus eliminating the need for underground storage.

Among the main types of proposed locations for storing carbon are the following:

* **Depleted oil and gas wells.** This technique has been used on a limited basis for decades in the practice called enhanced oil recovery, which has the economic—if not climatic—benefit of prolonging production from aging reserves (as the CO_2 is pumped down into the wells, it helps push out the remaining oil or gas).
* **Saline aquifers.** Found in basaltic formations and other types of rock, these typically hold salty water that's separate from the fresh water found in other aquifers. There may be enough saline aquifers to hold centuries' worth of CO_2.

Susan Rosenberg (left) and climate scientist James Hansen (center) were among those arrested at West Virginia's Goals Coal Company during a protest of mountaintop-removal coal mining on June 23, 2009. (RAN Field Photography/Wikimedia Commons)

There are two big pieces to the CCS puzzle that are only now starting to align. One is the industrial-scale testing of sequestration itself. To date, there have been only a few such projects. Two Norwegian natural-gas plants have injected millions of tons of carbon dioxide into aquifers beneath the North Sea, with a similar Algerian project doing the same underneath the Mediterranean. In North America, CO_2 from a synfuel plant in North Dakota is being piped to Saskatchewan and used to enhance recovery at an aging oil field. Meanwhile, a multiyear process was set to culminate in 2015 with £1 billion in capital funding from the U.K. government going to a competitively chosen project that would carry out CSS on an industrial scale.

Along with sequestration, the other puzzle piece is to develop the technologies to reliably and inexpensively separate out the CO_2 within power plants. This can be done in various ways. The **post-combustion** approach uses chemical processes to remove the CO_2 from the exhaust stream of a regular power plant. **Pre-combustion** CCS, by contrast, involves using a plant with gasification capability to turn the fuel into a mixture of CO_2 and hydrogen before any burning takes place. The CO_2 is then separated off and the hydrogen is used as the carbon-free energy source. Finally, the **oxy-fuel** approach uses special equipment to extract large quantities of pure oxygen from the air. The fuel is then burned in the pure oxygen (rather than the usual mixture of oxygen and nitrogen). That means the exhaust contains only CO_2 and steam, which can be easily separated.

Each of these approaches has its pros and cons. Post-combustion and oxy-fuel are attractive principally because, at least in theory, they could be retrofitted to existing power plants—which could be crucial considering the long operating lifespan of the world's existing and planned power stations. A small-scale oxy-fuel plant has been running for years in Germany, and many other trials of both these technologies are being developed around the world. Pre-combustion CCS, on the other hand, involves using specialist modern power stations, such as those based on **integrated gasification and combined cycle** (IGCC) technology, which was first explored in the 1980s. IGCC plants employ gasification but also use the waste heat to make steam, which drives a second turbine to increase efficiency. As of 2013 there were a handful of pilot IGCC plants in operation across the world, mainly in Europe, the United States, and China. Those numbers should increase substantially through the decade.

Perhaps surprisingly, the two big puzzle pieces—CSS and IGCC—have yet to come together as of this writing. Although it would be a fairly simple matter to capture the CO_2 from an IGCC plant, it has taken years longer than many people had hoped for the first such facility to take shape on an industrial scale, given the factors cited above. However, a number of projects in several countries now appear to have a shot at finally becoming the world's first industrial-scale IGCC-CSS plant. One is China's GreenGen, a multistage effort on the way to having a 400-megawatt plant sequestering more than 80% of its CO_2 emissions by 2018. Significantly, one of the key partners in GreenGen is United States–based Peabody Energy, the world's largest private coal company.

The United States had originally planned to have its own IGCC plant, the Illinois-based FutureGen project, up and running by 2011, with the help of an international set of funders and industry partners. In 2008, however, the project was cancelled by the U.S. Department of Energy, which blamed unexpectedly high construction costs—due in part, ironically, to the breakneck pace of conventional coal-plant construction elsewhere driving up the cost of key components. Later revived by U.S. president Barack Obama, FutureGen 2.0 is slated to begin construction in 2014, but this time as an oxy-fuel plant rather than an IGCC facility.

Three other U.S. sites have been vying to become the nation's first IGCC plant to store carbon. Construction was well under way in 2013 on a 582-megawatt IGCC plant, the Mississippi-based Kemper project, which was expected to begin operations in late 2014. However, this facility

Making sense of power

As you read through this chapter, you may find the following rough comparisons useful. Watts and their derivative terms (kilowatts, megawatts) are measures of the power being delivered or consumed; a watt is one joule of energy per second.

* **100 watts** = the power used by a strong incandescent light bulb (the non-energy-saving type that's now banned in a number of countries).
* **1000 watts = one kilowatt** = the order of magnitude of the average electricity demand of a typical household in the developed world (the actual demand goes up and down depending on the time of day).
* **1,000,000 watts = one megawatt** = the amount of power delivered by a relatively small commercial wind turbine (some of the newest ones provide 7 MW or more) = a thousand households' worth of electricity.
* **1,000,000,000 watts = one gigawatt** = the amount of power generated by a large coal-fired power plant = a million households' worth of electricity.

Kilowatt-hours (kWh) and similar terms describe the amount of energy delivered or consumed over a period of time. The world's total electricity usage in 2012, including both residential and commercial needs, was close to 19 trillion kWh. If consumed at a steady rate, that would represent about 2200 gigawatts of power at any one time.

will only sequester about 65% of its total CO_2 emissions. Two others—the 421-megawatt Hydrogen Energy California project and the 400-megawatt Texas Clean Energy Project—are expected to go into construction in the mid-2010s. Both of them are projected to tuck some 90% of their emissions into nearby oil fields. As a profitable by-product, both the California and Texas plants will each produce hundreds of thousands of metric tons of fertilizer each year.

Even with CSS lately regaining a bit of its lost momentum, some experts doubt that any of the sequester-oriented technologies can be refined and deployed globally soon enough to avoid major climatic trouble. After all, even if we did manage to store 90% of *all* fossil fuel plant emissions underground, the remaining 10% going into the atmosphere is no small matter. And not all environmentalists are thrilled with CSS as it's being put forth. The Sierra Club, for example, has opposed all three of the major IGCC-CSS initiatives in the United States, citing various concerns and more generally stressing the need to ensure that CSS testing is rock-solid before commercial-scale efforts are undertaken. Still, the coal industry is likely to face continued pressure to step up its CCS efforts. James Hansen, the eminent NASA scientist who helped trigger the first wave of climate change awareness in 1988, is among many who see coal at the heart of the greenhouse threat. Hansen caused a stir in 2007 when he called for a moratorium on all new coal plants that do not include functioning CCS systems.

Beyond fossil fuels: Renewables and nuclear

Cleaning up fossil fuel–based electricity sources is all well and good, but it's equally important to consider the alternatives. These fall into two camps: **renewables**, such as wind, solar, hydro, tidal, geothermal, and biomass, and **nuclear**. The degree to which each of these sources can meet our demand for low-carbon energy is among the most important and fiercely debated issues in climate change policy. It's an immensely complex area—an intimidating mix of economics, technology, and politics.

For one thing, it's not just the *amount* of energy each source can generate that counts. Equally important are various other factors, including the following:

* **Generating time.** Demand for electricity varies throughout the day and year, and storage is difficult, which can be a problem for solar, for example.
* **Reliability.** Wind power, for instance, has enormous potential but needs to be backed up by other sources during calm periods.
* **Public safety.** It's a perennial concern for nuclear power, not helped by Japan's Fukushima Daiichi disaster.
* **Roll-out time.** In an era of proliferating coal power, every year wasted is bad news for the climate.

As for **cost**, this is complicated by subsidies and tax regimes and the fact that each source becomes cheaper and more efficient as it becomes more widespread. Things are further clouded by the lack of definitive statistics on the total greenhouse emissions or maximum potential contributions of each energy source. "Nobody agrees about figures," U.K. commentator Simon Jenkins notes. "Energy policy is like Victorian medicine, at the mercy of quack remedies and snake-oil salesmen." On top of all this, there's often a heated debate about where the most emphasis and the most funding should go: either deploying current renewable technologies as quickly as possible or investing in research and development on the next generation of renewables. The latter could pay off handsomely over time, but at the risk of allowing greenhouse gas concentrations to pile up even further before the new technologies come online.

To scale up renewables in a serious way, major new infrastructure would be critical in order to transmit and/or store the vast amount of energy generated at different locations and times. There's also the key fact that some amount of fossil fuel will be needed to build, deploy, and maintain renewables such as wind and solar on the mammoth scales needed. And most experts see a long-term need for at least some amount of "baseload" power available around the clock—either from fossil fuels or nuclear plants—to help back up renewables as they make up an increasingly bigger piece of the electrical-generation pie. This isn't a linear problem: as renewables' portion of the total electric grid rises, the balance between variable and 24/7 power sources can change in unexpected ways. Still, it appears that an intersecting web of renewables could go a long way toward meeting the world's total energy needs, especially if vehicles go electric in a big way. The *Renewable Electricity Futures Study* (*RE Futures*), a major 2012 analysis by the U.S. National Renewable Energy Laboratory, found that with currently available technology, renewables could easily supply some 80% of the nation's electricity by 2050 (as opposed to the actual 10% in 2010) while keeping supply and demand balanced on an hour-by-hour basis. Despite the challenges of energy storage and transmission, "the upswing in wind and solar is a welcome trend and an important contribution to energy sustainability," notes the International Energy Agency.

One of the most detailed and optimistic analyses of how the United Kingdom and the planet as a whole might actually go carbon neutral for all of its energy needs was produced by engineer and energy guru David MacKay (University of Cambridge) in his 2009 book *Sustainable Energy—Without the Hot Air*. The task won't be simple, MacKay says: "To complete a plan that adds up, we must rely on one or more forms of solar power. Or use nuclear power. Or both."

With so much to consider, a full assessment of non-fossil energy is beyond the scope of this book, but following is a brief description of the pros and cons of each major source.

Solar power

Nearly all our energy ultimately comes from the sun. Wind, for example, is kicked up by the temperature contrasts that result when sunlight heats

Earth unevenly, and you can't have biofuels (or even fossil fuels) without the sunshine that makes plants grow. However, when people refer to **solar power**, they're usually talking about devices that convert sunlight directly into electricity or heat.

Solar energy took off in the oil-shocked 1970s, but its momentum sagged as the limits of that era's solar technology became clear and as oil prices dropped. Today, with the risks of climate change looming, the sun is rising once more on solar power. As of 2012, about 100 gigawatts of grid-based electricity was being derived from solar-based systems, a fivefold increase from 2009. Solar power now makes up about 2% of total global electrical capacity. Solar plants beat out wind farms for the first time in 2013 in terms of newly added capacity, although wind power is still more prevalent when all current plants are considered (see below).

Historically, solar power has cost several times more than coal-derived electricity, mainly because of the expensive raw materials that go into solar cells and concentrating systems (see below). The costs of solar technology have been steadily dropping, though, and in some locations, tax structures and other factors already make solar power roughly competitive with other sources. And for people who want to go off the grid—or who have no grid to draw from—solar power ranks near the top of the list. The costs are likely to keep dropping as solar power continues to grow.

Some solar technologies have been used for centuries, such as orienting windows toward the south (in the Northern Hemisphere) to allow warming sunlight in. These and other **passive** solar techniques are now commonly used in home and office construction. The higher-tech approaches fall into several camps:

* **Solar cells (also known as photovoltaic or PV systems)** use sunlight to split off electrons from atoms and thus generate electricity. A typical PV cell is about 10 cm × 10 cm (4 in. × 4 in.). Dozens or hundreds of these are combined into each solar panel, and panels can be joined to make an array. In small numbers, PV cells are ideal for small-scale needs such as patio lights. With a small array and a sunny climate, you can power an entire household; a large array can serve as a power plant. The standard PV panel, used since the 1970s, has been joined more recently by thin-film versions; these are more flexible and cheaper, but less productive (they can harvest roughly 10% of the solar energy reaching the panels, compared to nearly 20% for the most efficient standard PV panels).

Despite its northerly latitude, Canada boasted the world's largest PV plant at the start of the 2010s, with over a million panels providing a capacity of 97 megawatts. However, the latest PV plants now in design and construction are dwarfing those of just a few years ago. Arizona's Agua Caliente Solar Project was expected to reach a capacity of 397 MW by 2014, and two 550-MW plants were in the works in California. When running at peak capacity, each of the latter would provide roughly as much electricity as a large coal-fired power plant. India and China each have solar parks with a collective capacity of around 200 MW, and Europe's largest PV facility as of 2013 was Germany's Neuhardenberg Solar Park, a complex of 11 plants with a peak capacity of 145 MW.

✳ **Solar thermal systems** use sunlight to create heat rather than converting it directly into electricity. On a small scale, this usually involves using flat, roof-mounted "collector" panels to provide hot water in homes or commercial buildings. Tubes that carry water (or, in some cold climates, antifreeze) are sandwiched between a glass top and a back plate, which is painted black to maximize heat absorption. The fluid heats up as it flows through the collector and then proceeds to the hot-water tank. The whole process is typically driven by a pump, though there are also passive systems in which the heated water produces its own circulation.

✳ **On a much larger scale, concentrating solar power plants** use massive arrays of mirrors to direct huge quantities of sunlight onto tubes filled with a fluid, which becomes extremely hot and can be used to drive a steam turbine or Stirling engine. The world's second-largest concentrating system, California's vast 354-MW Solar Energy Generating Systems complex, which includes nine plants at three locations, is dwarfed only by the even larger Ivanpah Solar Electric Generating System (377 MW), also in California, that went online in 2014. Spain is a strong runner-up in concentrating systems, with a dozen recently completed plants that each have capacities of 100–200 MW.

Solar power is both limited and underused—limited because of the fixed distribution of sunlight around the world and around the clock, and underused because far more buildings could benefit from strategically placed solar systems than those that now have them. The developing world is a particular point of interest, because even a small PV array can power an entire household if the demand for electricity is modest (see chapter

Aerial view of a portion of Abengoa Solar's Solnova Solar Power Station in Spain. With a total of 250 MW of capacity, it ranked briefly as the world's largest concentrating solar power facility upon its completion in 2010. (Abengoa Solar/Wikimedia Commons)

18). At the other extreme, the world's most advanced showcases for energy efficiency make good use of passive and active solar systems.

The world's subtropical deserts are potential solar gold mines. David MacKay estimates that a square in the Sahara 1000 km (600 mi) on each side, covered with solar concentrating systems, would produce about as much power as the world's present consumption of all types of energy, including oil and coal. A plan called Desertec would develop huge solar power installations in North Africa and the Middle East to serve those regions plus Europe. The main challenge is storing and transporting the resulting energy. This could be done using molten salt tanks for storage and long-distance high-voltage power lines for transmission, though the upfront costs would be high—especially when the transmission distances are large. Hydrogen may also help with large-scale solar storage and distribution at some stage, but there are technical hurdles to overcome. As of 2013, the weak European economy and a split between Desertec's two main champions (the Desertec Foundation and the Desertec Industrial initiative) had left the future of the concept uncertain, but proponents remained upbeat. Even in the absence of a grand plan, solar power continues to blossom across North Africa. And in the wake of the Fukushima disaster, the Desertec Foundation signed an agreement with the Japan Renewable Energy Foundation to explore bringing the massive-solar-grid idea to East Asia.

Wind power

Although it's growing at a less rapid pace than solar power, **wind power** still represents a larger slice of the planet's renewable energy profile. As of 2012, wind accounted for about 280 GW of electric capacity, a more than fourfold increase in just three years. This occurred despite the recent blunting of wind power's growth in two big markets, China and the United States, because of uncertainty in federal policies and incentives. A similar slowdown may occur in Europe during the mid-2010s.

Averaging out the variability of wind speeds, wind power overall provides about 5% of the electricity generated globally. More than a third of the world's installed wind-power capacity is in Europe: wind power is especially popular in Germany, and it's grown quickly in Spain as well. After lagging other parts of the global for years, the United States now produces more wind power than any other nation—about 20% of the global total, as of 2013. California's Alta Wind Energy Center has a maximum capacity of just over 1 gigawatt, and the American climate and landscape lend themselves to proposals as grand as the Titan Wind Project in South Dakota. Though recently stalled, the full project would generate peak power of more than 5 GW, the equivalent of 10 coal-fired plants. The potential for continued growth in North American wind power is impressive, especially across the perpetually windy Great Plains from Texas to North Dakota and on to Saskatchewan in Canada. The National Renewable Energy Laboratory estimates that the 48 contiguous United States could conceivably generate more than 10 terawatt-hours of wind power per year. That's more than twice the total energy generated by all U.S. power plants of any type. The United Kingdom also has massive potential for wind energy, especially when it comes to offshore wind farms.

Once in place, wind turbines produce no greenhouse gases at all—the only petrochemicals involved are the small amounts used to manufacture and lubricate the units. Another bonus is that the giant wind turbines are typically centered at heights of 80 m (260 ft) or more, leaving the land below free for other uses. Many farmers and ranchers rake in substantial royalties for allowing turbines to be placed on top of fields where crops grow and cattle graze.

On the basis of all this, wind power seems like a no-brainer. Yet wind farms regularly come up against local opposition, especially in the United Kingdom and the United States, where some activists have fought against them with a vehemence approaching that of the anti-nuclear movement that peaked in the 1970s and 1980s. Mostly, the opposition isn't based on ecological or economic concerns but on aesthetics (some Brits have referred to wind turbines as "lavatory brushes in the sky") and noise (the distinctive whoosh-whoosh of the blades and the creaking that sometimes results when the blades are moved to take advantage of a wind shift). Another allegation is that wind farms are a danger to bird life. These are all contested claims. Many people enjoy the motion and symbolic cleanliness of wind farms, and proponents claim that you can hold a normal conversation while standing right underneath a modern turbine. A low-frequency hum may disturb some people within a kilometer or two, depending on the level of other ambient noise, but if wind farms are properly sited and built, these effects can be minimized.

As for bird deaths, much has been learned from California's Altamont Pass wind farm, where thousands of bird deaths have been reported each year. One of the oldest of the world's major wind farms, Altamont Pass is located on a migratory route, and its old-school blade design (with latticework that serves as bird perches) appears to exacerbate the problem. Better-sited facilities with slower-spinning blades can help reduce the risk to avian life. The American Bird Conservancy has recommended mandatory design standards to protect birds and expressed support for wind power "when it is bird-smart." Raptors, including hawks and eagles, are a particular concern, as their attention is typically focused downward—on potential meals scurrying across the ground—rather than on blades spinning across their flight path. Some wind turbines now feature cameras that can sense approaching raptors and trigger noise and light bursts to steer them away. Overall, it's worth noting that for each unit of energy generated, fossil fuel plants appear to kill many more birds than wind farms do—and few would consider these plants any more appealing to look at.

Building **offshore wind farms** is one way to minimize objections, at least in theory. The last few years have seen a boom in major offshore facilities, most of them in or near the North Sea. Off the southeast U.K. coast, the Greater Gabbard farm opened with a capacity of 504 MW in 2012, and the London Array—the world's largest offshore wind farm as of 2013—helps power greater London with a capacity of 630 MW. However, going offshore adds substantially to the upfront cost of a wind facility, and it isn't always enough to placate local protesters. For nearly a decade, a wide range of people living and working near Nantucket Sound in New England fought plans for the first U.S. offshore wind farm. The facility was finally approved in 2010, though lawsuits continued to slow the project for years afterward. Scheduled for construction in the mid-2010s, the Cape Wind project is designed with 130 turbines, each extending about 90 m (290 ft) from sea surface to turbine center, and a total capacity of 454 MW.

A wind farm on the plains of far northeast Colorado. (© UCAR)

Outside of the local controversies, most polls show strong overall public support for wind power, so at least for now, the tugs-of-war at specific sites appear unlikely to throw wind power off its course of vigorous expansion. In countries with lots of wind potential, a bigger long-term challenge could be intermittency. Despite claims by protesters, dealing with the inconsistency of wind power isn't generally a big problem when only a small proportion of a country's total electricity is produced from wind turbines. But above a certain threshold, intermittency can become a more serious constraint, hence the huge amount of attention currently focused on so-called **smart grid** technologies designed to more dynamically deal with electricity supply and demand.

Geothermal energy

The same forces that can blow the top off a volcanic mountainside can also generate electricity through geothermal energy [not to be confused with the use of ground-source heat pumps for heating and cooling buildings (see p. 432)]. Most geothermal power plants tap into the hot water produced where rain and snowmelt percolate underground to form pools that are heated by adjacent magma and hot rocks. Some facilities draw hot water and steam from more than a mile below ground, while others harvest steam that emerges directly at the surface. Either way, a small amount of naturally produced carbon dioxide escapes into the air—but it's less than

Iceland's low-carbon lifestyle

Thanks largely to its fortuitous location atop a geological hot spot, Iceland may point the way toward a cooler and cleaner future for the rest of the world. This island nation of about 300,000 people draws on a unique portfolio of energy, with a proportion of renewables that's the highest of any country on Earth—providing about 85% of the nation's total energy and nearly all of its electricity.

The two continental plates that play host to North America and Europe are separating beneath Iceland, which leads to the world's most concentrated zone of **geothermal** energy. The very name of the capital city, Reykjavik, derives from *reyka*—the plumes of steam visible from many points around the country. Steam and hot water from underground furnishes heat for some 90% of Iceland's buildings. Although depletion of geothermal energy is a concern in some other parts of the world, it's less so here: the magma, rocks, and steam beneath Iceland are estimated to hold many centuries' worth of energy, even without factoring in the potential for their recharge through future volcanism. Although some of Iceland's electricity comes from geothermal power plants, most of it is produced by **hydroelectric** plants. They've been a reliable source of power in the island's moist climate, but also a source of friction from environmentalists who want to limit the intrusion of dams and reservoirs into wilderness areas.

There's far more potential for geothermal power in Iceland than the country can use—in fact, it's been exploring the option of building the world's longest transmission line in order to export electricity to the United Kingdom. At the same time, the nation still imports oil for its cars and ships. These intertwining factors make Iceland a perfect test bed for **hydrogen**, as evidenced by a long-term project carried out by a public–private partnership that includes DaimlerChrysler, Shell, and NorskHydro. The first step was a mini fleet of three fuel-cell-powered municipal buses in Reykjavik, which drew power from the world's first public hydrogen filling station, followed by passenger vehicle tests.

10% of the amount emitted by a standard coal-fired power plant of the same capacity. **Binary geothermal** plants avoid even these minimal emissions of CO_2 by using water from underground to heat pipes that carry a separate fluid; the geothermal water returns to Earth without ever being exposed to the atmosphere.

More than 11 GW of electric capacity was available from geothermal sources as of 2010, although the technology's growth (about 20% from 2005 to 2010) has been much more modest than for wind and solar power.

Geothermal plants can run close to their capacity almost 24/7, since their power source doesn't vary like sunshine and winds do. The best sites for geothermal plants are where continental plates separate or grind against each other, especially around the Ring of Fire that surrounds the Pacific Ocean. Around a quarter of developed global geothermal capacity is located in the United States, where the world's biggest plant—The Geysers, in northern California—provides more than 1.5 GW. In the Philippines, geothermal sources provide more than 25% of all electricity. They're also a big part of the palette of carbon-free energy in Iceland, where geothermal energy provides direct heating as well as electricity.

A completely different approach to geothermal that's recently been attracting interest is known as **enhanced geothermal systems (EGS)**. In an EGS plant, dry hot rocks deep under the surface are fractured artificially using explosives, and water is piped down from the surface. This approach could, its advocates claim, open up vast new areas for geothermal exploitation, as it removes the requirement for particular under-surface water systems. On the downside, a given EGS site will only provide energy for roughly 30 years. It's also possible that the technique's reliance on underground fracturing could strike a nerve among people who are already opposed to the growth of oil- and gas-related hydrofracking. As of 2013, EGS plants were operating in France, Germany, and the Netherlands, with the potential for similar projects in the United States, Turkey, Australia, and various other countries. The world's first large-scale EGS plant, dubbed Habañero, may take shape in Australia's Cooper Basin. A successful test took place there in 2013, though it's unclear whether there's enough demand for electricity from Habañero's remote location to finance a commercially scaled site.

Hydroelectric, tidal, and wave power

Hydropower plants use dams to channel water through electricity-generating turbines. This approach currently provides far more energy than any other renewable source—around 1100 GW of electric-generating capacity, as of 2012, with the total increasing by several percent a year. In theory, hydropower capacity could be scaled up severalfold, especially in Africa and Asia, but actual growth is expected to be much more modest. One reason is the high start-up costs: many of the more easy-to-engineer sites have already been used up. Another is that the construction of dams can wreak havoc with river-based ecosystems far upstream and downstream. There are also social issues, as large dams can flood huge areas and displace many residents.

Most of the large-scale dams and hydropower plants now in planning or under construction are in China and Brazil. In both countries, environmental and societal concerns about hydropower have largely taken a back seat to rapidly developing economies. Opened in 1984, the 14-GW Itaipu Dam, which borders Brazil and Paraguay, is the world's second largest in electrical capacity. Brazil is now pouring more than US$150 billion into a new wave of dams and reservoirs set to come online through the early 2020s. This includes the massive and controversial Belo Monte Dam in Amazonia, scheduled to start operations in 2015 with an expected capacity of more than 11 GW. The potential impact of Belo Monte's reservoir on indigenous residents and wildlife has made it a cause for local and regional activists as well as world celebrities, including musician Sting and director James Cameron.

China's Three Gorges Dam—the largest on Earth—is an especially dramatic example of hydropower's outsized potential for both good and ill. Over the last few years, the associated reservoir has swallowed up entire cities and forced more than a million people to resettle. Yet at its opening in 2012, the dam brought 22.5 GW of capacity online, which made it the world's largest power station and the single largest human-created source of renewable energy on Earth.

While hydroelectric plants don't come close to coal-fired power plants as sources of greenhouse gas, they can still have a significant carbon footprint both during and after construction, especially at lower latitudes. In the tropics, vegetation decomposing in reservoirs can produce large amount of methane, especially in places where the water level rises and falls substantially each year during distinct wet and dry cycles. A 2011 study in *Nature Geoscience* led by Nathan Barros (Federal University of Juiz de Fora, Brazil) surveyed 85 reservoirs that made up about 20% of the global surface area covered by artificial bodies of water. The study estimated that Earth's reservoirs collectively put about 48 million metric tons of CO_2 into the atmosphere (less than 2% of the amount emitted from fossil fuels) and about 3 million metric tons of methane (roughly 1% of the total produced by human activity). The study didn't account for indirect impacts on the carbon budget, including the buildup of carbon in reservoirs through sedimentation as well as the displacement of forests that would otherwise be drawing CO_2 from the atmosphere.

There are other ways to generate electricity from the movement of water. Turbines driven by the **tide** and **waves** are both being developed. Various technological and economic barriers need to be overcome before either source could meet a significant proportion of the world's energy demands, but new innovations in design have triggered growing interest. Most of the recent effort has gone into tidal power, which shows particular promise in certain countries (including the United Kingdom and Canada, for example), as its timing is predictable and it could be harnessed offshore and out of sight using relatively simple technology—the equivalent of underwater wind turbines. The main challenge is building turbines that can withstand the enormous forces of fast tidal streams. South Korea is now leading the way in deployment, with two major projects off its northwest coast. The Sihwa Lake station opened in 2011 as the world's largest (its 254 MW capacity nudging just past the 240 MW of France's 50-year-old Rance station), and the even-bigger Incheon station is scheduled to provide 1.32 GW of capacity by its completion in 2017. Still-larger projects have been proposed, including several by Russia and the United Kingdom.

Biomass

While the practice of using plant-derived fuels as a substitute for oil has engendered both enthusiasm and controversy (see below), there's been comparatively little buzz about using biomass as a source of electricity, where it can displace carbon-belching coal. In 2011, about 72 GW of global electric capacity was derived from biomass, with a growth rate approaching 10% per year. About 15% of the world's biomass-produced electricity is in the United States, where it represents about 1% of the nation's total electric capacity. The sources range from pulp and paper to landfill gas, and even solid waste from municipal sewage systems. Though biomass is often burned directly, many of the same practices used to make coal plants more efficient, such as gasification via IGCC technology, are being explored. And since the root sources of biomass pull carbon dioxide from the air as they grow, there's the tantalizing possibility that a biomass power plant using IGCC and sequestering its carbon might actually exert a net reduction on atmospheric CO_2. However, many of the types of land-use problems that emerged with the growth of biofuels could also slow the growth of biomass-derived electricity. For example, activists have voiced concern about the threat of monoculture forests reducing the diversity of local plant and animal life.

Nuclear power

Just when it looked as if it might keel over from the weight of negative public opinion and high start-up costs, nuclear power got a new lease on life with the help of climate change. It's hard to dismiss a proven technology that produces massive amounts of power with hardly any carbon emissions, yet it's also impossible to ignore the political, economic, and environmental concerns that made nuclear power so contentious in the first place.

Though it's taken a somewhat lower profile since the 1990s, nuclear power never went away. The world's 400-plus nuclear reactors as of 2013 were providing some 370 GW of capacity and close to 15% of global electricity, with a much higher fraction than that in some countries—particularly France, which embraced nuclear power in the 1950s and never looked back. France now gets around 75% of its electricity from more than 50 nuclear reactors operated through the state-owned Electricité de France and a private partner, Areva. As of 2013, the United States had more nuclear reactors than any other nation—exactly 100. Construction was under way on four more, which would be the nation's first new ones since the Three Mile Island accident in 1979. China also had some 30 nuclear reactors under construction, with ambitious plans to install as much as 58 GW of capacity by 2020.

Global warming changed the nuclear equation in tangible ways during the first decade of the new century. In 2003, Finland's legislature voted narrowly in favor of building its fifth nuclear reactor—one of the world's largest, to be completed in the mid-2010s after extensive delays and huge budget overruns—partly because of the hope that it might help the nation meet its Kyoto goals. In 2006, France and the United States agreed to provide nuclear technology to India, which as of 2012 had 20 reactors online and more than 30 others under construction or in the planning phase. Also in 2006, U.K. prime minister Tony Blair announced that he'd "changed his mind" on nuclear, paving the way for approval of a new generation of British atomic-energy plants.

Much of this action came to a screeching halt—or at least a pause—in 2011, after the earthquake and tsunami that disabled Japan's Fukushima Daiichi nuclear plant and caused a major release of radioactivity. All of Japan's reactors were closed in response, and as a result, the nation was forced to suddenly and drastically cut its UN-pledged target for greenhouse gas reduction. Instead of slashing emissions by 25% from 2005 to 2020, Japan's new reduction target was a mere 3.8%, with nuclear power assumed to be zeroed out during that time frame. At least one of the closed plants was restarted in 2013, and it's possible that other existing plants will reopen if they're deemed safe, though public opinion in Japan is now firmly against a resumption of nuclear power. Elsewhere, the Fukushima disaster put a distinct chill on nuclear development and prompted close looks at existing facilities and sites. The industry has also had to deal with a still-struggling global economy and, in the United States, the ascent of natural gas. Most of the several dozen new U.S. reactors being proposed at the start of the 2010s had been cancelled by 2013; four older U.S. plants closed that year due to economic and technical challenges. In the United Kingdom, several aging plants have also closed, while four reactors are on tap to be built over the next few years. Even so, there's been little net change to the number of existing plants outside of Japan. The IEA's 2013 World Energy Outlook envisions total nuclear output resuming after the Fukushima-induced pause, with capacity about two-thirds larger than now by 2035.

Though a certain amount of CO_2 is released in the process of building a nuclear plant (plus mining and processing its fuel), there's no question that nuclear power is far kinder to the climate than fossil fuels are. That simple fact has engendered some powerful crosscurrents. Environmentalists who oppose an expansion of nuclear power run the risk of appearing callous about climate change. Thus, the anti-nuclear arguments have shifted somewhat in recent years. The unsolved problem of disposing of nuclear **waste** remains a key point, as does the risk of **accidents** like Chernobyl in 1986 and Fukushima in 2011. Recent heat waves have also underscored how nuclear plants themselves can be vulnerable to a warming climate. Reactors were temporarily closed in Germany in 2003, Michigan and Spain in 2006, Tennessee in 2007, and New York in 2012, in each case because lake, river, or sea water had warmed too much to cool down the reactors. On the other hand, industry advocates claim that the risks of accidents from modern, well-planned facilities are overstated. They point out that Chernobyl was a low-budget, Soviet-era construction and that Fukushima (also an older design) was sited without adequate protection from tsunamis. It's also true that coal, oil, and gas production lead to many deaths each year, including more than 100,000 fatalities attributed to respiratory ailments related to coal burning. Though nuclear waste is undeniably risky, the waste products from fossil fuels—especially greenhouse gases—pose an equally unresolved set of risks to ecosystems and society.

Nuclear power opponents have also invoked **economic arguments**, claiming that the massive costs of building and decommissioning nuclear plants—which are typically subsidized and insured in whole or in part by governments—drain money that's urgently needed to fund research on renewables and energy efficiency. According to Amory Lovins of the Rocky Mountain Institute, nuclear's "higher cost ... per unit of net CO_2 displaced, means that every dollar invested in nuclear expansion will worsen climate change by buying less solution per dollar." Terrorism also figures in the mix: the more nuclear plants we have, say opponents, the more risk there is of radioactive material slipping into the wrong hands or of an attack on a power plant that could kill many thousands.

Despite these critiques, a number of influential figures have spoken out in favor of increasing the world's nuclear power capacity in the light of the climate change threat. These include Gaia theorist James Lovelock; Stewart Brand, founder of the *Whole Earth Catalog*; and neo-environmentalists Michael Shellenberger and Ted Nordhaus. In 2013, a group of four esteemed climate researchers—Ken Caldeira, Kerry Emanuel, James Hansen, and Tom Wigley—issued a statement addressing "those influencing environmental policy, but opposed to nuclear power." The scientists asserted their belief that safer, next-generation nuclear options are critical for reducing global emissions. "We cannot hope to keep future climate within tolerable limits unless the future energy mix has nuclear as a large component," the scientists claimed. These and other advocates argue that nuclear is the *only* low-carbon energy source that's sufficiently developed and reliable that it could be rolled out on a big enough scale to take a meaningful chunk out of the world's greenhouse emissions in the short and medium term. In particular, nuclear plants could replace coal in providing a 24/7 power supply to pair with time-varying sources such as wind and solar power.

New types of commercial reactors have the potential to be safer and much more efficient than older designs. For example, while many reactors use less than 1% of the energy contained in uranium, the **integral fast reactor** can use

> ❝ The only technology ready to fill the gap and stop the carbon dioxide loading of the atmosphere is nuclear power." —*Stewart Brand, "Environmental Heresies,"* Technology Review

> ❝ Nuclear power has already died of an incurable attack of market forces, with no credible prospect of revival." —*Amory Lovins, Rocky Mountain Institute*

> ❝ Nuclear energy could make an increasing contribution to low-carbon energy supply, but a variety of barriers and risks exist." —*IPCC Working Group III, 2014*

almost all of the available energy, which increases its potential power output by more than a hundredfold. Russia has had a fast reactor for more than three decades, and the U.S. government operated another such reactor from the 1960s to 1990s before it was mothballed as part of a general rampdown of nuclear research. An integral fast reactor also has the big advantage of being able to draw on spent fuel that's been stored as nuclear waste, thus reducing the need to mine new uranium while also helping solve the waste-storage problem. There's also the possibility in future reactors of expanding the use of thorium, which is more abundant, less radioactive, and more resistant to proliferation than uranium. Of course, how policymakers and the public view the risks of nuclear power compared to the risk of climate change will be a critical determinant of where society heads, and that comparison is bound to continue evolving over time.

Making transport greener

Transporting people and goods by road, sea, and air is responsible for around a quarter of global greenhouse emissions, and this percentage is rising. If we're to tackle global warming, we'll have to develop lower-emission vehicles, not to mention promoting public transport and, some would argue, reducing the flow of people and goods around the world.

The easiest gains could come from simply improving the **efficiency** of conventional vehicles, something that is already happening but which could go much further. New fuel standards across the globe will make an increasing difference over the next few years. By model year 2025, U.S. vehicles (including cars, SUVs, and light trucks) will be required to get an average of 54.5 miles per gallon (mpg), more than double the 25-mpg average that was the standard in 2007. China and Europe will see standards for passenger cars around the 50-mpg mark by 2020.

One positive development is the growing popularity of cars such as the Toyota Prius that have highly efficient **hybrid** engines, which are fueled by gasoline but can also run from a battery that's recharged when braking. Some hybrids can manage as much as 70 mpg. Ironically, the hybrid family also includes hulking SUVs and executive cars, which improve upon their non-hybrid counterparts but still require plenty of fuel per mile. Fully **electric** vehicles, which draw their power from the grid and/or renewable home energy systems, can provide far greater efficiency. Another option is **plug-in hybrid** vehicles, which include a gas-fueled hybrid engine as well as an electric motor capable of running anywhere from 15 to 60 km (10–40 mi) on a single charge. A rapidly expanding array of electric and plug-in hybrid cars has come onto the scene this decade (see chapter 19), and sales are increasing by severalfold each year. One challenge for these vehicles is that electricity in many areas is still primarily produced by fossil fuel. "An electric car is only as good for the climate as the electricity used to power it," notes the U.S. nonprofit Climate Central. In a 2013 report, the group found that in states where coal and natural gas are the main sources of electric power, a high-efficiency gas-fueled hybrid might actually be *more* climate-friendly over a 50,000-mile period than an electric car, especially when considering the amount of energy required to build electric vehicles and their batteries. However, the electric ride does win out when the power source is cleaner (wind, hydropower, and/or home solar) and the car's lifespan is maximized. In many parts of the US, electric grids are on track to become significantly cleaner over a car's working life, which shifts the equation further.

In the meantime, engineers are refining ways to power vehicles through **biofuels**—such as ethanol and biodiesel—which are made from plants. The primary attraction of biofuels is simple: though they release plenty of carbon when burned, much of this will be soaked up by the next round of fuel crops (after all, plants grow by absorbing CO_2 from the atmosphere). How climate-friendly a biofuel is, then, depends on factors such as the amount of fossil-based energy used to fertilize, harvest, process, and distribute the crops.

Biofuels were once a favored child of climate change action. Political leaders used them in bipartisan fashion to win the support of big agriculture, and in the United States they represented a form of domestically produced energy that carried broad appeal in a time of international conflict. By the late 2000s, the EU had mandated that all liquid fuels in Europe be composed of 10% biofuel by 2020 (assuming that the biofuel produces a specified savings in greenhouse emissions). In the United States, the tellingly named Energy Independence and Security Act of 2007 mandated that biofuel production soar from 4.7 billion gallons in 2007 to 36 billion by 2022. The United States produces more than a third of the world's corn, and by 2013 almost half of the U.S. corn crop was going into ethanol production.

Only a few years after this international surge in enthusiasm, biofuels have fallen into some disrepute as a tactic for fighting climate change. Attitudes have soured among a wide variety of folks, from environmentalists to free-market champions. Perhaps the most contentious issue is land. It would take truly massive areas of cropland to turn biofuels into a major part of the global energy picture. This could squeeze food production, or require the conversion of virgin land into farms, something that has im-

How much can hydrogen help?

Hydrogen, the simplest and most abundant element in the universe, has been hyped as a solution to the greenhouse crisis in recent years, especially in the United States. In truth, hydrogen is no panacea for global warming. Indeed, hydrogen isn't even an energy *source*, as we have to use power from other sources to produce the pure hydrogen gas that's useful as a fuel. Hydrogen does hold potential as a clean and portable energy *carrier*, but with an array of long-term challenges to its use, it's looking increasingly likely that electric cars will leapfrog over hydrogen vehicles in the race for climate-friendly motoring.

On our planet, hydrogen is usually bound to other molecules, as in water (H_2O). But in its pure form, H_2, hydrogen is a lightweight gas (or a liquid at extremely low temperatures) that contains a great deal of energy. Hydrogen gas can be burned directly as a fuel, but the preferred means of exploiting it is through fuel cells, which convert hydrogen into electricity. This process produces zero emissions of carbon dioxide and other pollutants—the only by-product is steam, and though water vapor is a greenhouse gas, the net amount emitted is insignificant.

A few years ago, the potential for zero-emissions vehicles led to a good deal of buzz and some big-time investment, including the EU's European Hydrogen and Fuel Cell Technology Platform. In the United States, the Bush administration launched a program in 2004 aimed at bringing hydrogen-fueled vehicles into the marketplace, and Honda began rolling out its first mass-market hydrogen vehicle, the FCX Clarity, in 2008, albeit in small numbers—only 200 in the first three years.

The enthusiasm for hydrogen cars soon waned, however, as various problems were harder than expected to crack. There's much more to hydrogen mo-

plications not just for wildlife habitat but for the greenhouse effect itself, depending on where these land changes occur. If forests are chopped down to make room for biofuel crops, the shift could actually produce a net rise in carbon emissions. Research confirming this scenario helped prompt the EU to propose cutting the biofuel mandate by half. Should this rule change go through, it would effectively keep ethanol use at its current level.

The food-versus-fuel issue rose to prominence in 2007–2008, when world food prices spiked dramatically just as ethanol production was gaining momentum in the United States and, on a smaller scale, in Europe. Corn prices tripled in just three years, and many farmers who grew other crops began shifting to corn. Food riots rocked Mexico, Haiti, Bangladesh, and other emerging nations as the price of tortillas and other staples jumped.

toring than the vehicle itself. Before it can provide energy to fuel cells, hydrogen first has to be broken off from other molecules, stored (by compression or liquefaction), and distributed. Not only is each of these steps a daunting technical challenge in itself, but each one takes energy. One Swiss study found that hauling a hydrogen-filled truck 500 km (310 mi) would consume an amount of energy equal to almost half of the cargo. On top of that, the hydrogen itself is substantially more costly per unit of energy than gasoline. Such issues are partly why the Obama administration cancelled Bush's hydrogen-car initiative in 2009. Only a few dozen Honda Claritys were leased in the United States.

A new burst of interest came in 2013, though, as three automakers—Honda, Hyundai, and Toyota—unveiled new fuel-cell models, each to be introduced on a limited basis in the United States and/or Japan by the end of 2015. The latest selling points include convenience and range. While electrics and plug-in hybrids take hours to recharge and typically have an electric range of a few dozen miles at best, a fuel-cell car could be restocked with 300 miles' worth of hydrogen in just a few minutes at a dedicated fueling station. Such stations remain scarce, although California now plans to build at least 100 of them by 2024.

Hydrogen could conceivably be produced in a home unit, but even that technology (which is many years from being available) could signal a continued reliance on fossil fuels. Critics such as veteran researcher and blogger Joseph Romm, author of *The Hype about Hydrogen*, argue that the natural gas or renewables used to produce hydrogen would make a bigger and more immediate dent in the global greenhouse picture if they were used to replace coal-fired power.

However, analysts disagreed on how much of the price spike was driven by biofuels as opposed to other factors, including market speculation and high prices for oil (which goes into making fertilizer and other agricultural components). Whatever the reasons, as of 2013, the commodity prices for corn—as well as for other key crops, including wheat, soybeans, and rice—were still running substantially higher than before the 2007–2008 spike.

In response to these pressures, there's a growing push for **second-generation biofuels**, which are derived from cellulose (see below) and other biological materials outside the food stream. After years of development, some of these fuels are now entering the large-scale commercial realm. In addition, hundreds of facilities in Europe and a few in the United States are using animal waste to generate natural gas—or, in this case,

biogas. Bacteria feed on the animal waste and generate methane that can be burned directly or used to generate electricity. (In theory, human waste could be used in exactly the same way, though it's hard to imagine the idea taking off.) Even **algae** can be used as biofuel, and once the economics are sorted out, that could become the most useful biofuel of all, thanks to algae's ultra-speedy growth habits.

Ethanol

Ethanol (ethyl alcohol) is a plant-based substitute for gasoline that's quite common in some parts of the world. According to the Renewable Fuels Association, the total global production in 2012 stood at more than 105 billion liters (28 billion U.S. gallons), of which almost half was produced in North and Central America and about 20% in Brazil. Across the United States, most vehicle fuel is now sold in a standard mixture of 10% ethanol and 90% gasoline (E10), which most cars can accommodate. If people want to use higher proportions of ethanol, such as 85/15 (E85), they need a flexible fuel vehicle (FFV) that can handle any blend of the two, and they need to find a gas station that sells it. (In the United States, the handful of stations that offer 85/15 at this writing are mostly in the Midwest.) More than 35 million FFV vehicles have been sold, mostly in the United States and Brazil, though many American FFVs never see a drop of the fuel they're built to accommodate. Ethanol made from sugar cane provides more than half of Brazil's vehicular fuel by volume (all gasoline sold there is either 25% or 100% ethanol). A plus for ethanol use in Brazil is that its mild climate enables motorists to avoid the cold-start problems that can otherwise occur with high-ethanol blends.

There's been much debate over whether ethanol really helps protect the climate. Most U.S. ethanol to date is derived from corn, typically grown with fairly extensive use of petrochemicals. As a result, corn-based ethanol seems to provide some climate benefit, but not much. The University of California, Berkeley, has estimated that corn-based ethanol uses about 74 units of fossil fuel energy to produce 100 units of ethanol energy. That's somewhat like a charity spending $74,000 to raise $100,000. As for emissions, a vehicle running on ethanol does produce greenhouse gases, and it won't travel quite as far per gallon, but when you factor out the CO_2 absorbed by the plants it's made from, corn-based ethanol blends can add up to roughly 20%–30% less carbon to the atmosphere than standard gasoline for every mile traveled. The exact percentage depends on several factors, including the wetness or dryness of the source crops and the choice of fuels used for processing the biomass at the ethanol plant.

A more promising biofuel is just beginning to come into play: **cellulosic ethanol**. Standard ethanol comes from fermenting the sweet starches in corn kernels and sugar, but ethanol can also be brewed from the sturdy material called cellulose (as in celery). Cellulose is found in corn husks as well as a variety of other plants, such as fast-growing grasses, and even in human and animal waste. Because many plants high in cellulose can thrive on land that's marginal for regular farming, they're less likely to displace food crops, and the yield per acre can be twice or more that of corn. Deep-rooted perennials like switch grass also help stabilize soil and need little if any help from petroleum-based farming techniques. As a result, 100 units of cellulosic ethanol can require as little as 20 units of fossil energy to produce, according to the U.S. Department of Energy.

The main catch is that cellulosic materials don't break down very easily, so the cost of producing cellulosic ethanol is still considerably higher than that of the corn-based variety. Chemists have been bioengineering enzymes that could be produced en masse to break down the cellulosic materials more affordably. That might help level the playing field with corn-based ethanol over the next decade, although the economic downturn and other challenges pushed the development of cellulosic alternatives well behind expectations. Of the 36 billion annual U.S. gallons of biofuel specified by U.S. law to be produced by 2022, almost half—16 billion gallons—was to be derived from cellulose. However, the EPA was forced to slash interim targets drastically in the early 2010s. Production in 2012 was less than one million gallons, a minuscule amount next to the original mandate for that year of 500 million gallons. On the plus side, several large-scale plants were each on track to start producing cellulosic ethanol by the tens of millions of gallons per year in the mid-2010s. The industry could scale up rapidly from that point onward, so it's still possible that cellulosic ethanol will be a significant player by the latter part of the decade and into the 2020s.

Before that point, U.S. ethanol may face the "blend wall"—the need to raise the mix beyond 10% ethanol in order to employ the increasingly large amounts of ethanol now being produced to meet the national mandate. The unexpected recent drop in total U.S. vehicle miles traveled only adds to the crunch, since ethanol is being mixed into a smaller pool of gasoline than originally anticipated. Research by the U.S. Department of Energy suggests that many vehicles could handle an 20% ethanol/80% gasoline blend without serious trouble (though using this might run afoul of some car manufacturers' warranties). The nation's complex infrastructure for storing and transporting fuel would need some adjustments if it's going to handle more ethanol, as the biofuel could damage some existing storage tanks and pipelines.

Biodiesel

Diesel cars—which are very popular in Europe, though far less common in the United States—are typically more climate-friendly than the gasoline-based alternatives, since they burn their petroleum-based fuel about 40% more efficiently. (The flip side is higher emissions of nitrogen oxides and particulates, though in both the United States and Europe these pollutants have been greatly reduced by increasingly tighter emission standards.) Theoretically, diesel cars can be greener still when powered by **biodiesel** made from sustainably grown crops. This isn't a new concept. The term "diesel" originally described the engine type rather than the fuel, and its inventor, Rudolph Diesel, used peanut oil to drive the prototype that won him top honors at the 1900 World's Fair in Paris.

Biodiesel received a boost in the 2000s from tax breaks and government mandates on both sides of the Atlantic. By 2011, global production was just over 19 billion liters (5 billion U.S. gallons). As with ethanol, biodiesel is often added to fossil fuel in small percentages and sold as a way of helping engines run more smoothly while emitting fewer pollutants, but you can also buy 100% biodiesel in some parts of the United Kingdom and the United States. Country singer Willie Nelson has peddled his own blend, called BioWillie, in Texas, Oregon, and Hawaii.

At its most climate-friendly, biodiesel can be a convenient product of recycling, as in the United Kingdom, where it's often derived from used vegetable oil donated by chip shops and other commercial kitchens. In the United States and mainland Europe, however, it's typically produced commercially from canola, soybean, and other inexpensive vegetable oils. This allows for a far greater level of supply, but it also raises questions about the wider impacts of the burgeoning international market for oil crops. Huge areas of tropical rainforest have already been cleared to make way for soybean- and palm-oil cultivation, with disastrous impacts on both biodiversity and the climate. For this reason, biodiesel from crops (rather than from waste products) is just as controversial as ethanol, with some environmentalists seeing it as a problem rather than a solution.

Making buildings more efficient

Buildings and the appliances they contain are responsible for more than 20% of all global energy consumption, according to the U.S. Energy Information Administration, and that percentage is rising. In the United States, it's close to an eye-opening 50%, so no matter what energy source we use to power, heat, and cool buildings, it makes sense to do so as efficiently as possible. The main challenge here isn't developing new technologies—there are plenty to choose from already, with more on the way—but getting consumers, manufacturers, architects, and governments to choose (or require) efficiency. Because so many offices and homes are rented out, it's all too common for everyone involved to ignore the efficiency issue—especially when renters pay the utility bills but don't have the long-term incentive to improve a property that's owned by someone else.

The often-underappreciated role of buildings as contributors to the greenhouse effect is largely a legacy of the way offices and homes have been designed over the last few decades. With fuel for heating and cooling assumed to be cheap and plentiful, aesthetics and short-term costs often trumped energy efficiency and long-term savings. "Commercial and public buildings in the developed world have generally become sealed, artificially lit containers," says architect Richard Rogers of the U.K. design firm Arup. "The increasingly evident threat to the global environment posed by buildings of this sort cannot be ignored."

What's inside can be as important as the building itself. In the United Kingdom, overall home electricity consumption actually *increased* about 1% a year in the decade after 2000. One factor was the profusion of electricity-guzzling equipment, from computers to big-screen TVs, which cancelled out gains in other areas. An even bigger problem is persuading people to make substantial changes to their homes—such as insulating solid walls and changing doors and windows. These changes save money in the long run, but they require more upfront investment of time and money than many homeowners (and even more renters) are prepared to put in.

Most U.S. states now have energy codes for both commercial and residential buildings, as do many cities, but the country as a whole hasn't yet adopted stringent goals for building efficiency (apart from federal buildings). The lack of national policy is a serious concern, because American homes and offices are massive consumers of energy. Since the 1970s, the size of the average new U.S. home has grown by more than 50%, and the fraction of homes with air conditioning has risen from roughly a third to more than 60%. (About 90% of new U.S. homes are now air conditioned.) Amazingly, per capita energy use in U.S. homes has stayed relatively flat over the last four decades, as energy-hogging items such as refrigerators have become far more efficient. Countering this trend is the nation's relentless growth in population—now more than 310 million, and projected to top 400 million by 2050—as well as the tendency toward smaller families and more people living alone. If nothing else, this implies a continuing need to expand and refresh the U.S housing stock, with the accompanying chance to make it more efficient.

Low-power towers

Skyscrapers have always been more than mere office buildings. The best rank among the world's greatest architecture, and they often serve as icons for the companies they house. Now a new breed of eco-oriented towers is going up, aiming to impress with their low-carbon credentials as well as their high-rise aesthetics.

An award-winning, pickle-shaped tower, originally built for Swiss Re in 2004 and now known by its address—**30 St Mary Axe**, in London's financial district (see photo opposite)—was dubbed "**the Gherkin**" years before it was even built. The design can be seen as an elongated riff on Buckminster Fuller's geodesic dome, with the curvature provided by gentle turns in the frame. Love it or hate it, the building's provocative shape has a point. Wind flows easily around it, reducing the amount of support structure needed. Less obvious is a latticework of tunnels through which air can flow into and through the Gherkin's inner reaches, cutting heating and cooling costs by up to half. The tunnels also bring natural light to the building's interior. Also in London, the 43-story **Strata** apartment building made history in 2010 as the world's first skyscraper to incorporate wind turbines (a row of three is mounted on top).

In New York there's **Four Times Square**, also known as the Condé Nast Building. The 48-story structure, completed in 2000, was the first major U.S. skyscraper built with green credentials in mind. The building draws on a wide array of technologies, including fuel cells that kick in at night, state-of-the-art windows, plentiful solar panels, and a ventilation system that pulls in 50% more fresh air than comparable towers. New York also hosts the **58-story Bank of America Building**, which was the city's second tallest building at its completion in 2009.

The American Institute of Architects has launched a proposal called the 2030 Commitment, which includes the goal of halving the energy consumption of the typical new U.S. house by that year. However, the road to such targets is impeded by the fragmented nature of the U.S. construction industry and the lack of a federal mandate. Still another hurdle has been the collapse and slow recovery in U.S. construction from 2008 onward, brought about by the financial downturn and mortgage crisis. Progress may be more rapid inside U.S. homes. Since the 1990s, Americans have bought more than 4.5 billion products approved by the Energy Star program, which identifies and certifies the nation's most efficient appliances and electronics. But as in the United Kingdom, increasing the energy effi-

It's been billed as the world's most eco-friendly skyscraper, although a 2013 report by *New Republic* magazine found that its computer-laden, power-intensive trading floors put a major dent in the tower's green credentials. The building's sides are subtly canted to provide some of the streamlining effects of the Gherkin. Windows reflect ultraviolet light while letting in visible rays. The main power source is a 4.6-MW cogenerator drawing on natural gas, while a heat pump helps keep the building cooler in summer and warmer in winter.

With many of the world's tallest buildings now sprouting in Asia and the Middle East, it's no surprise that China has the first tower over 1000 ft to join the ranks of the world's green-

London's original eco-tower, 30 St Mary Axe. (Aurelien Guichard/Wikimedia Commons)

est designs. Completed in 2011, the 71-story **Pearl River Tower** in Guangzhou City (just northwest of Hong Kong) includes automatic blinds that can not only block sunlight but harvest it through attached photovoltaic cells.

ciency of appliances won't make a huge difference if the number of devices in use keeps rising. Local attitudes can play a role, too. In many U.S. neighborhoods, homeowner associations prohibit people from using a backyard clothesline in lieu of an energy-gulping dryer.

The quest to make large buildings—such as schools, factories, and offices—more energy efficient differs somewhat from the strategy for homes. In the United States, lighting and office equipment make up almost half of the energy consumed by commercial buildings. All those lights and computers produce lots of warmth, which means commercial buildings typically demand less heating but more air conditioning than homes. Given this energy profile, some of the most constructive steps to cut energy use in big buildings involve illumination. Along with **LEDs**—the highly efficient light-emitting diodes used in many traffic lights and stadium signs, and increasingly in homes and offices—large buildings can also make use of **hybrid solar lighting**, in which sunlight is piped into upper floors through optical cables feeding from a roof-mounted solar collector. Green roofs— gardens that spread across much or most of a large building—help keep the building cooler (and, as a bonus, whatever's planted absorbs a bit of carbon dioxide). Several high-profile skyscrapers have become showcases for a variety of energy-saving features (see sidebar, "Low-power towers").

The emissions that result from heating and cooling buildings can be slashed by drawing on the steady year-round temperature of soil through **ground-source heat pump** systems (also called geothermal heat pumps). More than a million such units were in place worldwide as of 2013; they're most common in the United States and northern Europe, particularly in Sweden. Whether ground-source heat pumps are cost-effective and car-bon-efficient depends on climate, household size, and other factors, including the fuel used to create the electricity that drives the pump. As a rule, they're quite expensive to install but require 30%–60% less energy than systems that burn fossil fuel directly. However, if installing a pump involves a switch from natural-gas heating to coal-based electricity, it could actually *increase* one's carbon footprint, as coal produces more than twice as much carbon dioxide per unit of energy.

Ground-source units rely on fluid flowing through pipes buried under-ground, either in flat arrays a little below the surface or vertical coils that can extend down more than 100 m (330 ft). The fluid returns at ground temperature, which may run anywhere from 10° to 16°C (50°–60°F) in midlatitude areas but which changes little through the year in any one location. A compressor is used to concentrate this low-level warmth, and the resulting heat is then distributed around the home (often via a sub-floor heating system), with the cold fluid returning underground to be warmed up once more. A major bonus in warm climates is that many ground-source heat pumps can also provide air conditioning that's up to 40% more efficient than conventional cooling systems (which are usually electricity-powered to begin with).

Air-source heat pumps are another option for many regions, especially in warmer climates. Although slightly less efficient than ground-source units, these are much cheaper to install and have the advantage of being suitable for existing homes that lack big yards or gardens. Both air- and ground-source systems must be installed and used correctly to provide their promised benefits.

Geoengineering: Practical solutions or potential disasters?

It's possible, if not probable, that the various strategies discussed so far in this chapter won't be implemented widely enough, or soon enough, to prevent the risk of the worst effects of climate change. But some visionaries think there's another way: using technology to tackle climate change head-on, either by sucking greenhouse gases from the air or changing the amount of sunlight that reaches Earth. These ambitious plans are often lumped under the heading of **geoengineering** or **climate engineering**—global-scale attempts to reshape Earth's environment. Although there are plenty of warning lights around the concept, it's become a topic of increasing research and even some tentative steps toward policymaking—a trend that's perhaps related to the lack of meaningful progress on a post-Kyoto plan.

Humans have long dreamed of controlling weather and climate. In the 1950s, the USSR pondered the notion of damming the Bering Strait that separates Alaska from Russia. The idea was to pump icy water out of the Arctic and make room for warmer Atlantic currents, thus softening the nation's climate and easing the way for ships. Famed computer scientist John von Neumann reflected the era's slightly wild-eyed optimism when he wrote in 1956 of spreading dark material over snow and ice to hasten its melting and warm the climate: "What power over our environment, over all nature, is implied!" Today's geoengineering schemes operate in a more sober realm, the idea being not to create the perfect climate but simply to stop the existing climate from changing too much.

Climate engineering plans fall into two main categories. One of them, **carbon dioxide removal**, is focused on reducing the amount of CO_2 in the atmosphere. Klaus Lackner (Columbia University) has prototyped "synthetic trees," devices through which CO_2 in the air reacts with an ion-exchange resin. It remains to be seen how much money—and, crucially, energy—it would take to produce, operate, and maintain systems like these. There's also the issue of how to safely store the vast quantities of captured CO_2, although one of the benefits of capturing CO_2 from the open air is that the collection devices can be placed anywhere in the world (with no need for a pipeline infrastructure). This enables the operator to select the most suitable geological storage sites. For example, some advocates of air-capture technology hold great hopes for the storage potential of basalt rocks underneath regions such as Iceland.

Another variation on this theme is to pull carbon from the air into the oceans. Experiments hint that spreading iron over parts of the ocean where it's limited, and where enough nitrogen and phosphorus are available, could produce vast fields of carbon-absorbing phytoplankton. A number of small-scale field studies over the last 20 years have shown that phytoplankton growth can indeed be stimulated by iron, but most weren't able to find any increase in carbon being deposited by the organisms into the deeper ocean. The many geopolitical challenges of geoengineering came to the fore in 2007, when the private U.S. firm Planktos launched plans to dump 100 metric tons of iron near the Galapagos, 10 times more than previous iron-fertilization studies. The eventually cancelled plan drew scrutiny from the International Maritime Organization and the global London Dumping Convention, which adopted a resolution stating that any such non-research activities "should be considered as contrary to the aims of the Convention and Protocol." Nevertheless, Planktos founder Russ George managed to deposit 100 tons of iron-enriched material off the coast of British Columbia in 2012 in a bid to bolster salmon production. The resulting phytoplankton bloom ended up covering 10,000 km² (3800 sq mi), and the action spurred an investigation by Environment Canada.

Among other ocean-based proposals for removing CO_2, Gaia theorist James Lovelock and Chris Rapley (director of London's Science Museum) have proposed dotting the sea with a set of vertical pipes that would allow wave action to pull up rich, deep water and promote algae blooms. Still another marine take on geoengineering is the idea of adding huge volumes of pulverized limestone to the oceans. This could theoretically react with atmospheric carbon dioxide to pull lots of carbon out of the air in a way that avoids ocean acidification—though the plan would only be worth trying if the CO_2 generated in the production of the lime could be captured and stored.

Yet another approach to removing CO_2 from the air—arguably not "geoengineering" as such—involves exploiting biological systems such as **plants and soils**. Perhaps the most visible recent proponent of this approach is maverick biologist Allan Savory. He's gained widespread attention with his proposal for "holistic grazing," which would mimic prehistoric ecosystems with great herds of cows that would tromp across vast swaths of semiarid land (rather than overgrazing in more confined territory). The idea is that the free-ranging bovines would stimulate soil and grassland health and produce a net loss of atmospheric carbon. Despite the appeal of Savory's plan (especially for carnivores), it's found little support in mainstream field studies, and the sheer volume of anthropogenic CO_2 emissions works against the claim made by Savory in a popular 2013 TED talk that holistic grazing could bring CO_2 back to prehistoric levels. A simpler strategy already being practiced is **no-till agriculture**, which minimizes the plowing up of land that can send CO_2 into the air (along with causing major problems with erosion). No-till approaches have been on a steady growth path in the Americas and Australia, though the technique is less common elsewhere. A 2009 survey showed that about 10% of U.S. croplands were using no-till techniques on a continuous basis, and about 25% more were being tilled only selectively.

There's also **biochar**—the technique of creating charcoal (which is a very stable and virtually pure form of carbon) by heating crop wastes, wood, or other sources of biomass in a simple kiln that limits the presence of oxygen. The resulting "char" can then be mixed in with agricultural soils. This practice—which appears to go back centuries in some parts of the world—not only locks the carbon into the soil for hundreds or even thousands of years, but also appears to significantly increase crop productivity. This virtuous circle could, its advocates claim, sequester huge amounts of carbon and increase the availability of food in poor areas. Field trials back up these claims, especially in the tropics, where soils tend to have lower carbon content. However, some environmentalists are cautious about biochar as a climate solution on the grounds that—if carried out on a large scale—the practice may create extra demand for wood, which in turn could boost deforestation and therefore carbon emissions.

The other main category of geoengineering concepts—**solar radiation management**—involves trying to reduce the amount of sunlight reaching Earth. This could be done by adding massive amounts of sulfates to the stratosphere, or deploying a colossal (and hugely expensive) array of mirrors or lenses far out in space that would deflect or refract sunlight before it reached our atmosphere. Astronomers Roger Angel and Pete Worden of the University of Arizona proposed an array of ultrathin lenses spanning an area the size of Western Europe that would sit about 1,600,00 km (1,000,000 mi) from Earth toward the sun. On a less grand scale, David Keith (Harvard University) has suggested testing balloon-borne systems that could deposit sunlight-blocking sulfuric acid at a tempo that could be sped up or slowed down as the effects become clear. A closer-to-Earth idea, first put forth by engineer Stephen Salter and physicist John Latham, is to deploy many hundreds of ships at sea, each with giant turbines that would stir up the ocean and send salt particles airborne. The salt would serve as nuclei to help brighten stratocumulus clouds—which extend across a third of the world's oceans—and thus increase their ability to reflect sunlight.

Even as research revs up, some observers worry that geoengineering is a less-than-ideal way to deal with the root problem. The late climatologist Stephen Schneider likened it to giving methadone to a drug addict. In the case of managing sunlight, especially, there's also the classic risk of unintended consequences. Even if enough sunlight could be reflected to compensate for the extra CO_2 in the atmosphere, for example, those two factors (less sunlight, more carbon) could torque regional weather patterns and produce other effects quite different than our present climate, especially for vegetation and ocean life. A 2013 study led by Simone Tilmes (NCAR) points to a risk of weaker seasonal monsoons and reduced global precipitation if solar radiation management were used to keep global temperature rises in check. As for sequestration, it's yet to be attempted on anything close to the scale needed for effective carbon removal. On top of these concerns, there are political ones. Nations might easily disagree on which geoengineering routes to take, and there's nothing to block a rogue country from embarking on a project unilaterally. Moreover, some of the researchers involved have financial stakes in companies that might benefit from geoengineering.

With all these question marks looming, science organizations and policymakers have become increasingly involved in geoengineering guidance. In a 2009 report, the U.K. Royal Society called for international development of a code of practice for geoengineering research. The society has also been working with the Environmental Defense Fund and other partners on a governance plan for solar radiation management. Both the American Meteorological Society and the American Geophysical Union adopted a 2009 position statement on geoengineering (later reaffirmed by both groups) that recommends enhanced research on the relevant science and technology, coordinated study of the various implications for society, and development of policy options that promote transparency and international cooperation while restricting "reckless efforts to manipulate the climate system." And the UN Convention on Biological Diversity, which includes nearly all of the world's nations, called in 2010 for a de facto moratorium on geoengineering plans (except for small-scale research projects) until the full array of risks—social, cultural, environmental, and economic—has been thoroughly vetted.

The IPCC's Working Group III report on climate change mitigation, issued in 2014, reflects how geoengineering has yet to take hold as a widely accepted option. Though the assessment does include some discussion of geoengineering, the topic is absent from the report's summary for policymakers. As the full report notes, "[G]eoengineering schemes to alter the planet's radiation balance have attracted particular attention; however, because they also create many risks that are difficult if not impossible to forecast, only a small but growing number of scientists have considered them seriously."

What Can You Do?

REDUCING YOUR FOOTPRINT AND WORKING FOR ACTION

Solar panels being installed on a home in Vleuten, Holland, October 2012. (Jeroen Komen)

CHAPTER SEVENTEEN

Getting Started
WHERE TO BEGIN

The gravity of our greenhouse predicament is enough to weigh anyone down, but the smart way to deal with climate change is to channel concerns and frustrations into constructive action. In this chapter, you'll find a wealth of ways to get started. Not all of them will suit your particular situation, but some probably will.

The most obvious way to take individual action on climate change is to reduce the size of your **carbon footprint**—the total amount of greenhouse emissions that result directly and indirectly from your lifestyle. Since this will include increasing your energy efficiency at home and on the road, you may even save some money in the process. The following chapters take a brief look at various ways to reduce your footprint.

While individual actions do make a difference, climate change is such an enormous global challenge that it's also worth **getting involved** by raising the issue with your political representatives, employers, and/or local communities, as described in "Making things happen."

Measuring your carbon footprint

Just as new dieters often keep a food diary, an excellent way to start reducing your emissions is by using carbon calculators. These simple tools, available online or in book form, allow you to calculate how much carbon

Making things happen

In addition to reducing your own carbon footprint, you can help move society as a whole toward a lower-carbon future. Joining with other like-minded citizens will enhance your own power.

* **Work and school.** Find out what policies exist on energy efficiency at your workplace or school. If there aren't any, or they seem half-hearted or inadequate, you can work for something better by relaying your suggestions to the powers that be and encouraging coworkers or fellow students to do the same.
* **Community.** Cities and towns vary hugely in how committed they are to solving the greenhouse problem. Contact your local government and ask what climate change measures they've adopted. Then see what you can do to raise awareness and make change happen—by contacting local media, for example, or by attending local-government open meetings.
* **Finances.** If you've got savings or investments, consider moving them toward a bank or fund that supports action on climate change. Generally this will be one that engages in "socially responsible investing"—that is, one that considers global warming as well as other social and environmental issues when deciding which companies to include in its portfolio. If you hold stock directly in a company that's behaving in a less-than-ideal way on climate change, look into introducing or joining a **shareholder resolution**. Typically, these require companies to report on their actions and plans regarding climate change and its impact on the company's bottom line.
* **Politics.** Climate change won't be solved without political will, and there won't be political will without pressure from voters. The Internet makes it easy to contact your political representatives. If you're in the United States, simply drop into **Congress** (www.congress.org) and enter your zip code. A good starting point is to ask what emission targets (if any) your legislators support and what they're doing to help ensure these targets are met.
* **Activist groups.** You could also consider joining one of the organizations working for political change around the climate issue. They vary in their policy focus and their attitudes toward issues such as globalization and nuclear power, but they're all working toward major emissions cuts.

each activity in your life generates and how your total compares to those of the people around you and elsewhere in the world. Carbon calculators vary by country, reflecting differences in the way energy is generated, priced, and taxed. They also vary among themselves in how they organize and categorize activities and how they handle uncertainty about the exact greenhouse impact of particular activities, such as flying. Sometimes the calculations in each step are explained in detail; in other cases they're not, which makes the site simpler but not necessarily more user-friendly. Note that U.S.–based calculators often use short tons (2000 lb each), which are about 10% less than a metric ton.

Carbon-offset companies (see below) offer calculators for specific activities, but to quickly assess your overall carbon footprint, try the following:

* **Carbon Footprint**: www.carbonfootprint.com/calculator.aspx
* **CoolClimate Network** (U.S. only): coolclimate.berkeley.edu/carbon-calculator

Both of these sites—unlike many others—cover emissions not just from the fuel and power we're each directly responsible for, such as heating and driving, but also less obvious sources such as the manufacture of the goods we buy.

CHAPTER 18
Home Energy
HOW TO SAVE MONEY WHILE SAVING THE PLANET

Saving energy at home will provide a dual satisfaction: less greenhouse gas in the atmosphere and, sooner or later, more money in your pocket. The following tips should help you streamline your household energy use. You may also want to check with your local government to see whether it offers free home-energy audits or energy-efficiency grants.

The Internet is a good place to find out more. For example, Home Energy Saver (**hes.lbl.gov**) allows U.S. residents to carry out a simple online audit of their household energy use. Of course, not everyone is in a position to make substantial changes to their home on the grounds of carbon and energy savings—even where grants are available. In particular, renters are often left frustrated that they cannot justify investing time or money in a building that doesn't belong to them. Other than appealing to the home-owner, there's often little that can be done in this situation.

Heating and hot water
Heating and cooling is a key area, accounting for the majority of emissions in most homes. Older houses are especially inefficient in this regard, though there's room for improvement in nearly all domestic buildings. (If

you're feeling really ambitious, and flush, consider building a Passivhaus, pioneered in Germany. Though expensive to construct, these ultra-efficient homes get by without any central heating or cooling source, thus cutting utility costs by as much as 90%.)

Insulation

Start small with weather-stripping—sealing up cracks around doors and windows—but be sure to consider beefing up your attic and wall insulation. The attic is a good place to start, as it's easier to access and less expensive to insulate. (If it's already been done but only to a depth of a few inches, consider adding to it.) Energy-efficient windows can also make a substantial difference, though they are typically a more expensive way of saving energy than insulating walls and attics. If you live in a hot climate, you can save on air conditioning by using bright, reflective window drapes and shades wherever sunlight enters and by using light colors or a reflective coating on your roof.

Thermostats and heating controls

Reducing your room temperature in the cool season by just a small amount can make a substantial difference to your energy consumption. You may find you sleep better, too. Throw on a sweater and try dropping your home's temperature toward 16°–18°C (61°–64°F), especially overnight. Inexpensive digital thermostats allow you to program different temperatures for different times of day, so you can maximize your utility savings by letting the house cool down when you're not at home. Zoned heating, which lets you specify the temperature of individual rooms within a central-air system, is more expensive, but it's another way to reduce your energy demands. If you're thinking about a space heater to keep select rooms warmer in winter, make sure it's a safe and efficient model; many space heaters are electricity hogs.

If you use air conditioning in the warm season, shoot for 25°C (77°F). Many utilities offer discounts if you're willing to install a switch that turns off your A/C for a few minutes every so often, which helps utilities balance their loads. Better yet, if you're in a warm, dry climate, try a whole-house fan to bring in cooler morning and evening air, or investigate swamp coolers, which can take the edge off summer heat while humidifying the air and using far less energy than an air conditioner.

Hot-water heating

An efficient hot-water heater makes a sensible long-term investment for many households. Modern versions can produce up to a third more usable heat from each unit of fuel than inefficient older models. For even greater savings, consider swapping your hot-water tank—if you have one—for a tankless "on-demand" system. These save energy by heating water only as it's needed. If you do stick with an older hot-water heater, though, be sure to insulate it. Blankets designed for this purpose are inexpensive and can save 25%–45% of the energy required to heat the water. (Many newer hot-water tanks have built-in insulation.) To save more energy, try bringing the temperature setting toward 120°F, which will still give you hot showers while reducing the risk of accidental scalding.

Pick your fuel

In general, natural gas is a more climate-friendly fuel for home heating and hot water than oil, coal, or regular grid electricity. If you're not connected to the gas network, wood could be your best bet (unless you can access wind power—see below), as long as the fuel is coming from forests that are being replenished as fast as they're being harvested.

Showers and baths

Everyone knows that showers use less energy than baths, and that shorter showers use less energy than longer ones. Less widely known is that a low-flow shower nozzle, which mixes air with the water flow, can reduce the amount of hot water needed by half. Shower heads made before the early 1990s had flow rates as high as 5.5 gallons per minute (gpm); new ones can't exceed 2.5 gpm, and some manage 1.5 gpm or less. If you're not sure about yours, take a bucket, turn on the shower, and measure how long it takes to collect a gallon of water.

Electricity supply

Renewable electricity

Many utility companies allow you to specify that some or all of the power you use is generated from renewable sources, most often wind farms. Typically with such plans, you're not literally getting the power generated by the renewables. Instead, the power company agrees to put renewable energy into the grid in an amount equal to your own consumption. One point to note: in certain U.S. states, electricity companies are already mandated to buy a certain percentage of their power from renewable sources, and the companies that exceed this quota can sell credits to those that fail to meet it. The upshot is that your fee won't necessarily increase the

Starting at the top

Besides hosting solar panels and micro wind turbines, roofs offer various opportunities for combating global warming. One option is to make your roof more reflective. A study by the Earth Institute at Columbia University estimated that a world full of entirely white roofs could add a full 1% to Earth's reflectivity, and U.S. energy secretary Steven Chu has promoted the concept. But you don't necessarily have to paint your roof white to make it more reflective. Several types of tile, metal, and shingle roofs can boost reflectivity while maintaining a splash of color. For large buildings in urban areas, another good option is a "green roof" with plants that can not only help with cooling but also absorb excess storm water (and CO_2).

In 2010, the U.S. Naval Legal Service Office in Norfolk, Virginia, unveiled this green roof. (U.S. Navy/John Land)

overall amount of renewable power generated, so it's important not to use a green energy surcharge to justify being less careful about minimizing energy waste. Check with local suppliers or environmental groups for more information.

Generating your own power

There are various ways to generate energy at home—but don't expect to be self-sufficient without substantial upfront investment. One sensible starting point is a rooftop solar panel to heat your hot water. Unlike the photovoltaic (PV) systems that generate electricity, these **collector panels** funnel heat directly into your hot-water system. Even in a place as cloudy and northerly as Britain, a collector panel—typically spanning about 2 m × 2 m (6 ft × 6 ft)—can provide up to half of the hot-water needs of a house with a south-facing roof and a hot-water tank.

Full-fledged solar **PV systems** can provide homeowners with half or more of their total power needs. The upfront expense for buying and installing a PV setup remains relatively high, but costs have been dropping dramatically in recent years, with the price of a full PV system down roughly 50% from 2010 to 2013, when you could expect to spend around $10,000 to install a system with a 2-kW capacity (enough to meet roughly 15%–30% of a typical home's electricity demand, depending on location). At current electricity prices, that kind of investment might take decades to pay for itself in terms of savings on energy bills, but government grants and tax breaks have made PV much more attractive financially in many countries and U.S. states than it used to be. One way to avoid the steep initial expense is to work with a third-party solar provider that owns and installs the panels on your home. Typically, you'll pay a much more modest starting cost, and your monthly electricity bills will be fixed at a rate that's usually a bit lower than the previous year-long average.

For people with enough land who live in a breezy spot, a **wind turbine** that feeds power into batteries or the grid may be another option worth considering. However, building-mounted models tend to be inefficient and expensive, and the towers can run afoul of local building codes—and even jeopardize your home's structural integrity. Stand-alone turbines are a much better bet if you are lucky enough to have a piece of windy land to put one on.

Appliances and gadgets

Lighting

Close to 95% of the power that drives an old-fashioned incandescent light bulb goes to produce heat instead of light (which explains why those bulbs are so hot to the touch). It's no wonder that incandescent bulbs are being

Look out for labels

Electrical devices vary enormously in their energy consumption, so when shopping for such items, be sure to look out for the various labels that give you the information you need to make a low-carbon choice. In the United States, these include the following:

* **EnergyGuide**: Standard on large appliances, this label shows the typical amount of energy consumed per year and how that product compares to the best and worst performers in its category, as well as the expected annual energy cost.
* **Energy Star**: Identifies products that meet various criteria for efficiency in more than 40 categories—including entire homes.

phased out across the world—a smart move in the long run, despite the political furor that erupted in some U.S. quarters. In 2007, Australia became the first nation to ban incandescent bulbs, effective in 2010. Major retailers in the United Kingdom agreed to phase out their sale by 2011, with the U.S. phase-out unfolding between 2012 and 2014. Some incandescents will still be eligible for retail sale in the United States, including specialty sizes and some three-way bulbs. Otherwise, if you're replacing or upgrading your lighting, you've got two main options.

* **Compact fluorescent (CF) bulbs** generate far less heat, enabling them to produce about four times more light per unit of energy. (The replacement for an incandescent 100-watt bulb typically uses about 23 watts.) These efficient bulbs also last around 10 times longer, meaning that they earn back their higher initial cost several times over. Though older compact fluorescents took a few seconds to reach full brightness, many are now "instant-on" and even dimmable in certain types of lamps. You can now find a CF bulb to replicate almost any type of light you desire, from the warmer, lower-wavelength hues put out by incandescent bulbs to the cooler, higher-wavelength light put out by the sun itself. To make sure you get the right bulb for the job, check its color temperature (usually expressed in Kelvins), which corresponds to wavelength. Note that compact fluorescent bulbs do contain tiny amounts of mercury, so when they eventually die, take them to a municipal dump to be disposed of properly.

* **LEDs (light-emitting diodes)** are now being used much more widely in homes after first gaining a foothold in traffic lights and outdoor displays. LEDs tend to be even more efficient than compact fluorescents, with the replacement for an old-school 100-watt bulb often using 20 watts or less. The cost of a typical LED bulb has plummeted in recent years, from as high as $50 in 2012 to as low as $10 by early 2014. Although LED bulbs don't contain mercury, it's still best to recycle them rather than putting them in the trash.

What about **halogen** lights? These are a subset of incandescents, typically around twice as efficient as typical incandescents. However, they're still half as efficient as compact fluorescents, and halogen light fixtures often take multiple bulbs, raising their overall energy consumption. Moreover, they burn much hotter than other bulbs, making them more likely to damage paintwork. If you have halogen lamps, look for CF or LED replacements, available from specialty lighting sources.

Whatever kind of lights you have, turning them off when you leave the room is an obvious energy-saver. Organizing your home to maximize the use of natural light—putting desks next to windows, for instance—can also help.

Invisible power drains

The proliferation of remote controls and consumer gadgets has come at a surprisingly high energy cost. In the typical modern home, anywhere from around 5% to 10% of all electricity is consumed needlessly by big TV sets, stereos, digital video recorders, and other devices that are supposedly turned off but are actually on **standby**. There are huge variations here: cell phone chargers don't use much when they're not charging, while some devices use almost as much energy in standby mode as when they're operating. To stop the waste, unplug items—or switch them off at the socket—when they aren't in use. A power strip with a switch can help if you don't have switches on your sockets. You can root out the worst offenders by plugging your gadgets into an inexpensive electricity usage monitor, such as Kill-a-Watt.

Computers vary widely in terms of the energy they consume in standby and screensaver modes. Turn them off when not in use, or dedicate their downtime to the fight against climate change (see sidebar, "Modeling for the masses: ClimatePrediction.net" in chapter 12).

Refrigerators and freezers

These are among the biggest users of energy at home, particularly because they're running 24/7. Models produced since the 1990s are far more efficient than their predecessors, so if yours is old, consider taking the plunge. A modern replacement could recoup its costs within several years and save plenty of energy right from the start. Be sure to dispose of the old one properly, as many older refrigerators contain potent ozone-depleting and planet-warming chemicals. Check with your local government or a home appliance dealer for more information.

If possible, locate your refrigerator or freezer as far away as possible from hot-running items such as ovens and dishwashers, and make sure it's set no lower than the recommended temperature, which is typically −18°C (0°F) for freezers and 1°–4°C (34°–39°F) for refrigerators.

Doing the dishes

Like refrigerators, dishwashers vary widely in their energy consumption, so be sure to consider efficiency when purchasing. As for how they compare with washing by hand, this depends on the individual machine (some newer models use fewer than four gallons of hot water per load), the efficiency of your hot-water heater and, most importantly, how economical you are when washing by hand—where it's possible to use huge amounts of hot water. An analysis in the book *How Bad Are Bananas?* found that a dishwasher actually wins out on energy in most cases, assuming you run it only when fully loaded (and assuming your hot water doesn't come from a renewable source). This is true even considering the energy used to make the dishwasher itself.

Washing clothes

To save energy (and minimize color fade) keep water temperatures as low as possible to get the job done well. Most clothes do fine in cold water. Even more worthwhile is letting clothes dry on a line or rack rather than using a dryer, as it takes a lot of energy to evaporate water—many times more than it takes to run the washing cycle. In the United States, many suburban condominium and housing associations, fearful that clotheslines will tag a neighborhood as déclassé, actually prohibit hanging clothes outside to dry. In such situations, an indoor rack may do the trick.

CHAPTER NINETEEN

Transport and Travel

TRAINS, PLANES, AND AUTOMOBILES

Personal transport accounts for around a third of the typical person's carbon footprint in countries such as the United Kingdom and the United States. Moreover, transport as a whole is the fastest growing major source of greenhouse gases. Thankfully, in the case of road transport at least, there's much you can do to make your comings and goings less carbon-intensive. As we'll see, air travel is rather more problematic: for most keen travelers, it makes up the single biggest element in their carbon footprint.

The car and what to do about it

Compared to alternatives such as train and bicycle, most cars are quite energy-inefficient, especially when carrying only one or two passengers. If you can't live without a car, you can at least reduce its greenhouse emissions by changing your driving habits (see below). Making fewer and shorter journeys also helps, of course, as can carpools and ride-sharing plans. Depending on where you live, you may even be able to run your present vehicle on ethanol or biodiesel (see p. 428).

If you're due for a new vehicle, be sure to opt for the most efficient model that fits your needs. This may be a **hybrid**, such as the Toyota Prius

How to drive green

Even if your vehicle's a gas guzzler, the following tips could cut its fuel usage by as much as 30%—and the more inefficient your car is, the more gas savings you'll get from that 30% improvement. Some of these tips are now incorporated in the tests that new drivers take in the United Kingdom, the Netherlands, and several other countries.

* **Drive at the right speed.** Most cars, including hybrids, are most efficient at speeds of 30–50 mph. Above 55 mph, cars typically gulp more fuel to travel the same distance—as much as 15% more for every additional 10 mph. Thus, saving a few minutes on a trip by cruising at extra-fast speed can result in higher fuel costs and more greenhouse emissions per mile, not to mention the safety issues associated with high velocity.
* **Lighten the load.** Keep heavy items out of your car unless you need them—you'll typically lose a percent or two in efficiency for every 100 pounds you haul. While you're at it, check the tire pressure often: rolling resistance goes up and efficiency goes down by as much as 1% for every PSI (pound per square inch) below the recommended pressure range. However, there's no benefit, and some risk, to driving with over-inflated tires. If your car manufacturer recommends a lower pressure than the maximum shown on the tire itself, go with the manufacturer's verdict. Be sure to check tire pressure at home, as the values can increase after driving as little as a mile or two.
* **Avoid idling.** Except when it's required (such as in stop-and-go traffic), idling is a wasteful practice, and it doesn't benefit your car, except perhaps in extremely cold conditions. Even five minutes of idling can throw more than a pound of greenhouse gas into the air. Anything more than about 10

or Honda's Civic Hybrid. These look, work, and fuel just like regular cars, but they feature an extra electric motor that charges up when you apply the brakes. Hybrids can achieve up to 70 miles per gallon, though SUVs and executive cars with hybrid engines are far less efficient because big, heavy, high-powered cars eat fuel much faster than smaller, medium-powered ones. If you're concerned about the **safety** of a smaller car, check the official safety data for each model—bigger isn't always better in this regard.

Even greener than hybrids in many cases are **electric** and **plug-in hybrid vehicles**. Both can be charged via a standard household power socket,

seconds of idling generates more global warming pollution than stopping and restarting would. Cooling your heels in a line of cars at a drive-in window may feel convenient, but parking, stretching your legs, and going inside is the greener (and healthier) option.

* **Use the A/C sparingly.** As you'd expect, air conditioning normally saps energy and cuts down on vehicle efficiency by a few percent. However, if you're on a long road trip and it's a choice between driving with the windows down and running the A/C, there may be little difference in fuel usage, according to some studies. That's because wide-open windows can increase the car's aerodynamic drag, especially at high speeds. If outside temperatures aren't too hot, try using the vents and fan but leaving the A/C off.

* **Starting and stopping.** Jackrabbit starts and stops not only put wear and tear on your car, but they also drain fuel economy. Accelerate gradually, and anticipate stops by starting to brake well in advance. If you have a manual transmission, the best time to change gears is between 1500 and 2500 rpm.

The rules are a bit different for **hybrids**, which has led to some confusion. With a standard engine, you're best off accelerating lightly no matter what the speed. In a hybrid, you should accelerate briskly until you get to the optimal in-between speed, around 30–34 mph, where the car is at its most efficient. A technique called "pulse and glide" driving—hovering in that optimal speed range through small accelerations and decelerations—can boost your hybrid's efficiency. When it's time to slow down a hybrid, brake slowly at first, then increase the pressure: this ensures that the maximum energy goes into recharging the battery, versus creating unusable heat in the brakes themselves.

though charging speeds are considerably faster with a dedicated charging station installed on a 240-volt circuit (the same type used for heavy appliances). Most fully electric vehicles, such as the Nissan Leaf and the Ford Focus, have a range of a few dozen miles. The big exception is the high-end Tesla S, which cost upward of $70,000 in 2013 but can go more than 250 miles on a charge. Plug-in hybrids, such as the Chevy Volt and Ford C-MAX Energi, are similar to regular hybrids, except that the electric engine can be charged and operated independently of the gas engine. Plug-in hybrids can be ideal for people who need a car mainly for commuting and short local trips but who still want the option of a gas-powered engine for occasional longer trips. The mileage capacity per charge varies from model to model, so make sure you pick a vehicle whose electric range is compatible with your most common trip lengths. Getting a vehicle that plugs in is also a perfect time to make sure that your home electricity is as green as possible (see above).

Air travel

Aviation is problematic from a climate change perspective (see sidebar, "Aviation: Taking emissions to new heights"). It takes extra fuel to move people at the altitudes and speeds of jet aircraft, and people typically fly much farther than they would ever travel by train, boat, or car. A single round trip in coach class between Los Angeles and London generates more than a metric ton of CO_2 emissions. The International Civil Aviation Organization provides a carbon calculator for flights between specific cities at **www2.icao.int/en/carbonoffset**. Moreover, because the contrails produced by aircraft also have important effects (see chapter 10), a plane's impact on the climate may be twice as much as its already significant CO_2 emissions would suggest.

Planes are slowly gaining in efficiency, with improvements of a percent or so each year owing to technological gains as well as fuller flights. The problem is that these gains are mostly translating into cheaper and more popular air travel, so that the sector's overall greenhouse output continues to grow. More substantial changes are on the drawing board, including the use of new aircraft designs that could improve fuel efficiency by 50% or more. Biofuels have been explored—Virgin Atlantic flew a jet from London to Amsterdam in 2008 using a mix of coconut and babassu oil for 20% of its fuel, and in 2009 Air New Zealand carried out tests using 50% jatropha oil—but these fuels raise their own thorny issues (see chapter 16), and even the most optimistic experts don't expect developments in fuel or engine technology to offset the massive growth in the numbers of flights. Ultimately, putting a plane into the air and keeping it there is an energy-guzzling task, so it'll be hard to make flying dramatically more efficient than it is now.

With all this in mind, cutting back on air travel is one of the most obvious steps for people concerned about global warming. Climate journalist Eric Holthaus made a splash in 2013 when he made a public pledge to stop flying altogether. For others, cutting back might mean taking fewer flights and making up for it by staying longer each time. It might mean choosing vacation venues closer to home more often, or it might mean considering alternative ways of traveling. With two or more people on the same itinerary, it can be more climate friendly to drive than to fly—especially for short distances, where planes are most inefficient per mile traveled. Better still are **trains and boats**, which are typically responsible for substantially fewer emissions per passenger mile than either cars or planes. To find out how to travel almost anywhere in the world by rail and sea, check out The Man in Seat Sixty-One (**www.seat61.com**).

When you do fly, consider offsetting your emissions (see below) and plan your itinerary with as few stops as possible, since takeoffs and landings are energy-intensive. Use the airline's website to see what model of plane you're booking yourself on. Sites such as **seatguru.com** can help you figure out whether your flight will be on a newer (and thus likely more efficient) aircraft. Especially on long-haul flights, avoid the perks of luxury seating. Thanks to the extra room they occupy, business-class passengers are responsible for about 40%–50% more emissions than those in economy, and first-class travel may generate up to six times more carbon. Another easy trick is to travel light. Two heavy suitcases can nearly double the onboard weight of a passenger, adding to fuel consumption.

Also, try to favor flights in the **daytime** and avoid red-eye trips. This can make a surprisingly big difference: aircraft contrails reflect no sunlight during nighttime, thus eliminating their ability to partially compensate for the climate-warming gases spewed by aircraft.

CHAPTER TWENTY
Shopping
FOOD, DRINK, AND OTHER PURCHASES

It's impossible to know the exact carbon footprint of all the items you buy, but a few schemes are working to fix that. The Carbon Trust, in partnership with the U.K. government, launched its Carbon Reduction Label in 2007. It's now used by hundreds of companies worldwide to show the grams of carbon emitted as a result of producing a given item (typically expressed in a per-serving format for food). The Swedish fast-food chain Max Burger began providing carbon data for each item on its menu in 2009. Some experts claim that these kinds of labels almost always underplay true carbon footprints of the items they appear on, for the simple reason that it's impossible to include all the virtually infinite economic ripples caused by the creation and use of a product. Nonetheless, such labels can still be a handy way for individuals to get a rough sense of how different products compare.

For products that aren't carbon-labeled, there are a few good rules of thumb to follow. First, you can consider the energy used in transporting an item. Heavy or bulky goods manufactured far away inevitably result in substantial carbon emissions, especially if they've traveled long distances by road (shipping is comparatively efficient, though not negligible). On the other hand, smaller items transported by air—such as perishable, high-value fruit and vegetables, and cut flowers—are far more greenhouse-intensive than their weight would suggest.

Another thing to consider is the material that something is made from. Steel, aluminum, concrete, and precious metals all require large amounts of energy to create. By contrast, wood from sustainable sources—such as local softwoods and anything certified by the Forest Stewardship Council—is practically carbon neutral. The same can't be said for tropical hardwoods such as mahogany or teak; demand for these helps drive the tropical deforestation that accounts for a large slice of the world's greenhouse emissions.

Food and drink

Cutting back on beef

Perhaps the simplest and most powerful way to reduce the emissions of your diet is to cut back on meat and dairy—especially beef. Cattle belch and excrete a substantial fraction of the world's methane emissions. Though we may picture cows munching contentedly on grassland that isn't being used for much else, much of Earth's rainforest destruction is driven by the clearing of land for grazing livestock or growing their feed. A 2013 report by the United Nations Food and Agriculture Organization found that factors related to livestock—from deforestation to fertilizer, feed, and cow flatulence—represent an astounding 14% of current greenhouse emissions, measured by carbon dioxide equivalent. About six-tenths of that total is related to the production of beef and cow's milk. It takes an estimated eight pounds of grain to yield a single pound of beef. Ratios are considerably lower for chicken and fish, but the same amount of protein can be obtained from a plant-centered diet with a lower carbon footprint, often more inexpensively.

Food miles

In general, food that's both local and in season is best for you and the climate. If there's a farmers' market or a home delivery system for fresh produce in your area, try it out, or consider starting your own vegetable garden. Keep in mind, though, that "food miles" alone aren't the whole story. Some foods—bananas, say—are relatively climate-friendly even if they are shipped thousands of miles. In one famous example, a 2006 study at Lincoln University found that, for U.K. eaters, New Zealand lamb has

The three Rs: Reduce, reuse, recycle

The old green mantra holds true for combating climate change. It's heresy in a capitalist society to suggest that downsizing might have its pluses, but a big part of reducing global emissions is taking a hard look at global consumption. One way to take action in this realm is to avoid unnecessary purchases, especially of products that take a lot of energy to manufacture and distribute. Second-hand stores are perfect for satisfying a shopping bug without enlarging your carbon footprint.

Equally, try to make use of the recycling facilities offered in your area. Almost all recycling helps to reduce energy consumption to some extent. For food waste, use a composter instead of the trash whenever possible; another option is home pickup services for compost, which are increasingly being combined with trash and recycled pickups. When food is buried in a commercial landfill, it decomposes anaerobically and generates the potent greenhouse gas methane. Landfills are, in fact, the source of about one-sixth of U.S. methane emissions, according to a 2013 estimate by the Environmental Protection Agency.

only a quarter of the total carbon footprint of British-raised lamb, mainly because the free-ranging Kiwi sheep require far less fuel to raise and are transported overseas using relatively efficient ships. The study omits trucking within national borders, and there are other hard-to-quantify elements in the comparison (which was, unsurprisingly, carried out by agricultural specialists in New Zealand). Such examples aside, favoring local and seasonal produce where it's available remains a good rule of thumb. When you do buy imports, such as coffee, choosing items marked with a **fair-trade** or rainforest-certified label will reduce the risk that carbon-rich forests are being chopped down to support your tastes.

Organics

Organic food eschews farming techniques that rely on petrochemicals and tends to result in slightly lower emissions per unit of food—though not all studies agree on this point. Whatever the truth, organic produce is rarely low-carbon enough to justify its being shipped (not to mention flown) over vast distances, so check the country of origin, and if climate is a key concern for you, then go with local food over organics when it's a choice between the two.

Weaning yourself from bottled water

Bottled water has made a tremendous splash among health-conscious consumers over the last several decades. More than 61 billion gallons were sold worldwide in 2011—that's more than 20 times the amount sold in 1990. All told, the United States uses more bottled water than any other nation, with a record 9.6 billion gallons consumed in 2012. Mexico leads the world in per capita use, with more than 65 gallons per year per person as of 2011. It's now expected that sales of bottled water will eclipse soft drinks by the end of the 2010s.

Drinking more water in place of sugary beverages is certainly a good move for one's health, but it would be hard to imagine a more gratuitously wasteful product than bottled water, in terms of both cost (it can be more than 1000 times more expensive than tap water) and greenhouse emissions. For starters, all those plastic bottles are made from petroleum, most likely at a factory that burns fossil fuels. Then there are the emissions involved in shipping the bottles long distances (water's quite heavy), keeping them refrigerated in many instances, and, finally, transporting them for recycling or landfill (yet another ecological impact).

Fortunately, there's a marvelous, time-tested alternative: the tap. There's no denying that a bottle of water can be convenient in certain places and at certain times. However, it's practically as easy and far less costly to buy a sturdy, refillable, washable bottle and keep it with you. (Avoid reusing bottled-water bottles, though; they may leach out harmful chemicals after only a few refillings.)

Buy basics—and in bulk

Some food-processing tasks require a surprising amount of energy, so buying ingredients rather than prepackaged meals is usually a good idea from a carbon perspective. Buying in bulk is a good idea, too, as it minimizes packaging. For fluids such as shampoo, look for shops that let you bring in and refill your own bottles.

Bring your own bag

Shopkeepers practically force store-branded plastic bags on us, but each of the estimated 1 trillion plastic bags that are used and tossed away each year adds to the world's waste load (not to mention killing thousands of creatures each year who consume plastic bags, thinking they're food sources). Each plastic bag carries a tiny carbon price tag, though some analysts have pointed out that a reusable bag can result in more net emissions than a

As for quality, the supposed superiority of bottled water has been vastly overstated. In most developed countries, bottled water is actually inspected less rigorously than tap water. In fact, many bottled-water brands are simply tap water from city sources, accompanied by a label that shows a mountain stream. As for taste, blind tests on the U.S. TV program *Good Morning America* found New York City tap water beating each of the three bottled alternatives by a clear margin. If water taste and/or quality is a concern in your own community, then try installing a filter on your tap or keeping a filtered pitcher in your refrigerator; you'll still save money over buying water in a bottle.

In the broader picture, many activists worry that the soaring popularity of bottled water—most of it distributed by a few huge corporations, including Coke, Dannon, and Nestlé—is paving the way for an increasing privatization of the world's water supply. In developing countries, the advent of bottled water as an alternative to poor public supplies only exacerbates the divide between haves and have-nots. In his book *Bottled and Sold*, water expert Peter Gleick takes note of a tragic irony: Western societies worked for safe municipal water in the 1800s, but now "we're afraid that public fountains, and our tap water in general, are sources of contamination and contagion." With climate change expected to make water supplies more variable in many parts of the globe, there's all the more incentive to make sure that everyone has access to water that's clean as well as climate friendly.

disposable one if it's of poor quality or used only sparingly—an example of how carbon footprints aren't always aligned with other environmental priorities. In any case, finding a good-quality washable, reusable bag and using it often (keeping it in the car you shop in, for example) will put you in the right direction. Aside from anything else, reusable bags are more comfortable to carry and much less likely to break. Some cities and nations are beginning to fight back against the plastic tide: Ireland cut its plastic-bag consumption by more than 90% after instituting a tax of about US$0.30 per bag.

CHAPTER TWENTY-ONE

Offsetting
PAYING A CARBON CLEANER

Carbon offsetting involves paying an organization to neutralize the climate impact of your own activities, thereby making those activities "carbon neutral" or "climate neutral." Some offset schemes focus on reducing future emissions by, for example, giving out low-energy light bulbs in the developing world or buying renewable electricity credits. Others focus on sucking CO_2 directly out of the atmosphere—by planting trees, for example.

Offsetting has become very popular in the last decade, and not just with individuals, but also with global corporations (HSBC and other office-based giants have gone climate-neutral) and celebrities (Pink Floyd, Pulp, and the Pet Shop Boys have all neutralized their tours). In 2007, Vatican City became the world's first climate-neutral state by planting trees in a Hungarian preserve. Some airlines, car rental firms, and other travel-oriented companies now make it possible for consumers to offset their emissions when they buy a ticket.

How offsetting works
Whether you want to cancel the carbon footprint of a single long-haul flight, a year of car trips, or your entire existence, the process is the same. First, you visit the website of an offsetting organization and use their **carbon calculators** to work out the emissions related to whatever activity

Tree planting: Help or hindrance?

It's one of the more clichéd approaches to tackling global warming, but tree plant-ing has also become one of the more controversial, especially in the context of carbon offsetting. It's true that trees soak up CO_2 as they grow. Two or three dozen can be enough to absorb an entire household's emissions. However, at snow-prone higher latitudes, trees can actually accelerate climate change, according to a landmark 2006 study led by Govindasamy Bala of the U.S. Lawrence Livermore National Laboratory. That's because their CO_2-absorbing benefit is outweighed by the impact of their dark color, which in wintertime absorbs sunlight that might otherwise be reflected to space by bright snow cover atop barren ground.

In snow-free warmer climates, and especially in the tropics and subtropics, the dark color of trees isn't a problem, but there are other catches. First, when a tree dies, much of its stored carbon returns to the atmosphere. Thus, the offset will only be permanent if each tree planted is replaced by another. Second, there are quicker, cheaper, and longer lasting ways to fight climate change, such as distributing low-energy technologies to displace fossil fuels.

For all these reasons, most offset schemes have switched from tree planting to energy saving. (The CarbonNeutral Company, for example, was the result of a rebranding of a scheme previously called Future Forests.) That said, unless you live in a snowy region, planting a few trees in your garden is still likely to be beneficial for the climate—at least for the crucial coming decades.

A separate, and altogether more pressing, approach is protecting the forests that are already standing. That's not so much because of the CO_2 that mature forests absorb. Rather, it's because deforestation, and in particular the destruc-tion of tropical rainforest, is one of the largest sources of greenhouse emissions (see p. 11). To help limit these emissions, the charity Cool Earth (**coolearth.org**) enables individuals and companies to sponsor areas of critically endangered rainforest in order to "keep the carbon where it belongs." The forest is purchased and given to a local trust, which protects it but allows sustainable harvesting of rubber, nuts, and other forest crops. You can even make sure your sponsored area is still tree-covered, thanks to satellite photos.

you want to offset. This will be translated into a fee which the offsetting organization will use to soak up—or remove the demand for—a matching amount of greenhouse gas.

Offsetting fees vary by organization and over time, but they're often calibrated against the price of carbon established in European emissions trading, which was in the neighborhood of US$7.00 for a metric ton of CO_2 in early 2014. If you pick a firm that charges an especially high rate per ton of CO_2, it doesn't necessarily mean your money is doing more good. As the Tufts Climate Initiative notes, "It is more important to invest in high quality offsets than to buy as many offsets as possible."

Offset debates

Despite their popularity, carbon-offset schemes are not without their critics. One argument leveled against offsetting is that it's just plaster on the wound—a guilt-assuaging exercise that hides the inherent unsustainability of carbon-intensive Western lifestyles. British environmental writer George Monbiot likened offsets to the medieval Catholic practice of selling indulgences to "offset" sinful behavior. The spoof website **cheatneutral.com** famously ridiculed carbon offsetting by offering a service for neutralizing infidelity: "When you cheat on your partner, you add to the heartbreak, pain, and jealousy in the atmosphere . . . CheatNeutral offsets your cheating by paying someone else to be faithful and not cheat."

Even the offset projects themselves tend to agree that offsetting emissions isn't as good as not causing them in the first place. That said, there's no reason why people can't buy offsets *and* make efforts to reduce their emissions directly. Indeed, the very act of digging into pockets for an offset fee may make consumers or businesses more conscious of carbon's larger cost.

A separate criticism is that offset projects may not make the swift, long-term carbon savings that are claimed of them. It's true that some projects are poorly conceived and/or executed. Tree planting (see box on previous page) may take decades to soak up the carbon you've paid to offset, which is one reason why many offsetting groups are moving toward sustainable energy projects instead of trees. As for whether the carbon savings are real, the better offsetting services are externally audited (see below) to address just this question. Some offsetters guarantee a given carbon reduction, so that if one project doesn't come through as expected, or if the project doesn't create "additionality" (carbon savings beyond those that would have otherwise occurred), then the offset goes to another project. Critics argue that, if we're to stand any chance of tackling climate change, then the kinds of projects invested in by offset schemes need to happen anyway, funded by governments, in addition to individuals and companies slashing their carbon footprints directly.

Still another point of contention is whether offsetting services ought to make a profit. According to the Tufts Climate Initiative, some offsetting firms use as much as 85% of their revenue for expenses before actually funding any emission reduction work. Some of the commercial firms that provide offsets don't reveal how much profit they earn, arguing that the good work they do justifies an unspecified return on their labor and risk. Other offsetters are bona fide charities, which means all of their income goes toward operations and the offset projects themselves. Even among these groups, some are more efficient at their work than others. Sites such as **charitynavigator.org** can help you find the offsetter that provides the most bang for the buck.

Ultimately, the benefits of offsetting are open to debate. If you pick the right plan, it's likely that your contribution will in fact help reduce emissions (especially if you pay to offset, say, two or three times the emissions of the activity in question). If you'd prefer to make a difference in another way, you could calculate an offset and send that money toward an environmental group or some other organization that's working through legislative or regulatory channels to reduce emissions. Some companies and individuals are switching from traditional offsets to alternative charity-like schemes that promise big environmental benefits rather than carbon neutrality.

If you do want to offset . . .

With dozens of offset schemes out there and no real regulatory framework, it can be hard to know which one to choose. One approach is to look for programs that adhere to voluntary standards, somewhat like the certifications created to identify organic or fair-trade food.

* **Verified Carbon Standard, www.v-c-s.org:** Billed as the most widely used voluntary standard, VCS has been applied to more than 1000 projects worldwide.
* **Gold Standard, www.cdmgoldstandard.org:** Supported by more than 85 nongovernmental organizations, this metric is designed to certify offsets that meet stringent Kyoto Protocol requirements.

Index

greenhouse gases (*continued*)
feedback processes with climate, 271–274
forest fires and, 216
goals to reduce emissions and mitigate climate change, 354–356
human-produced emissions, what happens to them, 43–46
hydrofluorocarbons (HFCs), 368
in ice ages, 281
in intense cold and warm periods, 276–277
in interglacial periods, 288
Kyoto Protocol reduction requirements, 373
levels of emission, and predictions of climate change, 16
longer-lived, in the troposphere, 246
measurement of rising levels of CO_2 in the atmosphere, 39–40
ocean acidification and, 167
other than CO_2, reducing, 359
outsourcing emissions, 52–53
prevalence and impact on enhancing the greenhouse effect, 30–34
shorter-lived, in the troposphere, 246
slashing emissions on scale needed, 22
stabilization level of, 21
Greenland, 98, 163, 203
Greenland ice sheet, 20, 21, 102, 114, 116–120, 121, 124, 145, 152, 154, 162
ice cores from, 280
Greenpeace, 330
Gregory, Jonathan, 145
Gressley, David, 89
grid points, 300
Grono, Switzerland, 64

groundnut, 225
ground-source heat pumps, 432
growing season (Northern Hemisphere), lengthening of, 4
Guardian, The, 342, 348
Gulf of Bothnia, 144
Gulf Coast, 175
Gulf of Mexico, 12, 175
Gulf Stream, 20, 161, 280

Hadley Centre (Met Office), 304, 305
half-life, 268
Hall, Timothy, 178
halochlorofluorocarbons, 37
Hansen, James, 61, 155, 238, 322, 340, 357, 386, 407, 421
Harriss, Robert, 52
Harvell, Drew, 169, 209
Hasnian, Syed Iqbal, 349
Hawaii, Mauna Loa Observatory, 39
Hayhoe, Katharine, 341
Heartland Institute, 243, 333
heat exchange between oceans and atmosphere, 13
heat, extreme, 59–76. *See also* heat waves
attributing the heat, 69–70
combined with pollution, 63
handling, 75–76
human cost of, 66–69
outlook for the future, 70–73
wildcard, Atlantic thermohaline circulation, 73, 75
heat waves, 64–69
defining characteristics, 74
duration and nighttime lows, 62
in cities, 7
in Europe, 2003, 59, 64, 65
future trends, 72
in Russia, 2010, 59, 61, 62, 64–66
Heathrow Airport, 64
heating and hot water (home), 443–444

oceans (*continued*)

sea surface temperature (SST),
242–243, 245

storms, 142

undersea volcanism, 275

warm ocean waters and tropical
cyclones, 173

warming of, 141

Oeschger, Hans, 284–285

offsetting, 463–466

offshore wind farms, 413

oil, 400–401. *See also* fossil fuels

synfuels, 402

Olson, Donald, 290

Oouchi, Kazuyoshi, 181

orbital cycles, 283. *See also* Earth's orbit

Orestes, Naomi, 335

organics, 459

Otto-Bliesner, Bette, 283

Our Changing Planet, 339

outlet glaciers, 116, 118

overfishing, 169

Overland, James, 104, 105

Overpeck, Jonathan T., 154, 267, 283

ozone, 32–33, 63

near-surface ozone and increased
CO_2, 222–223

ozone depletion, 8, 10, 37, 122

ozone hole, 14, 36–37, 120, 122, 156, 322,
328, 368

oxygen, 28, 36

oysters, 166

Pacala, Stephen, 361–363

Pacific Decadal Oscillation (PDO),
94, 157

Pacific Islands Applied Geoscience
Commission, 149

Pacific Ocean, 72, 94–95

El Niño and La Niña, 13, 159–160,
181

tropical Pacific and climate models,
310

tropical Pacific, effects on West
Antarctica, 124

paintings from the Little Ice Age,
290–291

Pakistan, 60, 75, 77, 78, 79

Paleocene-Eocene Thermal Maximum
(PETM), 276–278

paleoclimate proxy data, 265

Panama, Barro Blanco Dam, 376

Pan-Arctic Ice Ocean Modeling
and Assimilation System
(PIOMAS), 101, 105

Pangea, 260, 274

parasitic diseases, 209–210, 212–213

Paris, heat wave of 2003, 7, 64, 67–68

Parker, David, 239

Parmesan, Camille, 203

particulate air pollution, 63, 72

parts per million (ppm), 31

passive solar, 409

Patagonia, glaciers, 131

peak oil, 400

Pearl River Tower (Guangzhou City,
China), 431

Pederson, Louis, 133

percentile departures (temperature),
74

perfluorocarbons, 373

Perkins, Sarah, 62

permafrost, thawing in the Arctic,
109–112, 115

Permian-Triassic extinction, 277

per-person emissions, 43

personal solutions for climate change
getting started, 439–441
home energy, 443–450
offsetting, 463–466
shopping and purchases, 457–461
transport and travel, 451–455

Perth, 91, 93

Peru, 127, 130, 159

Petermann Glacier, 120, 121

Peterson, Thomas, 238